Introduction to
Nuclear Concepts for Engineers

Introduction to

Nuclear Concepts for Engineers

Robert M. Mayo

Department of Nuclear Engineering

The North Carolina State University

American Nuclear Society
La Grange Park, Illinois USA

Dedicated to
Bobby,
Jessie,
and Judy

Library of Congress Cataloging-in-Publication Data

Mayo, Robert M., 1962–

Introduction to nuclear concepts for engineers / Robert M. Mayo

p. cm.

Includes bibliographical references and index.

ISBN 0-89448-454-0 (alk. paper)

1. Nuclear engineering. I. Title.

TK9146.M36 1998 98–42552

621.48—dc21 CIP

ISBN: 0-89448-454-0
Library of Congress Catalog Card Number: 98–42552
ANS Order Number: 350019

555 North Kensington Avenue
La Grange Park, Illinois 60525 USA

Second Printing 2001

Printed in the United States of America

Contents

Preface

This text is the result of an accumulated experience over the past several years in teaching rising nuclear engineering students their first quantitative, theory-oriented course in nuclear principles with the emphasis placed on basic concepts and their quantitative usage. It has been derived from a set of detailed notes from a one semester course entitled the Fundamentals of Nuclear Energy required of second semester sophomore nuclear engineering students at the North Carolina State University. The objective and philosophy here is to provide the student with the timeless nuclear concepts, models, vocabulary, and problem solving skills essential for success in subsequent course work in reactor theory and engineering.

The material in this book is of broader interest, however, than just for those who wish to pursue a career in nuclear energy engineering. Without searching too far, one quickly encounters the ubiquity of nuclear topics throughout many engineering and science disciplines. Related fields such as nuclear medicine, health physics, and nuclear chemistry immediately come to mind. The topics developed here form the essentials in these sister disciplines, and are presented in a most general fashion, from a straightforward description of the basic concepts relevant to nuclear processes in a manner that appeals to one's intuition and understanding of basic physics concepts. This book may also appeal to those outside these fields wishing to satisfy a purely intellectual curiosity for a deeper understanding of the atomic nucleus, the radiation it produces, and how nuclear radiation interacts with the world around us. The practicing engineer too (especially nuclear) should find this a useful source of reference since it not only provides a description and equations for the basic nuclear concepts, but also some of the important data employed in nuclear calculations and references to sources of data not provided.

This text has been designed to be appropriate for a sophomore science or engineering student with a firm foundation in the basics of classical college physics and mathematics through ordinary differential equations. Concepts in modern physics (special relativity, quantum concepts, etc.) are introduced and developed as required in the presentation of the text material. Aside from a brief review section on special relativity in the first chapter, these subjects are applied to the problem at hand without diverting attention to them in and of themselves. The flavor of the book is a semi-rigorous, quantitative, and mathematical one without being intimidating. The objective is to present fundamental nuclear principles in a clear and understandable, yet physically sound manner. Most every result is derived from first principles without being overbearing, and is discussed vis-à-vis physical intuition. In doing so, we have occasionally needed to make assumptions to keep solutions analytically tractable. Every attempt is made to announce these assumptions as they are introduced and in many cases point the reader to more in-depth references for further reading. Finally, this book is not a substitute for texts in nuclear physics or nuclear engineering — both of which have very different objectives and audiences — nor does it cover the technology of fission and fusion energy production.

In chapter 1, we begin by discussing the role and importance of nuclear science in engineering and the world. Nuclear processes are extremely important in our world today, not just in the electrical power industry, but in many other ways that we have come to rely on. We are reminded of some of the indispensable medical uses of radiation and radioisotopes. Following this introductory discussion, some important concepts in special relativity are reviewed to refresh the reader's background in problems dealing with relativistic particles. The basic description and terminology of the atomic nucleus are also developed primarily for future use, but it is also helpful at this point to have a working model (or at least a mental image) of the entity under evaluation to appreciate scales and visualize processes.

The discussion of nuclear processes continues with chapter 2. Here, we deal exclusively with the discussion of nuclear reactions as general phenomena, the specifics of which are covered in later chapters in the context of particular reactions of interest; *i.e.* scattering, decay, capture, fission, etc. A general formalism is developed to describe the energetics of reactions, along with a detailed discussion of the necessary conditions for reactions to occur, both in the lab and center of momentum perspective.

In chapter 3, we move to some more specific nuclear reaction problems by considering scattering. Scattering and slowing down of neutrons are of paramount importance in nuclear reactors. Elastic neutron scattering leads perfectly to the implementation of the center of mass coordinate system formalism as a tool for describing interactions. It is also natural to discuss here the concepts of cross section and neutron particle flux, leading into a discussion of macroscopic systems, interaction probabilities, and attenuation.

Radioactive decay is covered in chapter 4, again as a specific type of nuclear reaction. The kinematics of radiation modes of primary importance in nuclear energy production and related disciplines are reviewed. Included in these discussions is the introduction of energy level diagrams and the nuclide chart, two very important tools in systematically describing nuclear processes. The radioactive decay law is given a great deal of attention as we develop decay systems and chains, and apply these to some practical problems.

The reader will find discussion of the nuclear reactions fission and fusion in chapter 5. The properties and energetics of the reaction processes are considered, rather than the engineering details of practical energy production in a reactor. The latter is better left to nuclear reactor and nuclear engineering design courses.

Radiation interactions with matter is covered in chapter 6, again from a semi-rigorous, mechanistic perspective. Consistent with our collision interaction theme developed earlier, charged particle (Coulomb) collision kinematics are discussed, as this is fundamental to the interaction of charged particles with matter. This chapter serves two important purposes. It reviews the fundamental interaction mechanisms of charged and neutral particles with material media, while it also serves as a prelude to the subjects of radiation protection and radiation detection, topics of subsequent and complementary study for nuclear and radiation professionals.

Each chapter contains a number of examples and exercises which illustrate concepts developed in the text. Input data, when required, can be found in the main body of the text or in the appendices. Every attempt has been made not to require outside sources of data except for the Chart of the Nuclides, which can be obtained from the General Electric Company, Nuclear Energy Operations as their publication, *Nuclides and Isotopes*, 15^{ed}, 1996.

I am indebted to a great number of very generous and hard working people for assistance and support in this effort. I have enjoyed and greatly benefitted from technical discussion with many, so many in fact that it is difficult to completely list. By omission, I mean no slight or disrespect, I thank you all. Most prominent are those who have been by me every day. I thank you especially, Kuruvilla Verghese, Douglas Peplow, Sven Bader, and Avneet Sood. Many thanks to those who have helped me with getting started on document and artwork preparation, and editing; Chris Barnes, Rebecca Carter, and Teresa Loignon. I'm also very appreciative of the patience and support provided by my family and friends while I toiled in this effort. But most of all, I owe thanks to my mentor, Professor Ward S. Diethorn (PSU). My interest in the subject grew from your teaching. The subject and style of technical presentation, I learned from you. As well, I cannot thank you enough for your help in writing the first chapter.

Robert M. Mayo
Raleigh, NC, 2001

Chapter 1

Basic Principles

1.1 Introduction

We live in a most interesting time, one that has been largely shaped by the technological achievements of the past century and a half. Although civilization has arguably existed for some one hundred centuries (10,000 years) or more, we have done and learned more in recent decades than in all preceding time. The ancients gazed at the heavens and pondered all sorts of majestic and fantastic theories about the moon and other celestial objects; what they might be, what are they made of, and what might be living there. We, in our lifetime, have gone to the moon and have answered some of these questions. We have explored the depths of the heavens, and have probed many of the mysteries of classical science. We have turned our discoveries and inventions into products and technological instruments that make our lives freer and less burdened by sustenance, allowing opportunity to pursue cultural and intellectual activities. We enjoy instant global communication made possible by the discovery of radio waves and further advanced by the invention of the transistor. Our society is healthier, safer, and more convenient than ever before due largely to such familiar technological achievements as electricity on demand, medical vaccines, X-ray machines, automobiles, and microwave ovens, to name a few, of which none were available before the late nineteenth century.

The ancients, too, pondered the minutest of objects. The Greek philosopher and mathematician Democritus (circa late fifth century B.C.) postulated the existence of the smallest bit of matter from which all macroscopic objects were comprised. This tiniest bit of matter, the fundamental building block of nature, was called the "atom," which comes from the greek *atomos* meaning undivided. We borrow from his verbiage and philosophy in our scientific language today with a slightly different connotation. Of course, we now realize that there are yet smaller particles than the atom that make up the atom and atomic nucleus, and still smaller particles comprising the nuclear particles called quarks. We'll discuss more about this first level of sub-atomic particles (*i.e.* neutrons, protons, electrons, ...) later in this chapter.

It should come as no surprise that many of the advances in our technological age have resulted from a better understanding of the atom and the atomic nucleus, and their properties. In fact many have called this the "nuclear" or "atomic" age. Despite the unfortunate association of this reference with nuclear weapons, it accurately portrays peacetime applications of nuclear technology equally well. It would be difficult for us to imagine a world without the medical diagnostic necessities such as the X-ray, *computed axial tomography* (CAT) scans, nuclear *magnetic resonance imaging* (MRI), and radiotracers. Nor would it be reasonable to overlook the nuclear contributions to our energy economy and its promise for the future. Nuclear techniques indeed touch many disciplines that affect our everyday lives, as we'll discuss in the next section.

A firm foundation in the basics of nuclear processes has led to these achievements and is essential to future advancement. To this end, we devote our attention in this book. It is exactly that collection of introductory ideas, scientific laws, vocabulary, and models of things nuclear that we call "nuclear concepts."

1.2 The Importance of Nuclear Technology

In the industrialized nations, nuclear technology in one form or another is both pervasive and diverse. To quantify the impact, several measures of economic importance are readily identified: jobs created, sales volume, corporate profits, and tax revenues are a few. Societal benefits, of course, also occur, but these are much more difficult to quantify and are frequently controversial, particularly when the trade-off between benefits and risks are considered. A list of social benefits would, however, certainly include convenience, increased efficiency, quality of life, reduced environmental impact, saving of lives, and relief of suffering. For the purpose of discussing economic and societal importance, the technology naturally separates into the two subdivisions of power production and non-power applications.

1.2.1 Power Production

At the close of 1995, seventeen countries depended on nuclear power for at least 25% of their electrical energy needs. In thirty countries, 437 nuclear power plants with 343 gigawatts of electrical capacity supplied 2228 terawatt-hours of electrical energy. This amounts to 17% of the world's total electricity production. Light water nuclear reactors (LWR), a technology pioneered in the United States, drive 80% of the nuclear plants. Moreover, nuclear electricity generated in 1995 exceeded all forms of energy consumption in 1958. Eighteen countries rely more on nuclear power for their electrical energy needs than does the United States. The top three are Lithuania (76%), France (75%), and Belgium (56%), followed by Spain (35%), Japan (33%), Germany (29%), and the United Kingdom (26%).

In the United States, 109 nuclear power plants generated 98.7 gigawatts of electric capacity and supplied 674 terawatt-hours, 22.5% of the country's electrical power in 1995. The first commercial U.S. nuclear power plant went into operation in 1957, producing 60 megawatts of electrical power. Nuclear is the second largest source of electricity and provides enough power for 65 million U.S. homes. Total U.S. production and sales of nuclear electricity in 1995 amounted to $12.8 billion and $47 billion, respectively. A merit index known as the capacity factor is one measure of the power industry's efficiency. Capacity factor is the ratio (in percent) of actual watt-hours generated annually to watt-hours ideally generated by round-the-clock operation 365 days per year. Based on the numbers above, the U.S. nuclear capacity factor for 1995 was 77%, four percentage points above the world capacity factor. One quality of life measure is the impact of the technology on the green house effect. CO_2 emission would increase by 21% worldwide if nuclear plants were replaced by fossil plants, assuming a 6%/77% mix of hydro to fossil generation.

A 1994 Nuclear Energy Institute (NEI) study[NEI94] provides some estimates on the impact of the nuclear power industry in the 1991 U.S. economy. Direct employment was estimated to be 98,000 with an amplification factor of 4.2. That is, the creation of an additional 319,000 jobs depended to some degree on the generation of nuclear electricity. Estimated sales of goods and services, and profits were $70 billion and $3 billion, respectively, which represents about 1% of the 1991 gross domestic product. Total federal, state, and local government tax revenues amounted to $3 billion in this economic model. Further, the 417,000 employment figure is 0.4% of the total U.S. work force for that year. As an independent gauge of the NEI's study job figures, one can readily estimate about 83,000 jobs in 1994 from data[NN96a] on personnel radiation exposures at nuclear utilities in that year. A second comparison comes from a Department of Energy report[NN96b] on employment related to nuclear power plant operations, in which professional engineering and physical scientist jobs in 1991 totaled about 44,000. Better agreement cannot be expected among these job estimates in view of inherent complexities and assumptions in studies of this kind.

In recent years, U.S. nuclear power plants as a whole have improved steadily, but the learning process has proved to be much slower than anticipated. Capacity factors are increasing, and production costs, refueling times, forced outages, personnel radiation exposures, and nuclear waste generation are declining. Work force safety compares favorably with the best in other major industries. The number of operating U.S. plants has reached a plateau; the most recent start-up occurred in 1993. A few plants are still under construction but their fates remain uncertain. Most plants now in operation were ordered during the 1965–75 period, and no new plants have been ordered since 1978. The situation is markedly different in the Pacific rim countries where nuclear power growth is very strong, paralleling the strong economic development in that part of the world, and strongly coupled to the absence of indigenous fossil fuels.

The world's natural uranium resources are sufficient with today's nuclear reactor designs to meet the planet's entire electricity needs for at least 100 years. Implementing the so-called breeder power reactor, a proven concept but nearly abandoned for the present, would extend this time limit for several millenia. There is, then, no technical reason why the United States could not expand its nuclear power program manyfold and perhaps double its present 22.5% over the next two decades. By contrast, a much smaller U.S. growth, if any at all, is expected over the next decade. If new plant orders are not forthcoming, the nation's nuclear electrical capacity will soon decline. Our present capacity will begin to drop off due to plant retirements starting in 2008 and fall to practically nothing by 2025. Reasons for the U.S. falling far short of its nuclear power potential range from simply the reduced demand for electricity forecast for the immediate future to the complexities of society's decision making processes when the public is confronted with the nation's long term energy needs.

Future technical innovations now in research and development will eventually expand the world's nuclear power options. Advanced U.S. nuclear reactor designs promise lower plant costs and inherently greater safety, but many years will pass before one of these new reactors is constructed. Lying much farther ahead is the promise of a fusion power plant. As with the fission reactors in use today, fusion too derives its power from the atomic nucleus, but operates on entirely different physical principles. We'll discuss much more on the subject of fusion in chapter 5. Because of the extremely challenging conditions required for fusion, a long research and development effort lies ahead (most probably well into the later part of the next century) before this power concept reaches a commercially viable position. Yet, conservative estimates place fusion fuel resources on this planet at an almost unlimited level (being derived from ordinary sea water), adequate to meet the world's electrical energy needs for an incredible 1–10 billion years!

Distinct from commercial nuclear electricity generation is the use of nuclear electricity for ship propulsion in the U.S. navy. Most large surface ships, as well as undersea vessels, are now nuclear powered, although the total megawatt capacity in service is far smaller than that of commercial nuclear power. The strategic and tactical advantages are enormous; higher speed and long deployment times are just a few. What's more, nuclear propulsion has practically eliminated the rationing of ship speed and fuel dependence on the ship's home port.

1.2.2 Non-Power Applications

This category of nuclear technology includes those products manufactured in the United States containing radioactive materials in some form, which find applications in research and commerce. Firstly, a definition is appropriate here. We need to distinguish between stable atoms and those that are unstable or "radioactive". Atoms in any of the common 92 elements in nature fall into either of these two categories. This subject will be covered in detail later (chpt. 4). For our purposes here, let it suffice to say that radioactive atoms (or radionuclides as they are sometimes called) emit particles of radiation whereas those that are stable do not. Examples of common stable/radioactive pairs include $^{1}H/^{3}H$, $^{12}C/^{14}C$, and the cesium pair $^{133}Cs/^{137}Cs$.

1.2.2.1 Radiotracer Techniques

Radioactive atoms have several properties that contribute to their usefulness. One is the radiation that is emitted when a radionuclide spontaneously transforms (decays) into a stable atom. (Two others are attenuation and energy deposition, and are important for applications discussed in the next two sections.) The three most common radiations are alpha particles (helium atom nucleus), beta particles (electrons), and gamma rays (photons). Radiation can be detected by electronic devices called "detectors" that are so sensitive that as few as 10^{14} radionuclide atoms can be detected (radioassayed). This property is the basis of a major radionuclide use known as a "tag" or "tracer". A minute number of tagged molecules added to a material permits an investigator to follow chemical and physical changes at both the macro and molecular levels without affecting the material's characteristic behavior. A material tagged with radioactive material is sometimes said to be "labeled". One example is water (H_2O) labeled with radioactive hydrogen (tritium), ^{3}H. Tritium replaces one of the hydrogen atoms in the molecule to form $^{3}H^{1}HO$. Another example replaces the stable ^{12}C with radioactive ^{14}C in the carbon dioxide molecule to form $^{14}CO_2$. In the automotive industry, piston rings are sometimes tagged with radioactive iron (^{59}Fe), which permits the study of engine wear dynamics by monitoring with a radiation detector the ^{59}Fe-labeled wear debris in the circulating engine oil. Almost every pesticide or insecticide registered by the Environmental Protection Agency (EPA) is synthesized as a tagged (^{14}C) analog to investigate its degradation products.

Dilution analysis is another powerful radiotracer application. A rather simple, but contrived, example best defines the technique. Suppose that a volume of water in an underground tank of limited access is required. A small sample of ordinary water containing a known amount of tritium label can be added to the tank. After the tank is mixed, a sample is taken and radioassayed for tritium. Because dilution has occurred in the tank, the new tritium concentration is lower. The before and after tritium concentrations and the initial volume of water added yields the unknown volume of the buried tank. This technique is much more general and can be applied to much more complex chemical systems.

Of the thousands of radionuclides known, only some 30 account for essentially all the applications. Most are produced in raw forms at federal facilities (usually nuclear reactors) and sold to manufacturers, who in turn provide commercial radioactive products: special forms (gases, liquids, solids), sealed radiation sources, and labeled materials. A few medical centers satisfy their needs by making their own radioactive materials in cyclotron accelerators. The economic benefits of the production, processing, transportation, and use of radioactive materials are indeed large. A 1994 study[ORRWS94] of the U.S. impact for the year 1991, invoking an examination of 80 industries and 475 occupations, found the following: $257 billion in total sales and services (4% of the 1991 gross domestic product), $11 billion in corporate profits, $45 billion in federal, state and local tax revenues, and 3.7 million jobs (3% of the total 1991 U.S. work force). This study attempted to evaluate both direct and indirect dollar and job benefits, and in so doing, included jobs that partly, as well as solely, depended on radionuclides. The job total includes a large number of semi- and low-skilled workers. Engineers, physical scientists, and medical and health professionals comprise in the aggregate only about 7% of the 3.7 million jobs. It is interesting to note that in comparing this study to the one above on nuclear power, sales and jobs in the radionuclide industry far exceed those in the power industry.

Radiotracers were first used in the early years of nuclear technology development during the second world war. In subsequent decades, a good share of research in chemistry, materials science, biotechnology, cell and molecular biology, genetics, and cancer benefitted from radiotracer use. As a result radiopharmacueticals and radiochemicals are now important commercial products. Many advances in biochemistry were based on the use of just two radiotracers: ^{3}H and ^{14}C. The importance of these two in unraveling the secrets of photosynthesis can hardly be exaggerated. Radiotracer techniques are often found to be indispensable because of their sensitivity, quantitativeness, and non-perturbing behavior in the system under investigation. Alternatives to radiotracers are frequently not available. For example, the organic compound thymidine is a precursor of DNA that must be duplicated in the living cell before the cell can divide. ^{3}H-labeled thymidine provides a unique tool in the study of cell renewal for this reason. Three additional observations further support the role of radionuclides in research. Eighty percent of all external biochemical research funded by the National Institutes of Health (NIH) is dependent on radiation. Secondly, three fourths of the Nobel Prize awards in physiology and medicine over the past twenty years were for research that utilized radioactive atoms. Finally, 90% of all new drugs approved by the Federal Drug Administration (FDA) require radionuclide-based tests.

1.2.2.2 Industrial Uses

Over the years, radionuclide applications have been well received in almost every U.S. industry. Table 1.1 lists some major ones, along with the radionuclides utilized. To provide a perspective, we offer a few details on each.

Table 1.1: Industrial Applications of Radionuclides

Use	Radionuclide
Industrial Irradiators	^{60}Co, ^{137}Cs
Well-Logging	^{137}Cs
Radiography	^{60}Co, ^{192}Ir
Static Eliminators	^{90}Sr, ^{210}Po
Smoke Detectors	^{241}Am
Nuclear Gauging	^{137}Cs, ^{147}Pm, ^{204}Tl, ^{241}Am, ^{252}Cf
Self-Powered Lighting	^{3}H, ^{85}Kr
Small, Remote Power Sources	^{90}Sr, ^{238}Pu

Irradiators are generally intense gamma sources capable of quickly delivering a rather large gamma radiation "dose" to a commercial product. The polymer industry uses radiation to process or cure certain products, such as protective coatings. Radiation is widely used to sterilize pre-packaged medical products like bandages, eye ointment, hypodermic syringes, sutures, tongue depressors, contact lens solutions, and tissue graft implants. Twenty-seven countries have approved the use of radiation to preserve some types of foodstuffs. Pathogens, molds, yeasts, and insects are eliminated, and shelf life is greatly extended. In some cases, refrigeration needs are greatly reduced, or even eliminated altogether. Approved U.S. foods for irradiation include spices, poultry, and fruit. This food application of radionuclides has not achieved anything near its potential in the United States because public acceptance of irradiated food products is tenuous. In agriculture, radiation sterilization has provided significant economic benefits by controlling screwworm, fruit fly, and gypsy moth populations. The basic idea here is to sterilize the male insect with radiation, release it to the wild where it mates, but no offspring are produced. One such screwworm control program in Texas is estimated to save cattle ranchers about $100 million per year.

Exploratory drilling for oil and gas is expensive and usually unsuccessful. A gamma emitting radionuclide, however, can be used during well-logging to assist the drilling evaluation. Scattered gamma rays observed with an *in situ* radiation detector helps identify rock composition, which, in turn, can be used to predict the likelihood of locating these fuels.

Radiography is the standard inspection method for welds in metal structures, such as pipes and bridges. To test the quality of a pipe weld, for example, during construction of a natural gas line, a gamma source is placed in the pipe, which produces an image of the weld on photographic film wrapped around the weld. U.S. airlines require radiographic inspection of all wing structures on domestic aircraft.

Buildup of static electric charge during continuous manufacturing processes can be a serious problem. Machining operations can be disrupted in printing and papermaking, for example. Sometimes, fires and explosions occur. Radiation sources strategically placed in the machinery ionize the surrounding air, promoting charge neutralization and leakage, and thereby reducing the buildup problem.

A common fire alarm system, the smoke detector, is widely used in U.S. homes and required in businesses and hotels. In this application, an alpha-emitting radionuclide ionizes air between two, closely placed, metal plates open to the atmosphere. A dry battery applies a small voltage across the plates. This assembly works like an ionization chamber and triggers an audible alarm when a smoke aerosol drifts into the chamber, abruptly reducing the small chamber current normally present.

Nuclear gauging is one of the largest industrial applications. Product density, thickness, and package content are frequently quality control issues in the manufacturing of sheet materials like glass, metals, paper, wood veneers, foils, etc. The basic principle is rather simple. A radiation source and detector are positioned on opposite sides of the moving sheet material. Changes in the sheet thickness, density, or composition alter the amount of radiation transmitted, and thus the strength of the detected signal. By knowing the scattering and attenuation properties of the material, automatic adjustments to the machinery in response detector signal changes can be used to return the manufacturing variables to normal. This principle is also used to inspect whether products or liquids in sealed packages moving along a conveyor belt meet their fill specifications before shipping.

Radionuclide-activated, glow-in-the-dark light sources are found in wrist watches, clocks, gauge dials, and instrument panels of all kinds, whenever compact, reliable, low-level illumination is required. The basic configuration is a mixture of a low-level, beta-emitting radionuclide with a chemical phosphor, which is painted on the relevant device area. Visible light is emitted when the beta particles strike the phosphor, deposit some of their energy, and excite the phosphor. Safety lights based on this principle have been used to mark emergency exits, light aircraft runways, and mark highway lines. A familiar example is the EXIT sign found in public places where operation during a power failure is imperative. This device is a sealed glass bulb coated on the inside surface with a phosphor and filled with the beta-emitting, radioactive gas tritium, $^3H_2$. The beta rays are completely absorbed in the glass so there is no beta radiation hazard to personnel from use of this device or the others listed above. Lights operating on this principle are said to be self-luminescent. No electrical power is required and maintenance is virtually eliminated. In a very cold climate like the arctic, conventional electric lights don't work well and the self-luminescent safety light is an obvious solution.

In small electrical power sources fueled by radionuclides, such as ^{90}Sr and ^{238}Pu, heat is liberated as radiation is absorbed to generate electrical power. Energy conversion is achieved by the thermoelectric effect, whereby a voltage source is created when heat raises the temperature of a bundle of thermocouples. A power range from watts to kilowatts is possible. Heart pacers use radionuclide sources to generate the small amounts of electrical power required. Other applications arise whenever remote needs make maintenance expensive or impossible. Navigational beacons are one example. Radionuclide power was used frequently in satellites and is a mainstay power source for some NASA deep space probes.

1.2.2.3 Medical Applications

Nuclear medicine is that field of medicine that uses radiation to diagnose and treat disease. The methodology includes radionuclide-labeled diagnostic and therapeutic drugs, laboratory radiochemical techniques, diagnostic nuclear imaging of body tissues and organs, and beam therapy. In the past two decades, this field has seen vigorous growth in the United States. Millions of nuclear medical procedures are now performed each year. Nuclear imaging procedures alone number about 12 million annually. As well, some

100 million laboratory tests involving radionuclides are performed each year on body fluids and tissues. One in every three hospitalized patients benefit from a nuclear medical procedure of some kind.

Nuclear imaging with radionuclides, a diagnostic tool, includes two basic approaches, *single photon emission tomography* or SPECT, and *positron emission tomography* or PET. We'll not discuss magnetic resonance imaging (MRI) here since the basic principle, interaction of proton spin with a pulsed magnetic field, although obviously nuclear in origin, does not involve radionuclides. MRI remains, as always, a premier diagnostic tool in medicine. SPECT and PET both evolved from the earlier X-ray imaging method CT or CAT. CAT is a form of radiography in which a three dimensional image is constructed by computer analysis of numerous, plane cross-sectional images made along the axis of the object. With CAT, a machine source of X-rays and an array of radiation detectors are rotated around the patient to provide the cross-sectional map of the body's interior. The image is usually viewed on a video monitor. In recent years, public awareness of X-ray CAT has been enhanced by its use in airport security systems to inspect for explosives in passenger baggage. Before CAT, the traditional use of X-rays in medicine gave only an overall map of superimposed, internal structures in the body. Imaging of the bone is good, but imaging of an organ and its functioning is unsatisfactory. The innovations called SPECT and PET utilize the CAT principle with its ability to eliminate the superposition problem and can image organ and tissue function, as well as pinpoint biochemical activity. Both rely on the medical practitioner's ability to place a radiation source (radionuclide) in the tissue to be imaged, a significant advance over X-rays alone or X-ray CAT. An arsenal of organ- and tissue-specific radiopharmacueticals (carriers) tagged with radionuclides are available for this purpose. Given orally or by injection, an appropriately selected carrier either metabolizes or accumulates in the organ or tissue of diagnostic interest.

The gamma-emitting radionuclides ^{67}Ga, ^{99m}Tc, ^{131}I, and ^{201}Tl are the most popular in SPECT. A good example of SPECT use is in bone scan. Cancerous areas in the bone are characterized by increased blood supply. In a bone scan, the patient is given a ^{99m}Tc-labeled carrier known as MDP, an organic compound, which preferentially goes to the bone matrix and concentrates where blood supply is elevated. Thus, the gamma rays from ^{99m}Tc decay "illuminate" the cancer in the three-dimensional bone image produced by the SPECT scanner. Cancers are detected sooner by SPECT than by X-ray methods. Other applicable carriers extend the method to the brain, liver, spleen, kidney, and heart. Detection of heart damage is another application for SPECT. The patient is first stressed to the maximum heart rate for his/her medical condition, and then a heart seeking carrier tagged with ^{204}Tl is injected into the bloodstream. Damaged muscle or partially blocked arteries in the heart are blood poor, and accordingly, the uptake of the carrier is low in these regions. A map of these abnormal regions shows up in the three dimensional heart scan.

PET is also based on the CAT principle and a tissue-specific carrier tagged with a radionuclide. Short-lived, positron-emitting radionuclides, however, are used rather than gamma emitters. Positrons are just like electrons except that they carry positive electrical charge. A positron annihilates when it meets up with an electron and produces gamma rays, which act as the imaging radiation source. Following its emission from the radionuclide, a positron always disappears (annihilates) very close to the decay event with the back-to-back emission of two identical gamma rays. These gammas are detected with a ring of detectors around the patient. A single pair of detectors at the end of the ring diameter responds to each pair of gamma rays and, thus, identifies where the positron originated in the body. Positron emitters used in PET include ^{11}C, ^{13}N, ^{15}O, and ^{18}F. They are usually produced in cyclotrons near the PET facility due to their short lifetimes. PET scanners can distinguish between metabolically active and inactive tissues, and thereby gauge the actual functioning of tissue. Cancer and heart disease have been diagnosed, and some success has been achieved in understanding brain function and mental processes, even the diagnosis of schizophrenia and Alzheimer's disease.

One of the more familiar medical uses of radionuclides is in cancer therapy. Deep cancers can be irradiated by an external beam therapy unit containing a large source of ^{60}Co or ^{137}Cs gamma rays. Hyperthyroidism and thyroid cancer can be treated by orally administering the gamma emitter ^{131}I. The patients are given the radionuclide in the form of water soluble potassium iodide. Iodine concentrates in the thyroid where the decay radiation from ^{131}I kills the diseased cells. Cancer therapy is also performed by placing small, sealed radionuclide sources (^{90}Sr, or brachytherapy with ^{192}Ir) in or on the patient. Other procedures include lung ventilation and blood flow studies with ^{133}Xe. Radioimmunoassay (RIA) is widely used in diagnostic testing, and is preferred where possible since it eliminates radiation exposure of the patient. RIA is an *in vitro* radionuclide procedure performed on a sample of body fluid for the purpose of assaying a particular chemical component. Relatively recently, a technique involving a tagged monoclonal antibody that takes radiation therapy directly to the cancerous cell, has been developed. A monoclonal antibody is a protein molecule that attaches itself to the outside of a cell membrane. In one approach to leukemia therapy, such an antibody tagged with the beta emitter ^{67}Cu and engineered to bind specifically to leukemia cells, is given to the patient in order to deliver radiation at the most fundamental level.

An entirely different type of cancer therapy utilizing the neutron, a charge neutral sub-atomic particle, is used when a rare type of brain cancer becomes inoperable. The technique is called *boron neutron capture therapy* (BNCT), which relies on the nuclear reaction that takes place when a boron nucleus captures a neutron, producing energetic, electrically charged helium and lithium atoms. The patient is given a specifically designed organic compound containing boron, which preferentially goes to the cancerous region in the brain. A machine source or nuclear reactor supplies the neutrons. The cancer killing agent in this case is the kinetic energy spent by the energetic helium and lithium atoms. Research has progressed to the point where aftereffects in the brain function are beginning to be minimized.

1.3 Dimensions, Physical Constants, and Symbols

Before embarking on our journey through the nuclear world, let's review briefly some basic definitions of units or dimensions, some physical constants, and introduce some shorthand notation (symbols). This will establish a common language with which we can build our nuclear concepts.

1.3.1 Dimensions

The physical quantities that are used to express the laws of nature require direct reference to the physical world and enable uniformity in measurement. We must all agree, for example, that my unit of length is exactly the same as yours, otherwise I'm likely to sell you a piece of rope that's much too short to dock your boat. A generally accepted system of measurement is then more than just a convenience, it is absolutely necessary for trade and commerce, and most every daily function. In science and engineering, in particular, it is imperative to adopt and maintain standards of measurement, since a very high value is placed on clarity and consistency.

Standards of measurement are manifest in an established system of units or dimension. In the United States, the British (or English) system is no doubt familiar to everyone. Having enjoyed an extremely durable public acceptance, it is the most popular system in this country, despite attempts at replacement. We are very comfortable in describing quantities of length, weight, and volume in units of inches, pounds, and gallons, for example. The situation becomes more cumbersome, even uncomfortable, however, if we for some reason have to communicate using idiosyncratic English units like fathom, furlong, rod, dram, square, grain, kip, gill, peck, etc. Further examples abound. There are twelve different "ton" units for example, defining such quantities as mass, volume, refrigeration capacity, and explosive force![Lord95]

Development of the metric system, early CGS (centimeter-gram-second), MKS (meter-kilogram-second), and the later version MKSA (adding the ampere), was a substantial improvement and quickly became popular in scientific and engineering circles, but not so in U.S. commerce where usage has been spotty. Even some early metric units raise eyebrows, the esoteric "stere" and "are", for example.

In this text, we use the SI[ANSI92] system of metric units, except in a few special nuclear contexts where the weight of custom and habit overrides cogent SI persuasions. This system of units evolved from its MKS and MKSA predecessors and is universally accepted in engineering and science. By 1870, the need to standardize units became evident and led to the first international meeting of the General Conference on Weights and Measures (CGPM). CGPM continues to meet every six years in Paris. The CGS and MKS metric systems were sanctioned in 1881 and 1900, respectively. In 1935, the next metric version, MKSA, was recommended following a 1901 suggestion by the Italian engineer G. Giosgi. CGPM added the degree kelvin and the candela units to MKSA in 1960 and gave the resulting system the title "International System of Units" (abbreviated SI from the French "Le Système International d'Unités"). By 1971, the seventh and last unit, the mole, was added to what is now a closed system of seven basic SI units.

Although legalized by the U.S. Congress in 1866, the infant metric system was not officially adopted by Congress until 1893. Feet, pounds, and other English units were defined in terms of their metric equivalents. Following further sporadic initiatives over the years, the next significant development occurred in 1975 when Congress passed the Metric Conversion Act promoting additional metric efforts. The Act created the U.S. Metric Board and gave the U.S. Department of Commerce the responsibility for implementing the SI system. No timetable for conversion to metric was mandated in the Act, however. In 1982, the Board, given no enforcement power by the Act, disbanded, effectively stalling metric conversion to a crawl in the United States. Finally, in 1988 Congress passed yet one more initiative, the Omnibus Trade and Competitiveness Act. This Act declared "...it is the policy of the U.S. (1) to designate the metric system of measurement as the preferred system of weights and measurements for U.S. trade and commerce, and (2) to require that each federal agency by the end of 1992 (if economically feasible) use metric in its procurements, grants, and business-related activities." The National Institute of Standards and Technology (NIST) represents the United States in CGPM activities and handles all international matters concerning the metric system. As of 1995, the United States is one of only three countries (Liberia and Myanmar are the other two) which has not yet implemented the metric system in commerce and trade.

Full acceptance of SI by the technical community contrasts with the slow growth of metric use elsewhere in the United States. The demands of international trade are essentially the only impetus to future growth. Arguments opposing further gains in metric use center on conversion costs, which are real and not easily dismissed by business, and an endemic, even xenophobic attitude toward change.

SI units simplify communication and arithmetic operations largely because there is one, and only one, unit for each physical quantity and a decimal relation between multiples and sub-multiple units. The system is also said to be coherent, wherein equations between units contain no numerical factors other than unity. There are four classes of SI units: base, derived, supplementary, and temporary. Many units have word names. Unit word names are never capitalized and follow the usual rules of English grammar. A shorthand symbol accompanying each named unit is lower-case, except when the name is a person's surname. Base units total seven and are the most fundamental (table 1.2). They are dimensionally independent and arbitrarily defined. That is, the meter, for example, is arbitrarily defined as the path traveled by light in vacuum during a time interval equal to 1/299 792 458 of a second. A space separating a continuous cluster of three digits is intentional here, as explained later. Likewise, the kilogram is a particular, arbitrarily-chosen platinum-iridium cylinder, which was designated the international archetype in 1899, and is stored in Sèvres, France. An excellent introduction to dimension standards and their measurement is given in Halliday and Resnick[Halliday81]. Conversion tables between the systems SI and English are provided in Appendix A.

Table 1.2: Base SI Units

Physical Quantity	Unit Name	Symbol
length	meter	m
mass	kilogram	kg
time	second	s
electric current	ampere	A
thermodynamic temperature	kelvin	K
luminous intensity	candela	cd
quantity of substance	mole	mol

Prefixes in table 1.3 provide multiples and sub-multiples of SI units. Except for the group of four in the center of the table, a factor of 10^3 separates each prefix from the one immediately below it. An attosecond, for example, is approximately the time taken by light to travel one atomic diameter in a solid. Typical atomic dimensions are often expressed in nanometers ($= 10^{-9}$ meters). Double prefixes are not permitted, so that while micromicroamperes is numerically equal to 10^{-12} amperes, the allowed SI unit is picoamperes. Kilogram is the only SI unit with a prefix in its name. The rule for multiples and sub-multiples of kilogram is to use the appropriate prefix with the word "gram" or "g".

Table 1.3: SI Prefixes

Value	Prefix	Symbol
10^{-24}	yocto	y
10^{-21}	zepto	z
10^{-18}	atto	a
10^{-15}	femto	f
10^{-12}	pico	p
10^{-9}	nano	n
10^{-6}	micro	μ
10^{-3}	milli	m
10^{-2}	centi	c
10^{-1}	deci	d
10^{1}	deka	da
10^{2}	hecto	h
10^{3}	kilo	k
10^{6}	mega	M
10^{9}	giga	G
10^{12}	tera	T
10^{15}	peta	P
10^{18}	exta	E
10^{21}	zetta	Z
10^{24}	yotta	Y

Derived SI units handle important quantities such as force, power, electrical resistance, frequency, etc., which are not in table 1.2 but can be derived from algebraic relationships connecting base units. These, in turn, are subdivided into named (with symbols) and unnamed (without symbols). Some named units are listed in table 1.4. Column four of the table defines the derived unit in the simplest possible way, *i.e.* by a formula. When the need arises, a formula not in base units can be so expressed. In table 1.4, entries 1, 4, and 9 are already in base units; the others are not. A derived unit can also be defined via the appropriate fundamental law of physics. For example, consider Newton's second law, $\mathbf{F} = m\mathbf{a}$. This law defines the unit of force, the newton, such that one newton (N) is the force required to give a one kilogram mass, one meter per second squared acceleration. Coherence is confirmed by the quotient (1 kg)(1 m/s^2)/(1 N) = 1. Many other derived units are also named after famous scientists in their honor. The SI unit of work ($\int \mathbf{F} \cdot d\mathbf{s}$) is the joule after the English thermodynamicist, James Joule. Formally one joule (J) is defined as the quantity of work done by a unit force (1 N) in moving a body a unit of distance (1 m) in the direction of **F**. Therefore, in base units J = N m = kg m^2/s^2. Likewise, V = W/A = kg m^2/(s^3 A). Further discussion of derived units and their measurement can be found in [Halliday81]. Examples of some unnamed, derived SI units are listed in table 1.5.

Table 1.4: Examples of Derived SI Units (named)

	Quantity	Unit Name	Symbol	Formula
1.	Force	newton	N	kg m/s^2
2.	Work, Energy, quantity of Heat	joule	J	N m
3.	Power	watt	W	J/s
4.	Electric Charge	coulomb	C	A s
5.	Electric Potential Difference	volt	V	W/A
6.	Electrical Resistance	ohm	Ω	V/A
7.	Magnetic Flux	weber	Wb	V s
8.	Magnetic Flux Density	tesla	T	Wb/m^2
9.	Frequency	hertz	Hz	s^{-1}
10.	Pressure	pascal	Pa	N/m^2

Table 1.5: Examples of Derived SI Units (unnamed)

Quantity	Unit
Velocity	m/s
Mass Density	kg/m^3
Area	m^2
Volume	m^3
Molar Energy	J/mol
Electric Charge Density	C/m^3

Supplementary units are a dimensionless sub-class of derived units which SI allows in expressions with derived units. There are just two of these: the radian with symbol, rd, and its three dimensional analog, the steradian with symbol, sr. The steradian will be defined in chapter 3.

The class of temporary units consists of seven units sanctioned for use with SI units, but only until such a time when their fates, acceptance or rejection, are finally decided by CGPM. All but two of these, the kilowatt-hour (kWh) and the bar (bar), are of nuclear origin. The remaining five are radiation related quantities and will be defined in the text as needed. In every case, there is a named, derived SI unit available to replace the temporary unit should rejection occur. With deep historical roots in nuclear technology, the nuclear temporary units continue to be intensely popular.

Another familiar unit to the atomic and nuclear spectroscopist is the angstrom having the symbol Å and being equivalent to 10^{-10} m. Although not an SI unit, the Å is quite popular and often appears in the literature.

Some do's and don'ts listed below (table 1.6) are a guide to good SI style and usage. When SI offers a choice, the preferred grammar is marked with an "$*$" and is the form used in this text.

1.3.2 Physical Constants

The physical constants give scale to the material world in accord with our chosen system of units. In a way, these can be thought of as little more than auxiliary derived units of our system. This view is not formally adopted in SI and is only useful as an analogy to the formal definitions presented earlier. Consider, for example, the mass of one electron, $m_e = 9.1094 \times 10^{-31}$ kg. It would be ridiculous for us to refer to this in this way on every occasion. Though this is the accepted value, it is much more convenient to say $1m_e$. Since it is very well defined, there will be no ambiguity in what is meant.

There is another significance to the physical constants; they are not alterable. They are indeed constants of the physical world. Every electron mass ever measured is $1m_e$ (to within experimental uncertainty) regardless of physical location, climatic conditions, or any other variable. The theory of special relativity (sec. 1.5) tells us that the mass of any body in motion relative to an observer is a function of the relative velocity between the moving body and the observer. In our discussion here, we refer to mass as that in a frame at rest with respect to the moving body, the rest mass. Some of the more important physical quantities in the SI system that are encountered in nuclear science are listed in table 1.7. A more extensive list can be found in [Cohen87].

Table 1.6: SI Grammar

Grammar	Guide
Capitalization	meter, not Meter hertz, not Hertz J, not j
Space	8 m, not 8m
Plural	7 J, not 7 Js or 7 J^s
Raised Dot	A s* or A·s
Solidis	m/s* or $m{\cdot}s^{-1}$
Exponent	$1\ mm^2/s = (10^{-3}m)^2/s = 10^{-6}\ m^2/s$
Period	... 7 W of power..., not ... 7 W. of power...
Mixture of Units	kilograms per cubic meter or kg/m^3, not kilograms/m^3, or kg per cubic meter
Prefix	megagram, not kilokilogram 1 pm, not 1 $\mu\mu$m km/s, not m/ms
Double Vowels	megohm, not megaohm kilohm, not kiloohm
Hyphen	newton meter* or newton-meter microgram, not micro-gram
Decimals	0.56 kg, not .56 kg 18.651 km, not 18651 m 654 321.123 456†, not 654321.123456 1234, not 1 234 0.1234, not 0.123 4

†Though this is strict SI grammar, contemporary usage relaxes this rule so that strings of more than five numbers are usually written without spaces, especially after the decimal.

Table 1.7: Physical Constants*

Quantity	Symbol	Value	Uncertainty†
Atomic Mass Unit (^{12}C scale)	amu $(= \frac{1}{12} m_{^{12}C})$	$1.6605402 \times 10^{-27}$ kg	0.59
Avogadro Constant	N_A	6.0221367×10^{23} mol^{-1}	0.59
Bohr Radius	a_o	$5.29177249 \times 10^{-11}$ m	0.045
Boltzmann Constant	$k = R/N_A$	1.380658×10^{-23} J/K	8.5
Electron Charge	q_e	$-1e$	
Electron Compton Wavelength	$\lambda_e = \frac{h}{m_e c}$	2.42631058 pm	0.089
Electron Rest Mass	m_e	$9.1093897 \times 10^{-31}$ kg	0.59
Elementary Charge	e	$1.60217733 \times 10^{-19}$ C	0.3
Gravitational Accel. (Earth Surface)	g	9.8 m/s^2	
Gravitational Constant	G	6.67259×10^{-11} m^3/(s^2 kg)	128
Molar Gas Constant	R	8.314510 J/(mol K)	8.4
Molar Volume (at STP‡)	$V_M = RT/P$	22.414×10^{-3} m^3/mol	8.4
Neutron Rest Mass	m_n	$1.6749286 \times 10^{-27}$ kg	0.59
Permeability of Free Space	μ_o	$4\pi \times 10^{-7}$ H/m (V s/A m)	
Permittivity of Free Space	$\epsilon_o = (\mu_o c^2)^{-1}$	8.8542×10^{-12} F/m (A s/V m)	
Planck Constant	h	$6.6260755 \times 10^{-34}$ J s	0.6
Proton Charge	q_p	$1e$	
Proton Rest Mass	m_p	$1.6726231 \times 10^{-27}$ kg	0.59
Rydberg Constant	R_∞	1.0973731534×10^7 m^{-1}	0.0012
Speed of Light in Vacuum	c	2.99792458×10^8 m/s	0.003
Stephan-Boltzmann Constant	σ	5.67051×10^{-8} W/(m^2 K^4)	34

*This list is a small subset of that found in [Cohen87]

†In parts per million (ppm).

‡Standard Temperature (273.15 K) and Pressure (1 atm = 101,325 N/m^2).

1.3.3 Mathematical Symbols

Let's take advantage of some shorthand notation. In many of the equations that follow in this and subsequent chapters, we'll have the need to, time-and-again, repeat certain phrases (mostly of equivalence) in derivations and explanations. We can greatly reduce the quantity of verbiage in our discussions by adopting these few simple mathematical symbols listed below.

$=$	equal to
$\neq$	not equal to
$\equiv$	defined to be
$\simeq$	approximately equal to
$\sim$	goes as, scales like
$\ni$	such that
$\Rightarrow$	implies
$\therefore$	therefore
$\lesssim$	less than or approximately equal
$\gtrsim$	greater than or approximately equal
$\ll$	much less than
$\gg$	much greater than

1.3.4 The Electron Volt Energy Unit

The SI derived energy units prove not to be very convenient when dealing with quantities of energy involving atomic and nuclear processes. Atomic and nuclear scales are quite small in energy, as well as in physical size. As an example, the formation of ordinary water, $H_2 + \frac{1}{2}O_2 \rightarrow H_2O$, releases 3.955×10^{-19} J, an extremely tiny quantity of energy. Chemical processes, of course, release respectable amounts of energy at the macroscopic level by virtue of the enormous number of molecules involved. The formation of a mere gram of water, containing some 3.3×10^{22} molecules, releases 13 kJ.

Describing individual reactions in our nuclear world, however, we'll need a more convenient scale. The most commonly used unit of energy in nuclear science is the "electron volt" (eV). Consider a particle of mass $1m_e$ and electrical charge $-e$ (*i.e.* an electron) held at rest in an electric field (perhaps created by conducting plates called electrodes connected to a battery). If we imagine holding our single particle in place at the cathode (negative electrode) until we establish the electric field, and then release it, our particle will gain energy of motion (kinetic energy) in the electric field as it is accelerated to the anode (positive electrode). Our particle, of course, gains energy at the expense of the electrical potential energy of the system. Work must be done by an external agent (battery) to restore the system to its initial condition. The electron volt is defined to be the energy equivalent of the energy of motion attained by a particle possessing a single quantum, e, of electric charge accelerated through a potential difference of 1 V in an electric field. Therefore, our electron in the example above gains KE (kinetic energy) equal to 1 eV in this process.

Mathematically, we can describe this process as follows. From the work-energy theorem, we know that change in kinetic energy is equivalent to the amount of work done on the particle by the external agent, the battery in this case,

$$\Delta \text{KE} = \int_o^s \mathbf{F} \cdot d\mathbf{s} \tag{1.1}$$

where KE is the particle kinetic energy, $\int_o^s \mathbf{F} \cdot d\mathbf{s}$ is the quantity of work performed by the external agent, $\mathbf{F}$ is the force experienced by the particle from the external agent, and $\mathbf{s}$ is the particle's vector trajectory as it progresses from point 0 to its final position s. The electric force is $\mathbf{F} = q\mathbf{E} = -e\mathbf{E}$ on an electron in an electric field, so that $\int_o^s \mathbf{F} \cdot d\mathbf{s} = q \int_o^s \mathbf{E} \cdot d\mathbf{s}$. Recalling the definition of electric potential difference, $\Delta V = -\int_o^s \mathbf{E} \cdot d\mathbf{s}$, so that in general

$$\boxed{\Delta \text{KE} = -q\Delta V} \tag{1.2}$$

For a $q = -1e$ particle initially at rest (initial velocity, $v_i = 0$) and ΔV of 1 V, we find

$$\text{KE} = \frac{1}{2} m v_f^2 = 1 \text{ eV}$$

The electron volt is not an SI unit. We can, however, relate it to the J derived unit. Since the magnitude of q_e is just $1e \simeq 1.6 \times 10^{-19}$ C and employing the definition of a volt as 1 J/C from table 1.4, we have

$$1 \text{ eV} \equiv 1.6 \times 10^{-19} \text{ C} \times 1 \text{ V} = 1.6 \times 10^{-19} \text{ J}$$

Although this is a minuscule quantity of energy by ordinary experience, being about equal to the amount of work required to lift one red ant 25 fm against gravity at the earth's surface, it is indeed a respectable quantity of kinetic energy for an electron. One eV kinetic energy corresponds to an electron speed of $v_e = \sqrt{2\text{KE}/m_e} \simeq 5.93 \times 10^5$ m/s. The keV, MeV, and GeV naturally follow and are common scales of atomic and nuclear energies.

The eV unit of energy is defined in terms of particles possessing electric charge. It is completely independent of particle mass, so that a very massive particle with the same net charge as an electron (say a +1 uranium ion which is some 460 thousand times more massive) is accelerated to the same 1 eV in a 1 V potential difference. It is also common to expresses energies in these units whether the particles involved are charged or electrically neutral, like the neutron or neutral atoms. We recognize that the acceleration of neutral particles, however, requires some means other than the electrostatic force.

Energy units are sometimes used to express the temperature of a fluid. This description naturally involves (and is only appropriate for) systems including a great number of particles since the concept of temperature has meaning only for a statistically large sample possessing a Maxwell-Boltzmann energy distribution. The energy/temperature connection is a natural one since we realize that a fluid possesses internal energy by virtue of random (thermal) motion of individual particles. By redefining the Boltzmann constant, we can rescale the energy/temperature connection to atomic dimensions as $kT = T(\mathrm{K}) \times 1.38 \times 10^{-23}\ \mathrm{J/K} \times 1\ \mathrm{eV}/1.6 \times 10^{-19}\ \mathrm{J} \simeq T(\mathrm{K}) \times 1\ \mathrm{eV}/11,600\ \mathrm{K}$. Therefore, a fluid at 11,600 K possesses the equivalent thermal energy per particle of 1 eV.

Example 1: Kelvin/eV Conversion
What is the eV equivalent of room temperature?

Solution:
Assume room temperature to be $\sim 20^o$ C = 293.15 K,
then $293.15\ \mathrm{K} \left(\frac{1\ \mathrm{eV}}{11{,}600\ \mathrm{K}}\right) = 0.0253\ \mathrm{eV}$

1.3.5 The Atomic Mass Unit

As we saw earlier, it is clumsy to continually refer to atomic and subatomic particle masses in multiples of the g. A more appropriate unit of mass on the atomic scale called the *atomic mass unit*, abbreviated amu and defined to be exactly $\frac{1}{12}^{\mathrm{th}}$ the mass of a neutral ^{12}C atom, has been adopted by international agreement. Therefore, $m_{^{12}\mathrm{C}} \equiv 12$ amu. Since the gram atomic weight of ^{12}C is 12 g, then $m_{^{12}\mathrm{C}} = 12/N_A = 1.99269 \times 10^{-26}$ kg. The kg equivalent of the amu is then

$$1\ \mathrm{amu} = 1.6605402 \times 10^{-27}\ \mathrm{kg} \tag{1.3}$$

The amu will be employed as the mass unit of choice in this text. A complete list of atomic masses in amu is given in Appendix B. In atomic mass tables, one will find all masses are of ground-state neutral atoms with the exception of the three most important sub-atomic particles: electron, proton, and neutron.

$$\begin{aligned} m_{\mathrm{e}} &= 9.1093897 \times 10^{-31}\ \mathrm{kg} = 5.485799 \times 10^{-4}\ \mathrm{amu} \\ m_{\mathrm{p}} &= 1.6726231 \times 10^{-27}\ \mathrm{kg} = 1.007276\ \mathrm{amu} \\ m_{\mathrm{n}} &= 1.6749286 \times 10^{-27}\ \mathrm{kg} = 1.008665\ \mathrm{amu} \end{aligned}$$

Atomic masses with a few exceptions are experimentally determined. Mass spectrometry is the primary method with a typical experimental uncertainty (one standard deviation) of 1–10 parts in 10^7, depending on the atom in question.

1.4 Fundamental Particles

All matter is comprised of fundamental particles. The first level in this microscopic world, as we have already discussed, is the atom. Of course, we now realize that the atom is not the smallest, indivisible physical entity as the ancient greeks first thought. Making up the atom in its first level of sub-division are some of the particles of greatest interest to applied nuclear science and engineering, the sub-atomic particles: proton, neutron, and electron. The structure of the atomic nucleus will be discussed in a later section (sec. 1.6). Although we expect little practical importance of still smaller particles, one should recognize their existence. Smaller than the sub-atomic particles are tiny particles called quarks, combinations of which make up our world of sub-atomic particles. Many scientists now argue that there are still smaller particles. Some even suggest that there may be no limit to this sequence. As we continue to probe further, we may indefinitely encounter smaller particles that combine to make up all entities at the next larger level.

Our interest here resides in the first sub-atomic level: protons, neutrons, and electrons. These are the entities of primary concern because they are directly responsible for the atomic and nuclear structure, and because they and their combinations comprise most of the nuclear radiation that is important in our applications. We have briefly mentioned these particles already in our discussion of units. Their physical properties can be found in table 1.7. Electrons and protons make up the electrically charged of the fundamental particles, possessing electrical quanta of $-1e$ and $1e$, respectively. Though the magnitude of their charge is identical, the proton is more massive than the electron by the ratio, $m_p/m_e \simeq 1836$.

Electrons come in two varieties, the more familiar negatively charged electron or negatron, and a positively charged anti-particle called a "positron" (*i.e.* $q_{e^+} = 1e$). Both the electron and positron can be products of nuclear decay and both have identical mass. The only difference being the sign of their charge. When anti-particles come into proximity, their interaction leads to annihilation, the complete destruction of both particles, leaving behind only energy in the form of electromagnetic radiation (sec. 1.4.1) called annihilation radiation (sec. 6.2.3). Protons too have anti-particle cousins ingeniously called "anti-protons" with the same mass as a proton but with opposite charge. Just as a proton and an electron can combine to form a hydrogen atom (sec. 1.6), an anti-proton and a positron can combine to form the anti-matter atom, "anti-hydrogen". An electron and positron, before annihilation, form the short-lived pseudo-atom "positronium".

The neutron is an extremely important particle in nuclear applications. It possesses about the same mass as a proton, but is electrically neutral. Along with the proton, it comprises the atomic nucleus. The neutron is responsible for many of the important and interesting features that one associates with the nuclear discipline, like induced radioactivity and the fission chain reaction. Both of these subjects will be taken up in great detail later in this book. The neutron, when it is not part of an atomic nucleus (*i.e.* when it is isolated as a "free neutron"), is an unstable particle. It undergoes radioactive decay (chpt. 4) into a proton and an electron with a mean life of 15.0 ± 0.3 min.

Massless particles round out our description of fundamental particles. Of interest to us are the photon and the neutrino. Both are expected to possess neither electrical charge nor mass. Since any physical effect of neutrino mass is extremely difficult to measure, mass is still an open issue for this particle and one of great importance to the future of the universe. Although the neutrino is probably very light, if it has mass at all (probably less than about 9×10^{-36} kg), there are so many of them that they could likely provide enough gravitational attraction in the universe to one day stop expansion and cause gravitational collapse. This entire process would take several billions of years if at all, so there is little for us to be concerned with in our lifetimes. Neutrinos are important to us in our studies here because they appear in the decay of certain radioactive atoms. There are six varieties of neutrinos in the world but only two are of relevance here, the electron neutrino and anti-neutrino, and they both may be treated identically for our purposes. The photon may also be the result of radioactive decay. Both particles travel at the speed of light, c.

1.4.1 Electromagnetic Radiation

In our future endeavors, we'll have many occasions to discuss various forms of electromagnetic radiation. Certain forms of this radiation are already very familiar to us. We recall that ordinary visible light is one such form. Radio waves and microwaves are others. More relevant to nuclear processes are higher energy forms called X-rays and γ-rays (gamma). These are all examples of the same phenomenon at different energies, namely electromagnetic rays propagating through free space.

In the wave theory of light, we attribute certain properties to these rays in direct analog to propagating fluid waves, the exception being that light waves do not require a material media to propagate in as do fluid waves. Light waves can travel in vacuum at the speed of light c, the most important of Maxwell's discoveries. As with other familiar sources of wave phenomena, electromagnetic waves possess identifiable features in this model, like wavelength, λ (m), and characteristic frequency of oscillation, ν (Hz). These are simply related through speed of propagation

$$c = \lambda\nu$$

the light speed in this case. (This simple relation holds strictly for light rays propagating in vacuum only, an adequate description for all of our purposes.) Recall that the linear frequency, ν, is related to the angular frequency, $\omega = 2\pi\nu$.

This simple wave theory of light successfully describes many features (*i.e.* diffraction, interference, scattering, ...) of this radiation very well, with one annoying problem, however. Light rays are just not this simple in all respects. They do not travel in an infinite wave train at constant amplitude as the wave theory tells us. In the early twentieth century, this picture was complemented by the "corpuscular theory". German physicist Max Planck suggested that light rays could be emitted as distinct, finite quanta of electromagnetic energy and not be incompatible with the wave theory. Hence, light rays can be quantized into individual packets traveling through space much like particles.

In his famous 1905 paper, Einstein cemented the corpuscular theory of light rays that he called "photons" by providing a beautiful explanation of the newly observed photoelectric effect (sec. 6.2.1). Experimentalists observed that when an intense light illuminated certain metal plates, electrons were emitted. The number of these "photoelectrons" increased as the light intensity increased. What's more, the emitted electron energy increased as the incident light's wavelength was decreased (frequency increased). If the light frequency is made too low, no photoelectrons are observed. These experimental phenomena could only be explained if light were quantized. There is no acceptable explanation in the wave theory. The photoelectric effect is the direct result of a single photon interacting with an electron in the metal plate. An increase in light intensity implies increasing numbers of photons incident on the plate, thereby increasing the number of electrons emitted. By increasing the photon frequency, the photon energy is increased as

$$E = h\nu \tag{1.4}$$

where h is Planck's constant (table 1.7). A higher energy photon transfers more energy to a single electron, thus producing higher energy photoelectrons in the interaction. Light of frequency below the cutoff for photoelectric emission is of insufficient energy to produce photoelectrons. Therefore, the corpuscular (particle) theory of light is successful. Light rays behave as if they possess both wave and particle qualities, and wave-particle duality is born.

Just as matter possesses linear momentum, so do photons. A photon's linear momentum, p, is related to its energy through

$$p = \frac{E}{c} = \frac{h\nu}{c} \tag{1.5}$$

This will be essential later in describing interactions involving photons. The quantized energy relation (Eq.(1.4)) also allows us to express a photon's wavelength directly as a function of energy.

$$\lambda = \frac{hc}{E} \simeq \frac{1.24 \times 10^{-6}\ \mathrm{m}}{E(\mathrm{eV})} \tag{1.6}$$

1.4.2 Matter Waves

To complete the wave-particle duality picture, matter must possess wave-like, as well as corpuscular qualities. This was the ingenious proposition of French physicist Louis de Broglie who, in 1924, postulated the existence of matter waves. The hypothesis of de Broglie was that the wave-particle nature of photon radiation should apply equally well to matter. In essence, matter and energy are synonymous, a view greatly strengthened by Einstein's postulates in special relativity (sec. 1.5). de Broglie proposed that for both matter and radiation, the total energy of the entity can be expressed by Eq.(1.4). The particles of matter must then possess wave characteristics so that a particle wavelength can be defined,

$$\lambda_{\mathrm{deB}} = \frac{h}{p} \tag{1.7}$$

the de Broglie wavelength. This was later confirmed experimentally by Davisson and Germer who discovered diffraction patterns for electrons, a definite wave characteristic. We then need both wave and particle models to independently describe either matter or massless particles like electromagnetic radiation.

Example 2: de Broglie Wavelength

Let's explore the importance of de Broglie waves on sub-atomic and macroscopic scales.

1. What is the de Broglie wavelength of an electron with 1 keV kinetic energy?

$$v = \sqrt{2\text{KE}/m_e} \quad \text{so that} \quad p = mv = \sqrt{2m_e\text{KE}}$$

then

$$\lambda_{\text{deB}} = \frac{h}{p} = \frac{6.626 \times 10^{-34}\ \text{J s}}{\sqrt{2(9.1 \times 10^{-31}\ \text{kg})\ 1000\ \text{eV}\ (1.6 \times 10^{-19}\ \text{J/eV})}}$$
$$\simeq 3.9 \times 10^{-11}\ \text{m} = 0.39\ \text{Å}$$

This wavelength is on the order of atomic dimensions, so de Broglie waves are an important part of atomic physics. (Recall that the Bohr radius of the hydrogen atom (table 1.7) is 0.53 Å.)

2. What is the de Broglie wavelength of a 50 g golf ball traveling at 50 m/s?

$$\lambda_{\text{deB}} = \frac{h}{mv} = \frac{6.626 \times 10^{-34}\ \text{J s}}{0.05\ \text{kg}\ \times 50\ \text{m/s}} \simeq 2.65 \times 10^{-24}\ \text{Å}$$

This is an extremely tiny wavelength, completely ignorable and undetectable even on nuclear scales and certainly in the macroscopic world.

1.5 Special Relativity and Mass-Energy Equivalence

In this section, we'll briefly review some of the postulates and consequences of the special theory of relativity. Those concepts (mass-energy equivalence, relativistic momentum and mass, ...) that will be utilized in our later discussions on nuclear reactions and interactions will receive the bulk of the attention. This is not intended to be a complete review on the special theory of relativity. An excellent introduction to the subject and many more facets of special relativity can be found in [Beiser81].

The special theory of relativity developed by Einstein in the early twentieth century has arguably had a more profound impact on science than any other single scientific theory, perhaps competing with only the quantum mechanics and general relativity in importance. Before the development of special relativity, it was unthinkable in all of science that any entity, including light rays, would travel with the same velocity, regardless of the motion of the observer, or that mass and energy were synonymous so that one might be converted into the other. All of these and many more amazing consequences stem from Einstein's two postulates of special relativity:

1. *The laws of physics may be expressed in equations having the same form in all frames of reference moving at constant velocity with respect to one another.*

2. *The speed of light in free space has the same value for all observers, regardless of their state of motion.*

These seem harmless enough, yet are quite revolutionary in that they lead to many profound and unusual predictions about the physical world. For instance, Einstein showed that our usual expression of Newtonian mechanics is incomplete for bodies moving at speeds approaching c. Though the differences are completely ignorable in the terrestrial world, they have dramatic influence in our sub-atomic one.

To gain an understanding of the changes required by special relativity, suppose that it were possible to provide space travel at enormous speeds, near the speed of light. Consider then a space traveler moving at a high rate of speed, v, with respect to a distant observer. Our space traveler carries a light clock that keeps time by recording light pulses traveling a fixed distance from source to recorder (Fig. 1.1). In the frame of reference at rest with the space traveler (shaded box), light pulses within the clock travel a distance ct_o before being recorded. The time t_o is called a "proper time" since events are witnessed without relative motion in this frame. An identical clock (not shown) at rest with the distant observer would behave the same in the rest frame of the observer. The distant observer, however, observes that light pulses emitted from the source in the traveler's clock are required to travel a longer path, ct, to reach the light recorder. The observer then measures this same event (light passing from source to recorder) to have required an apparent time interval t for which the space traveler has appeared to cover the distance vt. These distances are related by

$$c^2t^2 = v^2t^2 + c^2t_o^2$$

which reduces to the familiar time dilation expression

$$t = \frac{t_o}{\sqrt{1 - v^2/c^2}} \tag{1.8}$$

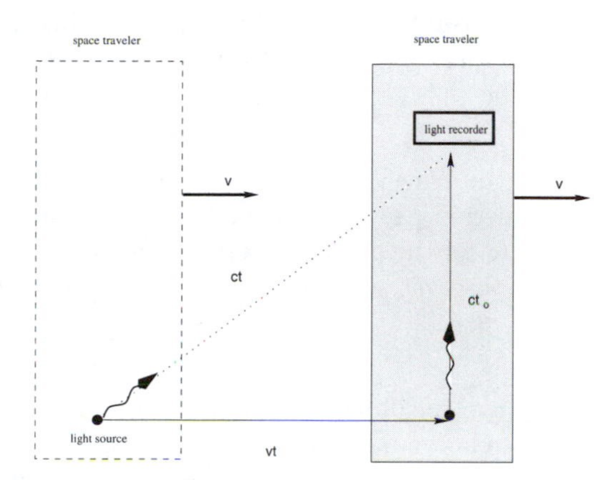

Figure 1.1: Time dilation of a relativistic traveler as viewed by a distant observer. The proper time, t_o, is measured with a clock in the traveler's frame (shaded box) by recording light pulses covering a distance ct_o. The same event is viewed in the frame of the distant observer to require the dilated time t.

The dilated time t is always greater than the proper time t_o. Hence, events which require a time interval, t_o, in the rest frame of the traveler, appear to require the longer time, t, to the distant observer. Since all inertial motion is relative, time dilation is reciprocal. The traveler sees identical time dilation of events occurring for the distant observer.

The low speed limit to this expression conforms to ordinary experience. When $v \ll c$, events in all frames are observed to require identical intervals, $t \sim t_o$. In this limit, light provides an instantaneous record of relative events. In the opposite extreme, when v approaches c, dilated time extends indefinitely. Light pulses can never catch up to a recorder moving at c. Of course, this scenario can never occur. As we'll see shortly, any material object is limited to $v < c$.

A direct consequence of time dilation is length contraction. Measuring distances in the direction of relative motion is subject to recording information carried at the finite speed of light. Consider making a measurement of the physical length of some object (in the direction of motion) on our same speeding space vehicle. The space traveler is at rest with the object in his frame. He may then measure the "proper length", L_o, with the aid of some measuring tool (*i.e.* a ruler). The distant observer, however, does not have this luxury. Instead the observer chooses a fixed (in his frame) reference point or marker in space. When the front edge of the object passes the marker the observer starts his clock. He stops his clock when the trailing edge of the object passes the marker a time t_o later. This is a proper time since both measurements have occurred at the same position in space in the observer's frame. The length of the object then appears to the distant observer to be

$$L = vt_o$$

Relative to the traveler, the same marker is moving with speed v in the opposite direction. As the traveler passes the marker, he measures a dilated time $t = L_o/v$ for the object to pass the marker. Reconciling this with what the distant observer sees, we find that the observer records a contracted length

$$L = L_o\sqrt{1 - v^2/c^2}$$

for the length of the object. The apparent length of the object is contracted, $L \ll L_o$, by relative motion.

The measure of an object's mass is also relative. Now, consider that our space traveler is flying in a trajectory that takes him past the observer at elevation $2h$ (Fig. 1.2). As he is approaching the location of the observer, our relativistic traveler tosses an object vertically downward in his frame. The object takes the path $\mathbf{v}_2$ in the frame of the observer. At the same time, the observer tosses an identical object vertically upward in his frame on trajectory $\mathbf{v}_1$. These two objects are identical in every way when measured at rest. In addition, they are released with the same speed when measured in each party's rest frame

$$v_1 = v_2{}'$$

where the prime notation indicates quantities measured in the travelers frame.

At altitude h, the objects collide elastically (kinetic energy is conserved), by which they rebound and return to their respective releaser with speeds identical to that with which they were tossed. Each party then witnesses the same event from different inertial reference frames. By Einstein's first postulate, momentum must be conserved in each frame. In the observer's frame, momentum conservation requires

$$m_1 v_1 = m_2 v_2$$

The speed v_1 is measured in the observer's frame by recording the proper time for object 1 to reach the collision point at h and return to the observer

$$v_1 = 2h/t_o$$

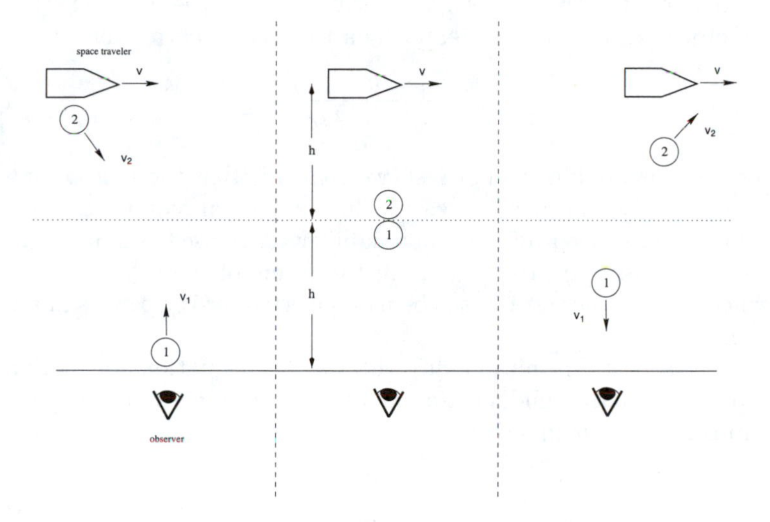

Figure 1.2: Elastic collision between objects 1 and 2, identical in their respective rest frames. In the three consecutive snapshots above, the collision is viewed from the frame of the observer, before, during, and after the collision. In his frame, the observer sees a dilated travel time for object 2. Hence, its mass appears greater in order to satisfy momentum conservation.

The traveler likewise measures $v_2{}' = 2h/t_o$ in his frame. Yet in the observer's frame, the flight of object 2 requires the dilated time $t > t_o$

$$v_2 = 2h/t = \frac{2h}{t_o}\sqrt{1 - v^2/c^2}$$

so that

$$m_1 = m_2\sqrt{1 - v^2/c^2}$$

When the vertical speeds v_1 and v_2 are small compared to v, we immediately recognize that m_1 is the mass of object 1 at rest in the observer's frame. We call this m_o, the "rest mass". It appears then, to the observer, that the mass of object 2, $m = m_2$, is greater by a relativistic correction

$$m = \frac{m_o}{\sqrt{1 - v^2/c^2}} \tag{1.9}$$

As these objects are identical at rest, we conclude that the relative speed v results in the apparent mass increase. This expression requires a retraction of the long held concept of mass invariability. A revised statement should provide that mass is invariable only in the frame of the moving body and only when none of the rest mass is being converted into energy as in nuclear reactions.

These relativistic results modify the way our equations of motion behave. Recall that the fundamental equation describing motion is the momentum conservation equation

$$\mathbf{F} = \frac{d\mathbf{p}}{dt} \tag{1.10}$$

where $\mathbf{p}$ is linear momentum ($= m\mathbf{v}$). In classical mechanics, this reduces to the familiar form of Newton's second law $\mathbf{F} = m\, d\mathbf{v}/dt = m\mathbf{a}$. However, the correct relativistic expression needs to include changes in mass with velocity so that

$$\mathbf{F} = \frac{d}{dt}(m\mathbf{v}) = m\frac{d\mathbf{v}}{dt} + \mathbf{v}\frac{dm}{dt}$$

and only reduces to the classical limit when $v \ll c$, so that $v(dm/dt) = v(dm/dv)(dv/dt)$ is small. We can use this result along with the work-energy theorem to assess relativistic particle energy. From Eq.(1.1) with motion along path $\mathbf{s}$, we have

$$\Delta\text{KE} = \int_o^s \mathbf{F} \cdot d\mathbf{s} = \int_o^s \frac{d\mathbf{p}}{dt} \cdot d\mathbf{s} = \int_o^{mv} \mathbf{v} \cdot d\mathbf{p}$$

Our chosen lower limit implies that the particle is beginning from rest so that $\Delta\text{KE} = \text{KE}$. Employing the relativistic mass expression (Eq.(1.9)) and a straightforward integration by parts, we arrive at Einstein's famous equation

$$\text{KE} = c^2(m - m_o) \tag{1.11}$$

which we rearrange to write

$$mc^2 = m_o c^2 + \text{KE}$$

We interpret this as the total energy of the isolated object, $E = mc^2$, made up of two contributions, the energy of motion (KE) and the residual energy when all motion ceases. This latter part is referred to as "rest energy," $E_o = m_o c^2$. This energy resides in the form of the rest mass of the object; *i.e.* contained in quantity of mass, m_o, there is the equivalent of $E_o = m_o c^2$ amount of energy. Since c^2 is a very large number,

$$c^2 = \left(2.997924 \times 10^8 \frac{\text{m}}{\text{s}}\right)^2 \left(\frac{\frac{1\ \text{eV}}{1.60218\times 10^{-19}\ \text{J}}}{\frac{1\ \text{amu}}{1.66054\times 10^{-27}\ \text{kg}}}\right) \approx 931.5\ \frac{\text{MeV}}{\text{amu}}$$

then E_o represents a tremendous amount of energy (on our nuclear, eV, scale). Most of this energy in our terrestrial world resides in atomic nuclei and can be released (at least in part) in nuclear reactions to do useful work.

Although the above formalism has been developed assuming a finite particle rest mass, it may be equally well applied to massless particles like photons. This is consistent with our concepts of wave-particle duality. Massless particles are not allowed in classical mechanics since objects must have rest mass to possess energy or momentum. Not so in relativistic mechanics. Consider the relativistic energy, $E = mc^2 = m_o c^2/(1 - v^2/c^2)^{\frac{1}{2}}$, and momentum, $p = mv = m_o v/(1 - v^2/c^2)^{\frac{1}{2}}$, expressions. When $v < c$, both E and p are identically zero for massless ($m_o = 0$) particles. But, when $v = c$, E and p need not be zero for $m_o = 0$, they can take on any value, *i.e.* massless particles are completely consistent with relativistic mechanics so long as they travel at the speed of light. Particles that possess rest mass ($m_o \neq 0$) must travel at $v < c$, always.

A generalized relativistic momentum-energy relationship can be developed. By squaring the expressions for E and p above, we can easily show that

$$E^2 - p^2 c^2 = m_o^2 c^4$$

or

$$E = \sqrt{p^2 c^2 + m_o^2 c^4} \tag{1.12}$$

For a massless particle, then, we must insist $E = pc$.

Example 3: Mass/Energy Equivalence
What is the energy contained in a particle with rest mass 1 amu? 1 electron? 10 amu?

Solution:

$$\begin{aligned}
E_o(1\text{ amu}) &= m_0c^2 = 1.66 \times 10^{-27}\text{ kg}(3 \times 10^8\text{ m/s})^2 = 931.5\text{ MeV} \\
E_o(1e) &= \frac{m_e}{m_{\text{amu}}} E_o(1\text{ amu}) = 0.511\text{ MeV} \\
E_o(10\text{ amu}) &= 10E_o(1\text{amu}) = 9315\text{ MeV}
\end{aligned}$$

Discussion:

1. About 1 GeV of energy resides in each proton or neutron.
2. This energy is liberated when an $m_o > 0$ particle and its respective antiparticle are mutually annihilated. An example is an electron/positron annihilation: $m_{e^-} + m_{e^+} \rightarrow 2\gamma$'s (0.511 MeV each).

Example 4: Classical Limit in KE
Show that we can obtain the classical expression for kinetic energy from the relativistic expression, Eq.(1.11), when $v \ll c$.

Solution:

$$
\begin{aligned}
\text{KE} &= mc^2 - m_oc^2 = \frac{m_oc^2}{\sqrt{1 - v^2/c^2}} - m_oc^2 \\
&= m_oc^2 \left[(1 - v^2/c^2)^{-1/2} - 1\right] \\
&\simeq m_oc^2 \left[1 + \frac{1}{2}\frac{v^2}{c^2} - O(v^2/c^2)^2 + \cdots - 1\right] \\
&\simeq \frac{1}{2}m_ov^2
\end{aligned}
$$

which is the classical limit, *i.e.* when $v \ll c$. We must consider a particle as being relativistic when its translational speed approaches that of light. We may also be interested in judging when a particle is considered relativistic based on its kinetic energy alone. From the above analysis, $\text{KE}/m_oc^2 \sim (v/c)^2$, when $v \ll c$. Therefore, an energetic particle need only be considered relativistic when its kinetic energy approaches its rest energy.

Mathematical Notes:

1. In this analysis, we have used the binomial expansion, $(1 \pm x)^n = 1 \pm nx + \frac{n(n-1)x^2}{2} \pm \frac{n(n-1)(n-2)x^3}{3} + \cdots$ for $(x^2 < 1)$.
2. The notation $O(x)^2$ means of order x^2. Since $x = v^2/c^2$ is a very small number, we can safely neglect terms in the expansion at this and all higher orders.

1.6 The Atomic Nucleus

We're all no doubt very familiar with what has become the accepted "working" model or picture of the atom. The atom consists of a very small but dense core called the nucleus possessing net positive electrical charge and surrounded by electrons which orbit the nucleus at much larger distances than nuclear dimensions. Electrons appear in exact numbers to balance the nuclear charge so that the atom remains electrically neutral. While simplistic, this elegant model results from the diligent, and in some cases lifelong, efforts of a number of prominent scientists too numerous to mention. The simplicity originates from only the bare essentials discussed here and ignores many important features only successfully described by sophisticated quantum mechanical analysis. It will do well for our intentions, however.

The atomic effects embedded in this model result from electron motion and are responsible for most of the chemical and mechanical properties (except mass of course) of macroscopic matter. The atomic nucleus, on the other hand, comprises nearly all the mass of the atom and is responsible for all nuclear effects like radioactive decay and nuclear energy. As well, it is the ability of atomic nuclei to exist in stable configurations with a variety of proton numbers that produce the multitude of chemical elements.

1.6.1 Basics and Terminology

Atomic nuclei consist only of an integer number of neutrons and protons densely packed at the core of the atom. Let us establish a convention that refers to an arbitrary atom with the symbol X. Since there is no chemical element with this symbol, we'll recognize this as a replacement for any such element in our general discussions. Similarly, we'll adopt the symbols N (called *neutron number*) and Z (called *proton number*) to represent the number of neutrons and protons, respectively, in the nucleus. Combined, protons and neutrons are known as *nucleons* as they are the only nuclear residents. It should be immediately recognized that the proton number, Z, is also the atomic number, which yields the quantity of orbiting electrons in an electrically neutral atom. An ionized atom is devoid of some or all of its atomic electrons and possesses net positive charge from $+1e$ to $+Ze$.

Symbolically then, we can describe an atom or nucleus as

$$^{A}_{Z}\mathrm{X}^{N}$$

This can refer to either an atom or nucleus because it says nothing about the extra nuclear electrons. Here, A is called the *atomic mass number* (or just *mass number*) and is defined to be the total number of nucleons in the atom, another integer.

$$A = N + Z$$

Usually, the neutron number, N, is omitted as being redundant such that ${}^A_Z$X completely represents an atom or nucleus of element X and having mass number A. The proton number, Z, is also redundant in the above symbolism, but is usually written since there is no simple scheme for recalling the proton number from the chemical symbol. Some examples of this symbolism are ${}^{14}_{6}$C (carbon-14) and ${}^{22}_{11}$Na (sodium-22).

In addition to the nomenclature developed above, some additional useful terminology is employed in the discussion of atomic nuclei and nuclear processes. The most important of these are listed below:

nuclide — An atom or nucleus identified by its proton and neutron numbers. A nuclide may be stable or unstable.

radionuclide — An unstable nuclide which spontaneously transforms into another nuclide. (These will be discussed extensively in chapter 4.)

isotopes — Nuclides with the same Z but with different N. (Note the "p" in isotope helps one remember that the proton number (Z) is the same.) One example of isotopes are those of the element hydrogen which has three:

$$\begin{aligned} {}^1_1\text{H} &\rightarrow \text{hydrogen (protium), H} \\ {}^2_1\text{H} &\rightarrow \text{deuterium, D} \\ {}^3_1\text{H} &\rightarrow \text{tritium, T} \end{aligned}$$

Some other isotopes frequently encountered in nuclear applications are: ${}^{235}_{92}$U and ${}^{238}_{92}$U, ${}^{12}_{6}$C and ${}^{14}_{6}$C. There are many, many such sets.

radioisotope — a radioactive isotope.

isotones — Nuclides with the same N but with differing Z. (The "n" in isotones should help one remember that the neutron number (N) is the same.) ${}^2_1$H and ${}^3_2$He are examples of isotones, as are ${}^3_1$H and ${}^4_2$He.

isobars — Nuclides with the same A, like ${}^3_1$H and ${}^3_2$He.

1.6.2 Nuclide Chart

The number of known nuclides is vast, totalling some 2930. Of these, only 266 are stable (non-radioactive), and only about 65 are naturally occurring radioisotopes. The remainder are man-made radioisotopes, most of which are produced in nuclear reactors.

As of this writing, the isotope containing the largest number of neutrons is ^{269}Hs with $N = 161$. The isotope with the largest proton number belongs to element 109, meitnerium (Mt). The former also claims the largest atomic mass number at 269.

To give order to this ensemble, a graphical presentation has been devised to systematically display all the known nuclei and some of their most important nuclear properties. This arrangement is called the "Chart of the Nuclides"[GE96]. As it is usually the first source of consultation on any nuclear related problem, the nuclide chart is an extremely valuable tool for nuclear scientists and engineers.

The chart is in essence a discrete two-dimensional plot of Z vs. N with a small square space reserved for each nuclide. Hence, isotopes and isotones appear on the chart as horizontal and vertical lines, respectively, while isobars are diagonal, top left to bottom right. Figure 1.3 shows a subset of the nuclide chart which includes only the stable isotopes. The dashed line displays $Z = N$ for reference. Within each square block (Fig. 1.4) resides a few of the most important pieces of data for each isotope. These data differ depending on whether the isotope is stable or unstable (radioactive). For all nuclides, the chemical element symbol and atomic mass number are listed on the first line. In the case of stable nuclei, the remainder of information includes the fractional abundance (in %), atomic mass (in amu), and any relevant absorption cross section data (sec. 3.3). The fractional abundance, usually expressed as a percentage and also known as percent abundance, f_n, is that number fraction of a particular isotope, n, relative to all stable isotopes of that element.

Data included in the chart for unstable nuclei are more involved. There is a further distinction of naturally occurring and man-made radionuclei, the former being represented in the chart by a black band at the top of the box. In the chart, information for radionuclei includes disintegration modes and energies, radiation energies, and half-lives (chpt. 4). Colors are used to indicate ranges of cross sections and half-lives for easy lookup. More detail on the information on the chart is provided on the chart and in the accompanying booklet[GE96].

1.6.3 Shape and Size of the Nucleus

To examine the shape and size of the atomic nucleus, experiments have been performed employing high energy electron and neutron beams. As a result of all such experiments, it has been found that nuclear radii are in proportion to $A^{1/3}$ and not a function of N or Z independently so that

$$R \sim R_o\ A^{1/3} \tag{1.13}$$

where $R_o \simeq 1.2 \pm 0.03 \times 10^{-15}$ m $\simeq 1.2$ fm. (In this context, the fm is sometimes referred to as a fermi, *i.e.* 1 fm = 1 fermi.) Further experiments involving sensitive spectroscopic measurements of atomic hyperfine structure reveal that all nuclei are, to first approximation, spherical in shape with

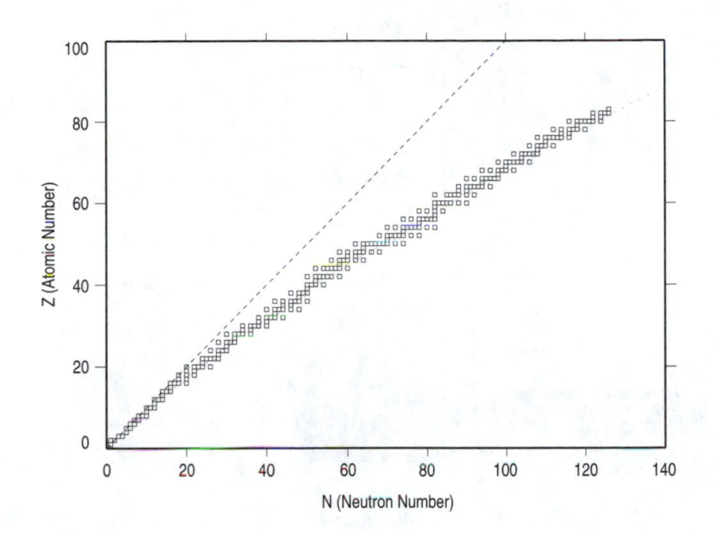

Figure 1.3: Location of stable nuclides in the nuclide chart. The dotted line is a plot of Eq.(1.24) while the dashed line displays $Z = N$ for comparison.

Example 5: Abundances

What are the stable isotopes of the element carbon? Given the percent abundances, how many atoms of each isotope are present in a natural sample of 10^{24} total atoms?

Solution:

From the Chart of the Nuclides, carbon has a total of 11 isotopes

$$^{8}_{6}\text{C},\ ^{9}_{6}\text{C},\ ^{10}_{6}\text{C},\ ^{11}_{6}\text{C},\ ^{12}_{6}\text{C},\ ^{13}_{6}\text{C},\ ^{14}_{6}\text{C},\ ^{15}_{6}\text{C},\ ^{16}_{6}\text{C},\ ^{17}_{6}\text{C},\ ^{18}_{6}\text{C}$$

However, only two of the isotopes are stable, $^{12}_{6}\text{C}$ and $^{13}_{6}\text{C}$. If a particular natural sample contains 10^{24} total atoms, and from the chart the fractional abundances are $f_{^{12}\text{C}} = 98.89\%$ and $f_{^{13}\text{C}} = 1 - f_{^{12}\text{C}} = 1.11\%$, respectively, then

$$\begin{aligned} n_{^{12}\text{C}} &= 0.9889 \times 10^{24} \text{ atoms} \\ n_{^{13}\text{C}} &= 0.0111 \times 10^{24} \text{ atoms} \end{aligned}$$

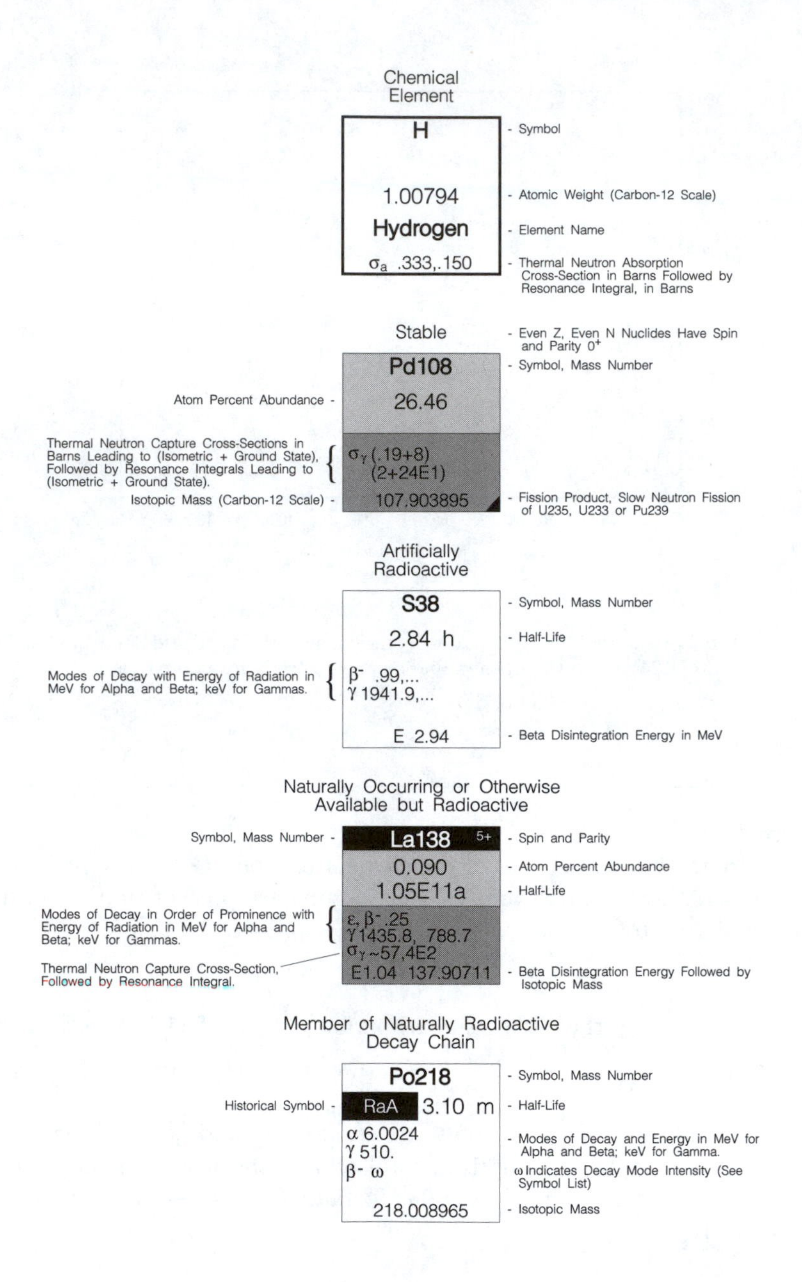

Figure 1.4: Some key features in the Chart of the Nuclides. The percent abundance, $f_n (= 100\% \times n_n / \sum_i n_i)$, refers only to that of the stable isotopes. Colors (shown here as dark shading) refer to range of half-lives (upper half of box) or neutron absorption properties (lower half of box). Consult your chart for more detail. (From *Nuclides and Isotopes*, Copyright©1996 by the General Electric Co. Reproduced by permission of the General Electric Co.)

a somewhat diffuse surface. In combination, these experimental results imply that the spherical nuclear volume ($\sim R^3$) should increase linearly with A, maintaining constant nuclear density. This suggests a close packing of nucleons so that increasing nuclear constituents increases nuclear mass and volume proportionately. This behavior bears close resemblance to incompressible liquids. And, indeed, a liquid drop model of the atomic nucleus, while crude, is successful in describing some nuclear features as we shall discuss (cf: sec. 1.6.6.1).

By contrast, the atomic electron cloud increases density with increasing electron numbers keeping the atomic size nearly constant at about 2–2.5 Å. The exceptions to this are the light atoms like hydrogen with a first Bohr orbital radius of $a_o \sim 0.53$ Å called the *Bohr radius*. The shape, however, may again be described as roughly spherical to first approximation, but with a much more "fuzzy" surface than the nucleus. This extreme diffuseness of the electron cloud surface is due to the electron's propensity to possess rather distorted, non-circular orbits that often take them quite far from the nucleus.

Because of the compactness of nuclear size, nuclear density is extremely high. Our spherical model argued above provides an estimate. Recall that since we've found $R \sim A^{1/3}$, the density is independent of A and we may choose any isotope for our calculation. As an example, let's select the nucleus of ^{1}H (1 proton) with a mass of m_p and a volume of $\frac{4}{3}\pi R^3$. Since $A = 1$, $R \simeq R_o$ and the nuclear density becomes, $\rho_n \simeq 2.3 \times 10^{17}$ kg/m^3 or about 230 megaton per cubic centimeter!

1.6.4 Mass Defect and Binding Energy

Consider the following interesting situation. The following particle and atomic masses have been measured very carefully:

Particle	Rest Mass (amu)	Rest Mass (MeV)
e^-	0.000549	0.511
p	1.007276	938.3
n	1.008665	939.6
$^1_1$H	1.007825	938.8
$^2_1$H = D	2.014102	1876.1

and we make the observations below:

1. $m_{\rm n} > m_{^1_1{\rm H}} > m_{\rm p}$.

2. $m_{\rm p} + m_{\rm e} = 1.007276 + 0.000549 = 1.007825 \simeq m_{^1_1{\rm H}}$ to very good accuracy. As we might expect, atomic hydrogen mass is almost com-

Example 6: Nuclear Density
Compare the nuclear radii and densities of $^1_1$H and $^{235}_{92}$U using the spherical nucleus model.

Solution:

- $^1_1$H has an atomic mass of $A = 1 \therefore R_{^1\mathrm{H}} = R_o A^{1/3} \simeq 1.2$ fm and $\rho_n \simeq 2.3 \times 10^{17}$ kg/m^3.
- $^{235}_{92}$U has an atomic mass of 235, so $R_{^{235}\mathrm{U}} = R_o A^{1/3} = (1.2\text{ fm})(235)^{1/3} \simeq 7.4$ fm and again a nuclear density of $\rho_n = \frac{m_{235}}{\frac{4}{3}\pi R^3_{^{235}\mathrm{H}}} \simeq 2.3 \times 10^{17}$ kg/m^3.

Discussion:
Nuclear density is very closely constant over a wide range of A. Let's look further at the comparative size of the atom and nucleus in our ^{1}H example. With a Bohr radius ~ 0.53 Å $= 5.3 \times 10^{-11}$ m, then

$$\frac{a_o}{R_{^1\mathrm{H}}} = \frac{5.3 \times 10^{-11}\text{ m}}{1.2 \times 10^{-15}\text{ m}} = 4.4 \times 10^4$$

Hence, the nucleus occupies only a very tiny fraction of the atomic volume. Most of matter is empty space.

pletely accounted for by the combined mass of one proton and one electron.

3. However, let's try the same for deuterium. If we add up the masses of the constituents of deuterium, we get

$$\begin{aligned} m_{\text{p}} + m_{\text{e}} + m_{\text{n}} &= 1.007276 + 0.000549 + 1.008665 \text{ amu} \\ &= 2.016490 \text{ amu} \\ &\neq m_{\text{D}} \end{aligned}$$

This is significantly greater than the mass of neutral atomic deuterium, certainly too large a discrepancy to be accounted for by experimental uncertainty. Since the combined mass of a proton, a neutron, and an electron does not equal the mass of a deuterium atom, something very significant is missing here.

In general, the sum of all constituent proton, neutron, and electron masses does not equal the atomic mass of the resultant atom:

$$Zm_{\text{p}} + Zm_{\text{e}} + Nm_{\text{n}} \neq m_{{}^A_Z\text{X}^N}$$

Therefore, rest mass is not conserved in the assembly process of the atom. We suspect that this must be true since, if it were not, there would be no difference between an assembled atom and the collection of sub-atomic particles. We know these entities are very different. Since there is a mass reduction or *mass defect* $(A - m_{\text{X}})$, we know through mass/energy equivalence that there is also an energy decrement associated with the assembly of sub-atomic particles into an atom. This energy of assembly is called the *binding energy*. More specifically, it is a nuclear binding energy, to distinguish it from the atomic binding energy, which binds electrons to the nucleus in the atom.

The nuclear binding energy is that energy that binds nucleons in the nucleus and would, therefore, be the energy required in order to break the nucleus of an atom into its individual nucleons. If some source of energy, external to the nucleus, were to provide this amount of energy and it were completely absorbed by the nucleus, the nucleus could dismantle into its sub-atomic constituents.

To examine the details of the assembly process, let's consider creating an arbitrary neutral atom ${}^A_Z\text{X}^N$ containing A nucleons and Z electrons at rest and initially far apart, such that,

$$Z\text{p} + Z\text{e}^- + N\text{n} \longrightarrow {}^{A'}_{Z'}\text{X}^{N'}$$

In doing so, we preserve the following quantities:

1. The number of protons is conserved.

2. The number of neutrons is conserved.

3. The number of electrons is conserved.

4. Mass/energy is conserved.

Preserving protons and neutrons independently implies that we are not allowing for the transformation of these species into others in the assembly process. From condition 1, we require that $Z' = Z$. With condition 1, condition 3 is identically satisfied for a neutral atom. From condition 2, $N' = N$ is required. Finally, from condition 4, we require that

$$\sum_j m_j c^2 + \sum_j \mathrm{KE}_j + \sum_j P_j = m_{\mathrm{X}} c^2 + \mathrm{KE}_{\mathrm{X}} + P_{\mathrm{X}} + E^*_{\mathrm{X}} \tag{1.14}$$

as a mass/energy conservation equation, where we have summed over all contributions from each of the nucleons and electrons and

$$\begin{array}{rcl} \mathrm{KE} & \longrightarrow & \text{kinetic energy} \\ P & \longrightarrow & \text{potential energy} \\ E^* & \longrightarrow & \text{nuclear excitation energy} \\ mc^2 & \longrightarrow & \text{rest energy} \end{array}$$

If we make some very realistic assumptions, we can greatly simplify this expression. Firstly, we can assume that all the reactants (constituents, j) are initially at rest so that $\mathrm{KE}_j = 0$ for all j. Similarly, it will not detract from our final result to assume that all reactants are initially at infinity (sufficiently well separated so that there is no interaction between reactants before assembly), then $P_j = 0$, all j. Since $\mathrm{KE}_j = 0$ (all j) the initial system possesses no linear momentum in our frame so, by momentum conservation, $\mathrm{KE}_{\mathrm{X}} = 0$. As well, if X is isolated in space after assembly, then there is no interaction with other entities and $P_{\mathrm{X}} = 0$.

Upon reducing and rearranging, we arrive at

$$E^*_{\mathrm{X}} = \sum_j m_j c^2 - m_{\mathrm{X}} c^2$$

This represents the excitation energy of the newly formed nucleus X and will subsequently decay by photon emission. Consequently, this is the amount of energy (photon or otherwise) that would need to be absorbed to dismantle the nucleus. As such, E^*_{X} is the energy binding the constituent nucleons, *i.e.* the nuclear binding energy.

$$\mathrm{BE} \quad = \quad E^*_X = \sum_j m_j c^2 - m_{\mathrm{X}} c^2$$

$$
\begin{aligned}
&= c^2(\underbrace{Zm_\mathrm{p} + Zm_\mathrm{e}} + Nm_\mathrm{n} - m_\mathrm{X}) \\
&= c^2(Zm_\mathrm{H} + Nm_\mathrm{n} - m_\mathrm{X}) + Z\mathrm{BE}_e^\mathrm{H} \\
&\simeq c^2(Zm_\mathrm{H} + Nm_\mathrm{n} - m_\mathrm{X}) \qquad (1.15)
\end{aligned}
$$

The left hand side of Eq.(1.14) represents the initial condition of the constituent sub-atomic particles before assembly, while the right hand side is the final state of the assembled atom. We are only considering here the initial and final states of the system and ask nothing about how these entities came to be in these conditions. The value of BE is unaffected (to order $Z\mathrm{BE}_e^\mathrm{H}$) by the route to nucleosynthesis so long as Z and N are both conserved. We may consider the assembly process in several steps or all at once as we have done. However, if we allow changes in either Z or N (or both) in the synthesis process (*i.e.* nucleons created, destroyed, or lost), the liberated energy will depend on the details of the synthesis route. This energy can be shown to always be greater than BE. Thus, BE as defined above, is the minimum energy route to synthesis.

The quantity BE_e^H in Eq.(1.15) represents the energy binding the atomic electron in the hydrogen atom, the atomic binding energy. The equivalence $m_\mathrm{H} = m_\mathrm{e} + m_\mathrm{p} + \mathrm{BE}_e^\mathrm{H} \simeq m_\mathrm{H} = m_\mathrm{e} + m_\mathrm{p}$ indicates that we are ignoring the atomic binding energy in the above expression. $\mathrm{BE}_e^\mathrm{H}/A$ is at most about 3 keV/nucleon, whereas BE$/A$ is on the order of 8 MeV/nucleon. Since $Z\left(\mathrm{BE}_e^\mathrm{H}\right) \ll \mathrm{BE}$, omitting atomic binding from the analysis is not a serious error in nuclear calculations. Ignoring electron masses in the above analysis would be, however. The electron rest mass must always be included since we are using "atomic" mass tables. Missing one electron rest mass produces an error of ~ 0.511 MeV. Since BE is usually on the order of a few MeV, this is a large fraction. A sure method for avoiding such problems is to always use neutral ground state atomic masses (as provided in the atomic mass tables) in these calculations. Here m_H should always be used, never m_p.

Listed below are some properties of BE, several of which refer to Fig. 1.5. This curve (solid line) is a plot of BE$/A$ for all stable isotopes. Superimposed on the data is a fit using the "liquid drop model" (dashed line) discussed in sec. 1.6.6.

1. $\mathrm{BE} > 0$, is always true since $\sum_j m_j > m_x$.
2. $\frac{d\mathrm{BE}}{dA} > 0$, in general (with some local variations especially at low A).
3. $2.2 \leq \mathrm{BE} \leq 1900$ MeV.
4. $\mathrm{BE(MeV)} = 931.5\,[Zm_\mathrm{H} + Nm_\mathrm{n} - m_\mathrm{X}]$, for all masses in amu.
5. BE must be absorbed by X in order to reverse this process.

Example 7: Nuclear Binding Energy
Determine the nuclear binding energy, BE, for deuterium.

Solution:

$$\begin{aligned} \mathrm{BE_D} &= c^2\left[m_\mathrm{p} + m_\mathrm{e} + m_\mathrm{n} - m_\mathrm{D}\right] \\ &\simeq c^2\left[m_\mathrm{H} + m_\mathrm{n} - m_\mathrm{D}\right] \\ &\simeq 931.5\frac{\mathrm{MeV}}{\mathrm{amu}}\ [1.007825 + 1.008665 - 2.014102]\ \mathrm{amu} \\ &\simeq 2.22\ \mathrm{MeV} \end{aligned}$$

Discussion:
Neglecting the electron rest mass in the hydrogen atom would represent approximately 23% under prediction of the nuclear binding energy in this case. However, since $\mathrm{BE}_e^\mathrm{H} \sim 13.6$ eV, neglecting this contribution only underestimates the nuclear binding energy by $(13.6/2.2 \times 10^6) \sim 6.2 \times 10^{-6}$, or about six parts in one million, completely insignificant.

6. $\frac{\mathrm{BE}}{A}$, binding energy per nucleon is a stability index. Stability is associated with high values of $\frac{\mathrm{BE}}{A}$ (greater mass defect implies a greater binding energy). The local peak at $A = 4$ in Fig. 1.5 represents ^{4}He, an extremely stable nucleus.

7. For $A > 60$, $\frac{\mathrm{BE}}{A}$ is a weakly decreasing, smooth function of A.

8. For $A < 20$, $\frac{\mathrm{BE}}{A}$ is a steep increasing function of A with large local variations, *i.e.* ^{4}He.

Figure 1.5 suggests several methods or routes to extracting nuclear binding energy and, thereby, ways to "put atoms to work". Consider the following three scenarios:

1. **Symmetric Fission**
   Imagine breaking apart (or fissioning) an atom of high atomic mass into two identical atoms of much smaller mass. If this occurs, then these new atoms may have a larger $\frac{\mathrm{BE}}{A}$, thereby having a greater mass defect than the original heavy nucleus. Energy is then liberated from the system. Symbolically we can describe this process

$$^A_Z\mathrm{X}^N \longrightarrow 2\,^{A'}_{Z'}\mathrm{Y}^{N'}$$

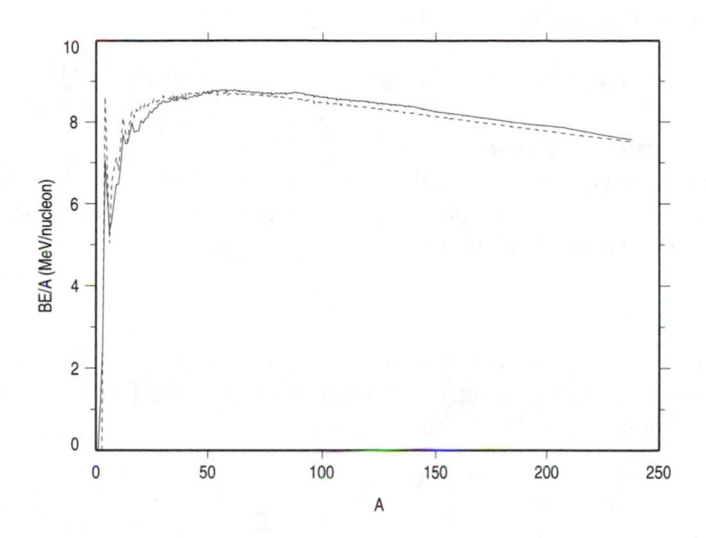

Figure 1.5: Binding energy per nucleon for the stable isotopes.

where X is the original heavy nucleus and Y is a new dummy symbol to represent the two lighter products. By conserving nucleons independently, the following assignments may be made:

(a) $Z = 2Z'$

(b) $N = 2N'$ and

(c) $A = 2A'$

Net energy can be released in this process if and only if $2\mathrm{BE_Y} > \mathrm{BE_X}$, or if

$$\frac{\mathrm{BE_Y}}{A'} > \frac{\mathrm{BE_X}}{A}$$

$$\frac{c^2\left[Z'm_\mathrm{H} + N'm_\mathrm{n} - m_\mathrm{Y}\right]}{A'} > \frac{c^2\left[Zm_\mathrm{H} + Nm_\mathrm{n} - m_\mathrm{X}\right]}{A}$$

Utilizing the conservation conditions above (a–c), this becomes after some rearrangement

$$\boxed{m_\mathrm{X} > 2m_\mathrm{Y}} \qquad (1.16)$$

Net energy is liberated in symmetrical fission, then, if this condition is satisfied, *i.e.*

$$\Delta m = m_\mathrm{X} - 2m_\mathrm{Y} > 0$$

so that the energy released is

$$E_{\text{released}} = \Delta mc^2 = (m_{\text{X}} - 2m_{\text{Y}})c^2 = 2\text{BE}_{\text{Y}} - \text{BE}_{\text{X}}$$

2. **Asymmetric Fission**
   Alternatively, we may consider asymmetrical fission as a nearly identical process as that above with the exception that the products of the process need not be identical

$${}^{A}_{Z}\text{X}^{N} \longrightarrow {}^{A'}_{Z'}\text{Y}^{N'} + {}^{A''}_{Z''}\zeta^{N''}$$

   where, again, X is the heavy isotope undergoing fission, but now Y and ζ are the two asymmetrical products. Again, we force independent nucleon conservation to find

$$\begin{aligned} N &= N' + N'' \\ Z &= Z' + Z'' \\ A &= A' + A'' \end{aligned}$$

   For net energy to be liberated (released) in this process, we must have

$$\text{BE}_{\text{Y}} + \text{BE}_{\zeta} > \text{BE}_{\text{X}}$$

   or

$$c^2\left[Z'm_{\text{H}} + N'm_{\text{n}} - m_{\text{Y}} + Z''m_{\text{H}} + N''m_{\text{n}} - m_{\zeta}\right] > c^2\left[Zm_{\text{H}} + Nm_{\text{n}} - m_{\text{X}}\right]$$

   Employing the nucleon conservation relations and upon some algebra, we arrive at the mass condition for net energy release from asymmetrical fission

$$\boxed{m_{\text{X}} > m_{\text{Y}} + m_{\zeta}} \qquad (1.17)$$

   for which the released energy is

$$E_{\text{released}} = \Delta mc^2 = (m_{\text{X}} - m_{\text{Y}} - m_{\zeta})c^2 = \text{BE}_{\text{Y}} + \text{BE}_{\zeta} - \text{BE}_{\text{X}}$$

3. **Fusion**
   Finally, we consider a third scenario for releasing nuclear binding energy, the joining together (or fusion) of atoms possessing low atomic mass into a single atom of higher atomic mass. If we could perform this synthesis in such a way as to provide a new larger atom with greater $\frac{\text{BE}}{A}$, this process would result in a greater mass defect, and

again, energy will be liberated. Here, we have only a single product Y possessing mass, formed by two lighter (and identical for argument sake) reactants X

$$2{}^{A}_{Z}\mathrm{X}^{N} \longrightarrow {}^{A'}_{Z'}\mathrm{Y}^{N'}$$

Once again, the following assignments can be made from nucleon conservation

$$\begin{aligned} 2Z &= Z' \\ 2N &= N' \\ 2A &= A' \end{aligned}$$

Energy will be released in this process if and only if $\mathrm{BE_Y} > 2\mathrm{BE_X}$, or

$$\frac{\mathrm{BE_Y}}{A'} > \frac{\mathrm{BE_X}}{A}$$

By this point the general condition should be clear. We must always move to larger $\frac{\mathrm{BE}}{A}$ in order to release energy. The mass condition for energy release in symmetric fusion then comes from

$$\frac{c^2\left[Z'm_\mathrm{H} + N'm_\mathrm{N} - m_\mathrm{Y}\right]}{A'} > \frac{c^2\left[Zm_\mathrm{H} + Nm_\mathrm{N} - m_\mathrm{X}\right]}{A}$$

which will reduce to $-\frac{1}{2}m_\mathrm{Y} > -m_\mathrm{X}$ so that

$$\boxed{2m_\mathrm{X} > m_\mathrm{Y}} \tag{1.18}$$

and the released energy is

$$E_\mathrm{released} = \Delta mc^2 = (2m_\mathrm{X} - m_\mathrm{Y})c^2 = \mathrm{BE_Y} - 2\mathrm{BE_X}$$

1.6.5 Separation Energies

Closely related to the concept of nuclear binding energy is separation energy. Suppose you would like to remove only one of the neutrons from the arbitrary nuclide X, as in the following process,

$${}^{A}_{Z}\mathrm{X}^{N} \longrightarrow {}^{A-1}_{Z}\mathrm{X}^{N-1} + \mathrm{n}$$

rather than completely dismantle it. Since there is always some positive BE associated with the attachment of any nucleon to any nucleus, then this

Example 8: Binding Energy: Fission

Determine the amount of energy released in the following asymmetrical fission process.

$$^{235}_{92}\text{U} \longrightarrow {}^{13}_{6}\text{C} + {}^{222}_{86}\text{Rn}$$

Solution:

$$\begin{aligned}
\text{BE}_{^{235}_{92}\text{U}} &= c^2\,[92m_\text{H} + 143m_\text{n} - m_{^{235}\text{U}}] \\
&= 931.5\,[92(1.007825) + 143(1.008665) - 235.043943] \\
&= 1783.9 \text{ MeV} \\
\text{BE}_{^{13}_{6}\text{C}} &= c^2\,[6m_\text{H} + 7m_\text{n} - m_{^{13}\text{C}}] \\
&= 931.5\,[6(1.007825) + 7(1.008665) - 13.003354] \\
&= 97.1 \text{ MeV} \\
\text{BE}_{^{222}_{86}\text{Rn}} &= c^2\,[86m_\text{H} + 136m_\text{n} - m_{^{222}\text{Rn}}] \\
&= 931.5\,[86(1.007825) + 136(1.008665) - 222.01761] \\
&= 1708.2 \text{ MeV}
\end{aligned}$$

so that

$$\text{BE}_{^{13}_{6}\text{C}} + \text{BE}_{^{222}_{86}\text{Rn}} = 1805.3 \text{ MeV}$$

and finally

$$1805.3 \text{ MeV} - 1783.9 \text{ MeV} = 21.4 \text{ MeV}$$

The nuclear binding energy of the products totals 21.4 MeV more than that of the original nucleus. This energy must be released in the synthesis process.

Example 9: Binding Energy: Fusion
A proposed (aneutronic) symmetric cold fusion reaction is:

$${}^2_1\mathrm{H} + {}^2_1\mathrm{H} \longrightarrow {}^4_2\mathrm{He}$$

Determine the energy that would be released in this process.

Solution:

$$\begin{aligned}
\mathrm{BE}_{^2\mathrm{H}} &= 931.5\,[1.007825 + 1.008665 - 2.014102] = 2.22\ \mathrm{MeV} \\
\mathrm{BE}_{^4\mathrm{He}} &= 931.5\,[2\,(1.007825) + 2\,(1.008665) - 4.002603] \\
&= 28.3\ \mathrm{MeV} \\
\therefore\ \mathrm{BE}_{^4\mathrm{He}} &> 2\,(\mathrm{BE}_{^2\mathrm{H}})\,, \quad \text{and } E_{\mathrm{released}} = 23.86\ \mathrm{MeV}
\end{aligned}$$

This symmetric reaction cannot occur in free space, *i.e.* without another body present to conserve momentum. A more common free space fusion reaction is:

$$\begin{aligned}
{}^2_1\mathrm{H} + {}^2_1\mathrm{H} &\longrightarrow \mathrm{T} + \mathrm{p} \qquad (50\%) \\
&\longrightarrow {}^3\mathrm{He} + \mathrm{n} \qquad (50\%) \\
\mathrm{BE}_{\mathrm{T}} &= 931.5\,[1.007825 + 2\,(1.008665) - 3.01605] = 8.48\ \mathrm{MeV} \\
\therefore\ E_{\mathrm{released}} &= 8.48\ \mathrm{MeV} - 2\,(2.22)\ \mathrm{MeV} = 4.04\ \mathrm{MeV}
\end{aligned}$$

Challenge: Show why the first reaction cannot occur in free space. We will answer this later when we discuss nuclear reactions. See if you can resolve it now.

process requires energy (usually in the form of photons). A mass/energy balance yields

$$m_{\mathrm{X}^N}c^2 + S_\mathrm{n} = m_{\mathrm{X}^{N-1}}c^2 + m_\mathrm{n}c^2$$

where S_n is the energy required to separate an outer shell neutron from the nucleus, the neutron separation energy,

$$S_\mathrm{n} = c^2\left[m_{\mathrm{X}^{N-1}} + m_\mathrm{n} - m_{\mathrm{X}^N}\right] \tag{1.19}$$

also called the "binding energy of the last neutron". In this sense, S_n is completely analogous to the first atomic ionization energy. It is exactly the energy required to move the last outer shell bound neutron to infinity. As a function of the binding energy BE, we can write

$$S_\mathrm{n} = \mathrm{BE}_{\mathrm{X}^N} - \mathrm{BE}_{\mathrm{X}^{N-1}} \tag{1.20}$$

We can also define a proton separation energy (S_p) in a very similar way

$${}^{A}_{Z}\mathrm{X}^N \longrightarrow {}^{A-1}_{Z-1}\mathrm{Y}^N + \mathrm{p} + \mathrm{e}^-$$

such that

$$S_\mathrm{p} = c^2\left[m_\mathrm{Y} + m_\mathrm{H} - m_\mathrm{X}\right] = \mathrm{BE}_\mathrm{X} - \mathrm{BE}_\mathrm{Y} \tag{1.21}$$

Trends in the separation energies vs. A, N, and Z, can help identify features in nuclear structure. For example, nuclear stability is favored for nuclei with even numbers of Z and N, something that we'll later call "paring." This paring situation is marked by much higher separation energies for nucleons in these nuclei so that they are much more difficult to remove.

1.6.6 Nuclear Models

The pictorial model of the atomic nucleus that has been presented thus far is useful in providing a qualitative description. Its real utility is in assembling a mental image of the physical entity under examination. To obtain a mechanistic physical understanding of the nucleus, we need more quantitative models. The ultimate goal would be to provide explanations for observed phenomena and make predictions regarding the future state of the nuclear system. While it is well outside the scope of this text to describe such models in all their glorious detail, a brief introduction serves to provide some insight into the essential physics, make some simple predictions, and pave a path toward further study in these subjects.

The two models we'll briefly discuss are the liquid drop and shell models. The former piece of vocabulary has already been introduced in the context of nuclear density (cf: sec. 1.6.3). We'll also exploit it to provide a quantitative prediction for binding energy in the next section and later a heuristic description of the nuclear fission process (chpt. 5). The shell model, on the other hand, is much more involved. Even to make the simplest of quantitative use, it requires mastery of the quantum mechanics. We'll mention it here only for a bit of completeness and to suggest that more sophisticated descriptions of the atomic nucleus are required to mechanistically describe some important phenomena like nuclear stability.

1.6.6.1 Liquid Drop Model

As we have earlier discussed, nucleons in the atomic nucleus appear very densely packed. This was principally borne out in experimental studies showing nuclear volume increasing as A, allowing little or no empty nuclear space. Moreover, the identification that stable nuclides are not rare in nature and that nuclear binding energy is large, indicate that nucleons are somehow strongly attracted to each other. This force of attraction is called the "strong nuclear force" and is, in fact, the strongest force known.

The strong nuclear force acts to bind nucleons rather tightly over only a small distance. It is a short range force and allows nucleons to interact only with their nearest neighbors. Recall that BE/A is largely independent of A (except at small A), so that the nuclear bond is essentially "saturated." This situation is rather analogous to incompressible liquids where cohesive atomic forces act to bind liquid molecules. These tend to act over only a relatively small distance so that liquid molecules essentially interact only with others close by. The combined effect of many such interactions in a water droplet results in the phenomenon we call surface tension, and uninfluenced by external forces, tends to produce roughly spherical droplets.

Further nuclear evidence that supports this connection includes the tendency for isobars to have similar BE/A values. Hence, nuclear bonds are at least similar, regardless of the n,p mixture. We can then treat all nucleons as being nearly identical in the model. Let's then treat the nucleus as a continuous and homogeneous charged drop of incompressible liquid. At the same time, we recognize that the nucleus is not really a liquid and that we may need to incorporate some quantum mechanical, non-fluid properties to complete the picture.

Our goal will be to use the liquid droplet analogy of the atomic nucleus to build a model that predicts nuclear binding energy as a function of A, Z, and N. Predictions based on this model can then be used as a gauge for its validity. This will be a semi-empirical model that we'll also expect to bring some order to the nuclide chart and provide some basis for the prediction of nuclear phenomena like radioactive decay and fission. (Semi-empirical models incorporate basic physical effects primarily to determine parameter dependencies while relying on guidance from experimental data to supply unknown coefficients. These empirical coefficients in turn provide the relative magnitudes of often several physical mechanisms combining to describe a single phenomenon.)

To derive the semi-empirical binding energy equation, we need to consider all contributions to BE implied by the model. For convention, we'll adapt as a rule that all contributions to energy liberated are positive, and all contributions to energy absorbed are negative. The first of these is the combined affect of nuclear attractive force within the entire nuclear volume, *i.e.* the volume energy. Each nucleon-nucleon bond is associated with some potential energy, P, shared between two nucleons. Each nucleon then "possesses" a volume binding energy of $\frac{1}{2}P$, but in a perfectly packed sphere is surrounded by 12 neighbors for a total of $6P$. If we assume that all nucleons are in an identical situation, then the total volume contribution is $6PA$, usually written

$$\mathrm{BE}_v = a_v A$$

However, not all nucleons are so surrounded. Some are near the surface and have fewer neighbors. A second contribution that incorporates the surface effect must be included. Since the surface area for a sphere $\sim R^2 \sim R_o^2 A^{2/3}$, the number of nucleons with fewer than 12 bonds is proportional to $A^{2/3}$. This is a negative contribution to BE since every additional nucleon added to the nucleus enlarges the surface area, which requires work to be done on the surface $\sim R^2$. The result is a contribution

$$\mathrm{BE}_s = -a_s A^{2/3}$$

This is exactly that contribution which yields the spherical shape in analogy to the liquid drop. By conforming to a sphere, the drop minimizes its surface to volume ratio and maximizes its potential energy.

Electrostatic forces contribute another negative component. Adding only positive charge to a nucleus that already possesses protons causes Coulombic repulsion, which fights against nuclear binding. The net work that must be done against the Coulomb force of Z protons in a nucleus yields this third contribution to nuclear binding energy

$$\mathrm{BE}_c = -a_c \frac{Z(Z-1)}{A^{1/3}}$$

The remaining two contributions are non-fluid effects and require some quantum physics. It so happens that the n-p interaction is slightly stronger than that of n-n or p-p. If $Z \neq N$, there will be, on average, more n-n and p-p interactions than n-p. Therefore, non-identical values of Z and N reduce BE. Nuclides with $Z = N$ are more stable for a given A. This is called the asymmetry effect and contributes

$$\mathrm{BE}_a = -a_a \frac{(N-Z)^2}{A}$$

to the nuclear binding energy.

Finally, there is a contribution from the quantum pairing effect which favors stability for evenly paired combinations of both N and Z. This can yield a net positive or negative effect on BE, depending on the pairing. Symbolically we can describe this contribution as

$$\mathrm{BE}_p = a_p \frac{\delta}{A^{3/4}}$$

where

δ	N	Z
1	even	even
0	even	odd
0	odd	even
-1	odd	odd

Combining all contributions, we arrive at the semi-empirical nuclear binding energy equation

$$\mathrm{BE} = a_v A - a_s A^{2/3} - a_c \frac{Z(Z-1)}{A^{1/3}} - a_a \frac{(N-Z)^2}{A} + a_p \frac{\delta}{A^{3/4}} \tag{1.22}$$

The empirical part, though, is still to be done. The five coefficients must be chosen to give the best fit to the atomic mass data. The model itself does not yield these values. Furthermore, no one set of five values gives a good fit to the data over the entire range of A. (We expect the model to be most troublesome at small A. When the number of nucleons is small, the entire nucleus is involved in the binding forces of every nucleon, a violation of the liquid drop premise.) The set that gives the best fit over the largest range of A is [Marmier69]

$$\begin{aligned} a_v &= 14 \text{ MeV} \\ a_s &= 13 \text{ MeV} \\ a_c &= 0.6 \text{ MeV} \\ a_a &= 19 \text{ MeV} \\ a_p &= 34 \text{ MeV} \end{aligned}$$

Through our BE definition (Eq.(1.15)), we can use the semi-empirical model result to predict atomic masses.

$$\begin{aligned} m_{\mathrm{X}}(\mathrm{amu}) &= \frac{1}{931.5}[931.5(Zm_{\mathrm{H}} + Nm_{\mathrm{n}}) \qquad (1.23) \\ &- a_v A + a_s A^{2/3} + a_c \frac{Z(Z-1)}{A^{1/3}} + a_a \frac{(N-Z)^2}{A} - a_p \frac{\delta}{A^{3/4}}\Big] \end{aligned}$$

This expression, however, should be considered a guide for understanding trends only. It is much more inaccurate than the atomic mass tables and should never be used to find atomic masses when the latter are available. The value of the semi-empirical binding energy equation is in its ability to combine in a compact, systematic way the influence of A, Z, and N on BE and m_X trends. One such trend is the dependence of m_X on Z for constant A (isobars). Since from Eq.(1.23) m_X is quadratic in Z, we expect a minimum or maximum in $m_X(Z)$. Let $Z^* = Z$ where $\frac{dm_X}{dZ} \to 0$. This can be shown to be a minimum mass and that

$$Z^* = \frac{\frac{A}{2} + \frac{a_c}{8a_a} A^{2/3}}{1 + \frac{a_c}{4a_a} A^{2/3}} \tag{1.24}$$

Differentiating $m_X(Z)$ implies that we are treating the discrete quantity Z as a continuous variable. This is conceptually puzzling, but causes no practical problems. As well, the expression Eq.(1.24) rarely yields integer values for Z^*. One should always round to the nearest integer. Fractional values make no sense. For small A, $Z^* \sim A/2 = N$. Equation (1.24) yields the isobar with the smallest rest mass or maximum BE. In the nuclide chart, these cluster to a band near but below $Z = N$ as depicted in Fig. 1.3, indicating that stable nuclides cluster around this minimum mass trajectory called the "valley of stability". We assume, then, that higher mass isobars are unstable and transform to more stable nuclides by losing rest mass. This transformation process we call radioactive decay.

1.6.6.2 Shell Model

In the development of the liquid drop model, we have made as the chief assumption that each nucleon interacts exclusively with its nearest neighbors. While this model was successful in describing some nuclear phenomena like some features of the BE/A curve, it fails in predicting other empirical data. Meanwhile, other models have been successful in describing additional observations in a rather complementary way to the drop model. One such model is the "shell model"[Lawson80], which derives its name from close analogy to the familiar shell model of atomic electrons. Here, the main assumption essentially reaches the opposite extreme as with the drop model. The basic assumption in the shell model is that each nucleon interacts independently with a force field produced by all other nucleons. Models of this type are sometimes called "individual nucleon models"[Evans55]. In addition, nucleons are forced to obey the exclusion principle and other quantum mechanical constraints which nicely yield Z and N periodicity.

One of the major triumphs of the shell model is its prediction of the so called nuclear "magic numbers". As with the electron shell model, there exist special stable nuclear configurations at closed shells in which all allowed energy states are filled for a given principle quantum state. The nuclear closed shells occur at magic numbers of 2, 8, 20, 28, 50, 82, and 126 nucleons. The increased abundance of atoms possessing these number of nucleons over other nuclei of similar mass is empirical evidence for the stability of their nuclear structure.

The shell model is able to explain some other important nuclear phenomena like the tendency for nuclear stability in evenly paired Z and N nuclei, what we earlier called pairing. Nuclear angular momentum is also correctly predicted in the shell model. Collectively, both the shell and drop models account for most of the observed nuclear properties. Combined in the "collective model," they constitute a rather complete nuclear description. Such higher level features as non-spherical shape and nuclear quadrupole moments, centrifugal distortion of rotating nuclei, and a consistent accounting of nuclear level spacing are well reproduced.

References

[ANSI92] ANSI/IEEE, *American National Standard for Metric Practice*, Std. 268, 1992.

[Beiser81] Beiser, A., *Concepts of Modern Physics*, 3^{ed}, McGraw Hill, Inc., NY, 1981.

[Cohen87] Cohen, E. R., and Taylor, B. N., *Reviews of Modern Physics*, **59** (1987) 1121.

[Evans55] Evans, R. D., *The Atomic Nucleus*, McGraw Hill, Inc., NY, 1955.

[Firestone96] Firestone, R. B., Shirley, V. S., Baglin, C. M., Chu, S. Y. F., and Zipkin, J., *Table of Isotopes*, 8^{ed}, John Wiley & Sons, Inc., NY, 1996.

[GE96] General Electric Co., *Nuclides and Isotopes*, 15^{ed}, 1996.

[Halliday81] Halliday, D., and Resnick, R., *Fundamentals of Physics*, 2^{ed}, John Wiley & Sons, Inc., NY, 1981.

[Lawson80] Lawson, R. D., *The Theory of the Nuclear Shell Model*, Oxford University Press, NY, 1980.

[Lord95] Lord, J., *Sizes*, Harper Perennial, NY, 1995.

[Marmier69] Marmier, P., and Sheldon, E., *Physics of Nuclei and Particles*, Academic Press, Inc., NY, 1969.

[NEI94] Nuclear Energy Institute, *Economic and Employment Benefits of the Use of Nuclear Energy to Produce Electricity*, Washington, DC, 1994.

[NN96a] Nuclear News, **39**, No.6 (1996) 22.

[NN96b] Nuclear News, **39**, No.10 (1996) 38.

[ORRWS94] Organization for Responsible Low-Level Radioactive Waste Solutions, *Economic and Employment Benefits of the Use of Radioactive Materials*, Washington, DC, 1994.

Web References

By now you know that "everything" is on the Web. Nuclear data is no exception. Exploiting the Internet for scientific information is its most worthwhile application. Besides, it's a fast, efficient, and exciting way to get data. Below is a short list of some of the more prominent sites that I've used to get nuclear and physical data. They also appear to be the most stable, but I don't make any guarantees. Net content changes at light speed. These sites are either chock full of data or links to other sites, or both. So bounce around until you find what you need. The *Table of Isotopes*[Firestone96] can be found at "toi" sites, and order information for the Chart of the Nuclides is found at the GE site. Have fun!

Physical Data

physics.nist.gov/
www.shef.ac.uk/chemistry/web-elements/web-elements-home.html
www.cs.ubc.ca/cgi-bin/nph-pertab/periodic-table

Nuclear Data

www.nndc.bnl.gov/
neutrino.nuc.berkeley.edu/NEadm.html
epicws.epm.ornl.gov/NUCDATA.html
www.dne.bnl.gov/ burrows/usndn/usndn.html
t2.lanl.gov/data/ndviewer.html
t2.lanl.gov/
www-rsicc.ornl.gov/rsic.html
www.fysik.lu.se/nucleardata/
wwwndc.tokai.jaeri.go.jp/index.html
www.iaea.or.at/
nucleardata.nuclear.lu.se/Database/ndgloss/

Table of Isotopes/ Isotope Explorer / Nuclide Chart

www.ge.com/nuclear/nucchart.htm
ie.lbl.gov/isoexpl/isoexpl.htm
www.fysik.lu.se/nucleardata/toi_.htm
isotopes.lbl.gov/isotopes/toi.html
www.dne.bnl.gov/CoN/index.html
csa5.lbl.gov/ fchu/toi.html
csa5.lbl.gov/ fchu/ip.html

Problems and Questions

1. A particle accelerator is used to create high energy electron beams. In a particular case, electrons are allowed to be accelerated through a 300 V potential difference.
   (a) Determine the electron's kinetic energy, total energy, and velocity.
   (b) Repeat your calculations for a proton instead of an electron.
   (c) Repeat your calculations for both the electron and the proton if the accelerating potential is increased to 1 MV.
2. An electron in the first Bohr orbit of a hydrogen atom ($a_o = 5.3 \times 10^{-11}$ m) has a kinetic energy of 13.6 eV. Express the de Broglie wavelength for this electron in multiples of the atomic circumference.
3. Following the procedure outlined in sec. 1.5, derive the mass/energy equivalence expression (Eq.(1.11)).
4. Determine the kinetic energy at which a proton attains $0.7c$. Express this energy as both an absolute energy in eV and relative to the proton rest energy.
5. Find the number of neutrons and protons in the nuclei of the following atoms: (a) $^{9}_{4}$Be, (b) $^{132}_{54}$Xe, (c) $^{197}_{79}$Au, and (d) $^{222}_{86}$Rn.
6. Define the following: (a) electron volt, (b) isotope, (c) isotone, (d) isobar, (e) nucleon, (f) nuclide, (g) radionuclide, (h) atomic mass unit.
7. Write down all the isotopes, isotones, and isobars for ^{7}Li and ^{40}Ar.
8. Using your knowledge of nuclear dimensions and densities, estimate the mass of an object that is about the size of a basketball ($\approx$10 inch diameter sphere) and is comprised of only nuclear matter.
9. Define the mass defect. Why is it important?
10. Determine the binding energy (in MeV) per nucleon for: (a) $^{4}_{2}$He, (b) $^{12}_{6}$C, (c) $^{59}_{27}$Co, (d) $^{148}_{62}$Sm, (e) $^{238}_{92}$U, and (f) $^{239}_{94}$Pu.
11. Complete the following nuclear reaction equations by conserving nucleons:
    (a) $^{239}_{94}\text{Pu} + ^{1}_{0}\text{n} \longrightarrow (?)$
    (b) $(?) \longrightarrow ^{234}_{90}\text{Th} + ^{4}_{2}\text{He}$
    (c) $^{16}_{8}\text{O} + ^{1}_{0}\text{n} \longrightarrow (?) + ^{1}_{1}\text{H}$
    (d) $^{207}_{82}\text{Pb} \longrightarrow (?) + ^{1}_{0}\text{n}$

12. Calculate the binding energy and the neutron separation energy for $^{7}_{3}$Li.

13. Calculate the binding energy per nucleon for deuterium, $^{56}_{26}$Fe, $^{115}_{49}$In, and $^{235}_{92}$U in MeV.

14. Draw a graph that shows the variation of the binding energy per nucleon with mass number. Explain the significance of this graph with respect to fusion and fission.

15. Calculate the binding energy and binding energy per nucleon for $^{238}_{92}$U and $^{56}_{26}$Fe. What is the significance of the difference between the two results?

16. Determine the nuclear binding energies for $^{236}_{92}$U and $^{118}_{46}$Pd and use this information to demonstrate that there is net energy liberated in the symmetric, thermal neutron-induced fission of $^{236}_{92}$U with $^{118}_{46}$Pd as a product (*i.e.* write the energy release as a function of the nuclear binding energy of the products and reactants).

17. Consider the following isotopes: ^{2}H, ^{4}He, ^{6}Li, ^{14}N, ^{35}Cl, ^{56}Fe, ^{110}Cd, ^{126}Te, ^{209}Bi, and ^{238}U.

    (a) Calculate the nuclear binding energies and binding energy per nucleon.

    (b) Plot the binding energy per nucleon (BE/A) as a function of A for these, and discuss the trends you observe with respect to the general trends discussed in the text.

18. Find the mass condition for energy release from a three-product asymmetric fission reaction.

19. Determine the neutron separation energies for the isotopes of Li and C, and discuss the trend(s) you observe in S_{n} with respect to the effect of odd and even combinations of Z and N on the pairing energy.

20. A nuclear scientist desires to perform experiments with stable ^{56}Fe. Determine the amount of energy that she will need to provide in order to:

    (a) remove the last neutron,

    (b) completely dismantle the ^{56}Fe nucleus into its sub-atomic components, and

    (c) fission it symmetrically (into two fragments).

21. Determine the separation energies (in MeV) of the last proton in the following nuclei: (a) ${}^{3}_{2}$He, (b) ${}^{51}_{23}$V, (c) ${}^{206}_{82}$Pb, (d) ${}^{210}_{84}$Po, and (e) ${}^{241}_{95}$Am.

22. Demonstrate that the neutron and proton separation energies, expressed as functions of nuclear binding energies, are given correctly in Eqs.(1.20) and (1.21), respectively.

23. Use the semi-empirical binding energy expression (Eq.(1.23)) to estimate the atomic masses of: (a) ^{4}He, (b) ^{6}Li, (c) ^{59}Co, and (d) ^{238}U. Compare these to the accepted values from the atomic mass tables and comment.

Chapter 2

Nuclear Reactions

2.1 Introduction

The defining characteristic of the nuclear sciences is the processes of nuclear reactions. Simply put, any processes involving the nucleus of the atom may be considered a nuclear reaction. When we need a more precise definition, we can state that nuclear reactions are *those interactions between the nucleus of an atom and a sub-atomic particle or the nucleus of another atom (or the decomposition of a single atomic nucleus) that result in the production of two or more particles consisting of atomic nuclei or sub-atomic particles.* The impact of such processes cannot be underestimated. We, in fact, owe our very existence to nuclear reactions that have taken place over the eons deep within the core of long since dead stars. The reactions of fusion, nuclear combining of light nuclei, in stars are responsible for the formation of all chemical elements heavier than helium. Terrestrially, nuclear reactions are, of course, responsible for the energy produced in nuclear reactors, and the plethora of industrial and other applications described early in chapter 1.

Our objective in this chapter is to describe the kinematics of nuclear reactions in both the laboratory and center of mass frames (a frame in which the net linear momentum of the system is identically zero). The utility of the latter approach will become apparent upon application, as it provides a description of nuclear processes from another perspective, one which may appeal to our physical intuition. Though well beyond the scope of this book, the center of mass system also provides a reference frame in which reaction probabilities are either independent or only weakly dependent on the emission angles with which particles leave nuclear reactions. This may greatly simplify the description of quantities related to interaction probabilities called cross sections (chpt. 3).

Many, many interesting and useful possibilities exist among the vast multitude of nuclear reactions. Those of practical interest in energy production are limited to fission and fusion. (These are so extremely important that they warrant detailed discussion in chpt. 5.) There exist a myriad of neutron absorption reactions available for radionuclide production or the detection of neutrons, not to mention the variety of reactions that we may exploit to study the structure of the atomic nucleus. Below are a few examples.

Reaction equations showing a few examples of some important nuclear reactions:

$$\mathrm{n} + {}^{235}\mathrm{U} \xrightarrow{\text{fission}} \text{fission products}$$

$${}^{2}\mathrm{H} + {}^{3}\mathrm{H} \xrightarrow{\text{fusion}} {}^{4}\mathrm{He} + \mathrm{n}$$

$$\alpha + {}^{9}_{4}\mathrm{Be} \xrightarrow{\text{neutron production}} {}^{12}_{6}\mathrm{C} + \mathrm{n}$$

$$\mathrm{n} + {}^{7}\mathrm{Li} \xrightarrow{\text{tritium production}} {}^{3}\mathrm{H} + {}^{4}\mathrm{He} + \mathrm{n}$$

$$\mathrm{n} + {}^{14}\mathrm{N} \longrightarrow {}^{14}\mathrm{C} + \mathrm{p}$$

(An α particle is the nucleus of the ^{4}He atom. The last reaction is the cosmogenic source of radioactive ^{14}C by cosmic ray produced neutrons in the upper atmosphere.)

A bit of terminology is necessary in describing such processes. The particles residing on the left hand sides of the above expressions comprise the *reactants* in the system. These reactants interact to produce the *products* of the reaction (right hand sides). In almost all cases of interest, a reaction will involve only two reactants. These are called binary reactions. Reactions involving three reactants are possible but extremely unlikely. For many such binary reactions, there are only two products. These two-product binary reactions comprise the remainder of our discussion in this chapter. Occasionally, however, important reactions involve more than two products. Specific examples include fission, as well as certain forms of radioactive decay, subjects to be taken up later.

We may further assign identity to the individual reactants and products. In many instances, nuclear reactions are induced by generating a beam of energetic light particles as one of the reactants. This reactant then takes on the role of *projectile* and is made to interact with the usually stationary and often heavier *target* reactant. In the middle example above, the α particle (^{4}He nucleus) takes on the role of projectile, while the ^{9}Be nucleus provides the stationary target for the interaction. Often, the products are of vastly differing mass and are labeled as heavy and light products.

For a general nuclear reaction, we'll adopt the following terminology

X, Y	heavy nuclides
x, y	light particles
m_j	rest mass of species j

so that we can write the general two-product binary reaction

$$x + \mathrm{X} \longrightarrow \mathrm{Y} + y$$

The shorthand notation

$$\mathrm{X}(x, y)\mathrm{Y}$$

is also in wide acceptance. One would then express the $\alpha + {}^{9}\mathrm{Be} \longrightarrow {}^{12}\mathrm{C} + \mathrm{n}$ reaction as ${}^{9}\mathrm{Be}(\alpha, \mathrm{n})^{12}\mathrm{C}$. This latter notation clearly represents this as a neutron production reaction of the (α,n) type.

In the next two sections, we'll discuss some of the classification schemes for the many varieties of nuclear reactions before developing the kinematics of binary two-product reactions. This formalism sets the stage for much of the remainder of the text as most nuclear interactions, including radioactive decay, can be viewed as specific examples of this type of reaction.

2.2 Classification of Nuclear Reactions

Because of their great variety, nuclear reactions can be classified in many ways. Effective classification can exploit features of projectile type, projectile energy, target atomic mass, product type, and others. A complete treatment of all the various classification schemes and rationale is well beyond the scope of this book. For further reading on this subject, see the discussion in [Kaplan63].

The beginning nuclear professional should be conversant in certain general classifications. For example, the projectile energy range can dictate the nuclear interaction mechanism. At high energies ($\gtrsim$ 50 MeV), an incident projectile is likely to interact only with a few or even only a single nucleon in the target nucleus. This is qualitatively justified since, at these energies, the projectile de Broglie wavelength is on the order of a single nucleon dimension. Such interactions are called *direct reactions* since there is interaction directly between the projectile and the target nucleon(s) while the remainder of the nucleus serves as a passive observer. Often, these interactions occur at the nuclear surface and are, thus, sometimes called *peripheral processes* [Krane88].

The energies required to produce direct interactions are much higher than those usually encountered in nuclear applications. At the low energy extreme, nuclear reactions are much more likely to proceed via the *compound nucleus* route proposed by Bohr in 1936. Since this is the most important energy range for our purposes, the next section will be devoted to the compound nucleus interaction mechanism. At intermediate energies, *resonance reactions* become important, wherein discrete energy levels representing "quasi-bound" states have a high probability of being excited.

Some familiar nomenclature comes from projectile and light product type classifications. *Transfer* reactions, for example, are direct interactions in which one or two nucleons are transferred between target and projectile, as in (d,n), (α,d). (Here, d is a *deuteron*, the nucleus of a deuterium atom.) If, in our general description, an interaction proceeds in which X = Y and $x = y$, we call this a *scattering* reaction. When the products of such a scattering interaction are left in their ground nuclear states, the scattering is *elastic* (n,n) so that kinetic energy is conserved in this process. Otherwise, when an excited state remains, scattering is *inelastic* (n,n$'$). Here, there is a reduction in kinetic energy of the system, which appears as nuclear excitation energy and decays quickly by photon (γ) emission. In certain interactions, the projectile itself is emitted in the company of one or more target nucleons. Such interactions are called *knockout reactions*, (n,2n), (n,3n). When the light product of an interaction is a photon (γ), the reaction is said to be *radiative capture* or just capture. *Neutron capture* (n,γ) occurs when the projectile is a neutron and the light product is γ. This is an extremely important reaction in nuclear technology since it represents the principle means of producing radioactive materials. A reaction in which the projectile is a γ is said to be a *nuclear photoeffect*, (γ,n), (γ,p).

2.3 The Compound Nucleus

At low energy ($\lesssim$ 50 MeV), compound nucleus formation is the principal mechanism responsible for nuclear reactions. In this energy range, the projectile (typically n, p, d, α) de Broglie wavelength is substantially larger than the characteristic nucleon scale. It is then much more likely that the projectile will interact with the entire nucleus rather than just a few nucleons as in direct reactions. The incident particle kinetic energy and the binding energy of this particle in the new nucleus are "shared" among all the nucleons. Initially, no one nucleon possesses enough individual energy to escape the nucleus. The new *compound nucleus* thus exists as a unique entity in an agitated (excited) state. This state can exist for a time on the order of $10^{-15} - 10^{-14}$ s, a relatively long time compared to the *natural nuclear time*, $10^{-21} - 10^{-17}$ s. The latter is also referred to as the nuclear crossing time since it is the time required for the energetic particle to travel across the nucleus, the characteristic duration of direct interactions. The excited levels in the compound nucleus are called *virtual states* or *virtual levels* since they can decay by nucleon emission, in contrast to the bound states or levels which must decay only by γ emission.

The shared energy in the compound nucleus is continuously redistributed during its lifetime by "random" interactions among nucleons. While on the average, no one nucleon possesses enough energy to escape the nucleus, eventually these intranuclear interactions have some probability of concentrating more than the average energy on a single or a few nucleons, which may then have enough energy to escape, *i.e.* the separation energy. This process is very much like that occurring in the evaporation of hot liquids, and is sometimes called the *evaporation model.* This situation is illustrated pictorially in Fig. 2.1.

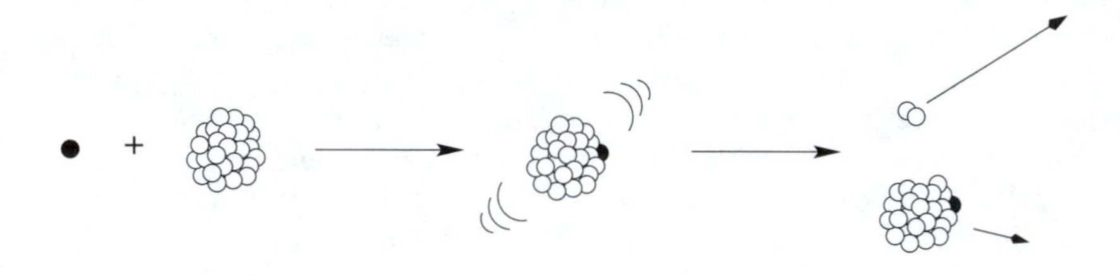

Figure 2.1: Conceptualization of a compound nucleus reaction in various stages showing before, during, and after the compound state.

Compound nucleus reactions are, therefore, two step processes; formation of the relatively long lived compound nucleus, and its subsequent disintegration releasing all its excitation energy. In our binary reaction language, we can write

$$x + \mathrm{X} \longrightarrow (x \ + \ \mathrm{X})^* \longrightarrow \mathrm{Y} + y'\mathrm{s}$$

where the asterisk refers to a nuclear excited level.

Since the lifetime of the compound nucleus is relatively long, it loses all memory of the formation details and of the incident projectile. It "forgets" how it was formed. The decay products are then independent of the formation process and depend only on the total energy given to the system and statistical rules. Two such examples are illustrated below

$$\left.\begin{array}{l} \mathrm{p} + {}^{63}\mathrm{Cu} \\ \alpha + {}^{60}\mathrm{Ni} \end{array}\right\} \longrightarrow {}^{64}\mathrm{Zn}^* \longrightarrow \left\{\begin{array}{l} {}^{63}\mathrm{Zn} + \mathrm{n} \\ {}^{62}\mathrm{Cu} + \mathrm{n} + \mathrm{p} \\ {}^{62}\mathrm{Zn} + 2\mathrm{n} \end{array}\right.$$

$$\mathrm{p} + {}^{27}\mathrm{Al} \longrightarrow {}^{28}\mathrm{Si}^* \longrightarrow \begin{cases} {}^{24}\mathrm{Mg} + \alpha \\ {}^{27}\mathrm{Si} + \mathrm{n} \\ {}^{28}\mathrm{Si} + \gamma \\ {}^{24}\mathrm{Na} + 3\mathrm{p} + \mathrm{n} \end{cases}$$

The set of paths for end reaction products are called *exit channels.* For certain reactions, many such exit channels may exist. An excellent example is fission which, for some fissionable nuclei, may have more than one hundred exit channels.

Example 10: Compound Nucleus
For the ${}^{28}\mathrm{Si}^*$ compound nucleus formed by a 10 MeV proton incident on an ${}^{27}\mathrm{Al}$ atom, calculate the excitation energy of the compound nucleus, and discuss this relative to the average nucleon energy in the compound nucleus and that required in the three massive particle exit channels listed above.

Solution:
The compound nucleus excitation energy is equal to the projectile kinetic energy + the binding energy of the projectile in the compound nucleus, the separation energy.

$$\begin{aligned} S_{\mathrm{p}} &= c^2 \left[m_{{}^{27}\mathrm{Al}} + m_{\mathrm{H}} - m_{{}^{28}\mathrm{Si}}\right] \\ &= 931.5\,[26.981538 + 1.007825 - 27.976926]\ \mathrm{MeV} \\ &= 11.58\ \mathrm{MeV} \end{aligned}$$

Therefore, the excitation energy is 21.58 MeV, making the average energy among the 28 nucleons in the compound nucleus just 0.77 MeV/nucleon. The separation energies are 9.98, 17.2, and 43.0 MeV, respectively, for the three exit channels leading to α, n, and 3p + n emission. The third would only be possible with a higher energy incident proton since the required energy is greater than the total available to the system. The α and n channels are energetically allowed with a 10 MeV proton, but require much more than the average nucleon excitation energy channeled to just a few nucleons in accord with the compound nucleus model.

The compound nucleus model works best when the projectile energy is low ($\lesssim$ 50 MeV) so that the projectile has little chance of escaping the nucleus with most of its original energy. Also, the target nucleus should be of medium or heavy atomic mass so that the target interior can absorb all of the projectile energy without the average excitation energy per nucleon exceeding the emission threshold. Thermal neutron-induced fission is an ideal situation for the compound nucleus mechanism.

Additional evidence supports the compound nucleus model. Since the compound nucleus lifetime is long, the decay process is independent of formation, and product emission should be nearly isotropic (angle independent). This is confirmed by directional detection experiments. As well, the evaporation analog suggests that as more energy is added to the system, additional nucleons can be ejected. Figure 2.2 shows the relative probability for multiple neutron emission in (α, an) compound nucleus reactions ($a = 1, 2, 3, \ldots$) on a heavy target as a function of α particle energy. The trend for higher probability of emitting larger numbers of neutrons as energy is increased is in agreement with the evaporation model.

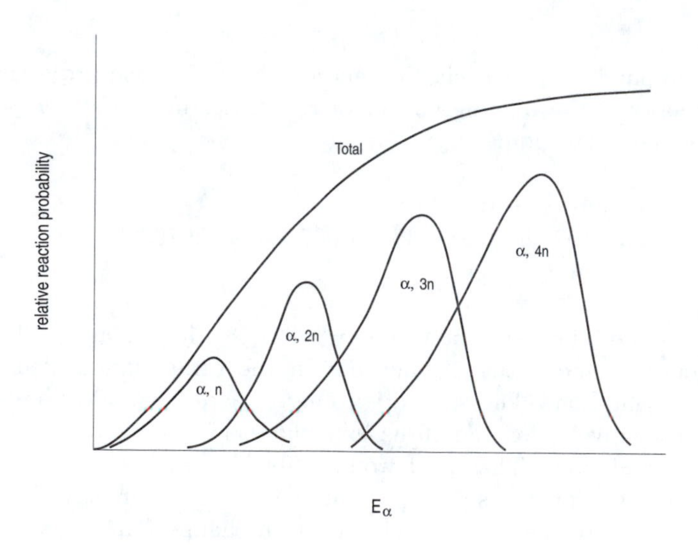

Figure 2.2: Relative probability for (α, n), $(\alpha, 2n)$, ..., emission from a heavy nucleus. The curve marked "total" represents the envelope of all individual reaction probabilities. (From *Introductory Nuclear Physics* by Kenneth S. Krane, Copyright©1988 by John Wiley & Sons, Inc. Reprinted by permission of John Wiley & Sons, Inc.)

2.4 Conservation Laws

Our goal in the remainder of this chapter is to describe the kinematics of binary two-product nuclear reactions. Most often unknown in such problems are the final product energies, which are found as functions of the known reactant masses and energies. Predicting product energies has obvious usefulness. Besides the possibility of elucidating features of nuclear structure, product energies may be important in designing radiation detectors, predicting the degree of materials damage, and determining the types of reactions the products may subsequently induce.

In order to conduct the algebraic formalism required to make these predictions, we need to describe some conserved quantities. These conserved quantities allow us to connect properties before the reaction (reactant properties) to after the reaction (product properties). Such conservation principles should be very familiar from mechanics and fluid dynamics where the independent conservation of mass, momentum, and energy completely describe the evolution of the system. As it must be in any physical system, linear momentum is always independently conserved in nuclear interactions. Beyond this, however, the analogy breaks somewhat. We saw earlier in the discussion of nucleosynthesis that mass and energy are not independently conserved in nuclear systems. Instead, they are conserved as the combined mass/energy (or total energy) of the system. Indeed, it is just this property of converting rest energy to other forms that make nuclear processes useful.

For nuclear reactions below about 50 MeV, nucleon numbers (both Z and N) are independently conserved. From our general binary two-product nuclear reaction

$$^{A_x}_{Z_x}x^{N_x} + \,^{A_\mathrm{X}}_{Z_\mathrm{X}}\mathrm{X}^{N_\mathrm{X}} \longrightarrow \,^{A_\mathrm{Y}}_{Z_\mathrm{Y}}\mathrm{Y}^{N_\mathrm{Y}} + \,^{A_y}_{Z_y}y^{N_y} \tag{2.1}$$

we can obtain nucleon conservation expressions

$$\begin{aligned} N_x + N_\mathrm{X} &= N_\mathrm{Y} + N_y \\ Z_x + Z_\mathrm{X} &= Z_\mathrm{Y} + Z_y \end{aligned} \tag{2.2}$$

and, of course, these imply that the atomic mass number is also conserved

$$A_x + A_\mathrm{X} = A_\mathrm{Y} + A_y \tag{2.3}$$

The only real utility in these expressions comes in identifying unknown reaction products. For example, consider the tritium production reaction

$$\mathrm{n} + \,^6\mathrm{Li} \longrightarrow \mathrm{Y} + \,^3\mathrm{H}$$

By balancing nucleons, one readily finds that the unknown isotope Y must consist of two protons and two neutrons for an atomic mass number of four, *i.e.* it must be a ^{4}He atom.

Electric charge is also conserved in nuclear reactions. This provides a useful reminder to balance electrons in mass/energy conservation. In addition, charge conservation will play an important role in completing nuclear decay reactions (chpt. 4).

2.4.1 Mass/Energy Conservation

In all nuclear reactions, except for scattering, rest mass is not conserved, $m_x + m_X \neq m_Y + m_y$. The total energy or mass/energy is, however, always conserved. Consider a binary two-product reaction (Eq.(2.1)) in which all reaction participants (before and after the reaction) are neutral, isolated, and in a ground nuclear and atomic state. Then, a mass/energy balance can be expressed in the following way

$$\mathrm{KE}_x + m_x c^2 + \mathrm{KE}_\mathrm{X} + m_\mathrm{X} c^2 = \mathrm{KE}_\mathrm{Y} + m_\mathrm{Y} c^2 + \mathrm{KE}_y + m_y c^2 \qquad (2.4)$$

where, as usual, KE_j and m_j represent the kinetic energy and rest mass of particle j, respectively. (Species j may be a photon in which case $\mathrm{KE}_j \to E_j$.) By reorganizing this expression, it can be concluded that a mass difference implies a change in the kinetic energy of the system. This change, $\mathrm{KE}_\mathrm{Y} + \mathrm{KE}_y - \mathrm{KE}_x - \mathrm{KE}_\mathrm{X}$, represents the energy released or liberated in the reaction and is given the symbol Q, called the *Q-value* of the reaction.

$$Q \equiv c^2(m_x + m_\mathrm{X} - m_\mathrm{Y} - m_y) = \mathrm{KE}_\mathrm{Y} + \mathrm{KE}_y - \mathrm{KE}_x - \mathrm{KE}_\mathrm{X} \qquad (2.5)$$

Since we'll most often encounter the situation where the target is at rest before the reaction (*i.e.* $\mathrm{KE}_\mathrm{X} = 0$), then without loss of generality,

$$Q = \mathrm{KE}_\mathrm{Y} + \mathrm{KE}_y - \mathrm{KE}_x$$

In the case of a non-stationary target, we may always transform our reference frame to one in which the target is stationary. The laws of physics are invariant under any such transformation to a frame moving with constant velocity, an inertial frame.

Note that we have not considered the intermediate, compound nucleus state during the interaction when determining Q. Though this would involve many fascinating details of the nuclear processes, it would contribute nothing to our kinematic discussion of the final state of the reaction products. The Q-value, and hence the product energies, are fully determined by the isolated conditions before and after the reaction.

There are three possibilities for the Q-value:

$Q > 0$	exoergic (exothermal) reaction
$Q = 0$	scattering reaction
$Q < 0$	endoergic (endothermal) reaction

When the combined mass of the reactants is larger than that of the products, kinetic energy is released and $Q > 0$, an *exoergic* process (the analog of the chemical exothermal process). This energy is liberated at the expense of reactant rest energy via Einstein's mass/energy equivalence, *i.e.* $\mathrm{KE} = c^2 \Delta m$. The liberated energy initially takes the form of kinetic energy shared between the reaction products which may then be put to practical use (in propulsion, electricity generation, ...).

In situations where the mass inequality is reversed ($m_\mathrm{X} + m_\mathrm{x} < m_\mathrm{Y} + m_\mathrm{y}$), the mass gain in the system comes at the expense of initial projectile kinetic energy. In this reaction, $Q < 0$ and the process is *endoergic* (the analog of chemical endothermal processes). Kinetic energy is not liberated in this reaction. Instead, energy in kinetic form is required for the reaction to proceed. We must provide this energy in a quantity of at least $-Q$ from an external agent to satisfy the total energy balance. There is then a *threshold* of at least $-Q$, below which this reaction cannot proceed. Reactions of this type cannot be spontaneous; they cannot proceed without intervention. Radioactive decay is a spontaneous reaction, and therefore must be exoergic. Conversely, γ induced reactions ($x = \gamma$, nuclear photoeffect) must be endoergic since there is no rest mass in excess of the target mass available to the system. To see this, consider the rest mass definition of the Q-value. For an $x = \gamma$ reaction this is

$$Q = c^2[m_\mathrm{X} - m_\mathrm{Y} - m_y]$$

For this to be exoergic, $m_\mathrm{X} > m_\mathrm{Y} + m_\mathrm{y}$ which yields $\mathrm{KE}_\mathrm{Y} + \mathrm{KE}_\mathrm{y} > E_\gamma$. Allowing $E_\gamma \to 0$ implies this reaction may proceed without requiring the presence of a second reactant. It would then be a spontaneous transformation (decay reaction). Hence, nuclear photoeffect must be endoergic. In a complementary view, the nuclear photoeffect is exactly that interaction scenario we envision when introducing the nuclear separation energy. Since nuclear separation requires energy from an external agent (*i.e.* a projectile photon in our reaction terminology), the interaction must be endoergic.

The remaining possibility is $Q = 0$. These are scattering reactions and are marked by no net change in mass of the system. Scattering reactions are neither endoergic nor exoergic and serve only to redistribute kinetic energy among reactants. Reactants retain their identity such that

$$\begin{aligned} m_x &= m_y \\ m_\mathrm{X} &= m_\mathrm{Y} \end{aligned}$$

There are two varieties of scattering reactions, elastic and inelastic. In elastic scattering, the composite KE is conserved so that $\mathrm{KE}_x + \mathrm{KE}_\mathrm{X} = \mathrm{KE}_\mathrm{Y} + \mathrm{KE}_y$, but individual particle kinetic energy is not

$$\mathrm{KE}_x \neq \mathrm{KE}_y$$

$$\text{KE}_\text{X} \neq \text{KE}_\text{Y}$$

The initially stationary target is given exactly that kinetic energy lost by the projectile. In inelastic scattering, one or both of the products may be left in an excited nuclear state which subsequently decays by photon emission. The kinetic energy of the system is reduced by the excitation energy.

2.4.1.1 Neutron Production Reactions

Availability of neutrons is important in many areas of basic nuclear and quantum physics research, as well as being indispensable in applications of nuclear technology such as radioisotope production and activation analysis. For many such situations, only a nuclear reactor, or perhaps a spontaneous fission source (chpt. 5), will provide a sufficient quantity of neutrons. In some instances however, smaller and perhaps portable sources may be sufficient and even desired (laboratory sources, detector calibration, field experiments, ...), or high neutron energy may be required. For such circumstances reaction-produced neutron sources are more appropriate. Some of the more important of these are listed in table 2.1 below along with their respective Q-values and typical sources. In a few, the advantages of portability and monoenergetic (single energy) emission are mentioned.

2.4.1.2 Mirror Reactions

All of our binary two-product reactions

$$x + \text{X} \longrightarrow \text{Y} + y$$

have a mirror image

$$y + \text{Y} \longrightarrow \text{X} + x$$

That is, the reaction may be reversed providing mass/energy is satisfied. If the reaction yields Q in the forward direction, then it must yield $-Q$ in the reverse direction. If Q is positive, we must supply $-Q$ to allow the reverse reaction. For example, consider the familiar $^9\text{Be}(\alpha, \text{n})^{12}\text{C}$ neutron production reaction. In the forward direction

$$Q = 931.5\frac{\text{MeV}}{\text{amu}}[9.012182 + 4.002603 - 1.008665 - 12] = 5.70\ \text{MeV}$$

an exoergic reaction. (Note that just as in calculating the nuclear binding energy, BE, it is imperative that we consistently use only neutral atomic masses in the calculation of Q-value!) In the reverse direction

$$\text{n} + {}^{12}\text{C} \longrightarrow {}^9\text{Be} + {}^4\text{He}$$

Table 2.1: Some Important Neutron Production Reactions

Type	Reaction	Q-value (MeV)	Source
(α,n)	^{4}He + ^{9}Be $\longrightarrow$ ^{12}C + n	5.70	Portable: decay αs from ^{210}Po, ^{238}Pu, ^{241}Am
	^{4}He + ^{7}Li $\longrightarrow$ ^{10}B + n	-2.79	
(d,n)	^{2}H + ^{3}H $\longrightarrow$ ^{4}He + n	17.59	Accelerator, Thermonuclear Fusion
	^{2}H + ^{2}H $\longrightarrow$ ^{3}He + n	3.27	
(p,n)	^{1}H + ^{7}Li $\longrightarrow$ ^{7}Be + n	-1.64	Accelerator: nearly monoenergetic
(γ,n)	γ + ^{9}Be $\longrightarrow$ ^{8}Be + n	-1.67	Nuclear Photoeffect: nearly monoenergetic

yielding $Q = -5.70$ MeV. We need to supply this quantity of energy (in the form of kinetic energy of the projectile) to the system for this reaction to proceed. Also note that the reverse reaction is endoergic. As a rule, if a reaction is exoergic in the forward direction, then it is endoergic in the reverse direction, and vise versa.

2.4.1.3 Q-value and Binding Energy

From the BE definition (Eq.(1.15)), we can express an arbitrary atomic mass (m_j) as a function of the constituent nucleon masses and the nuclear binding energy

$$m_j = Z_j m_{\mathrm{H}} + N_j m_{\mathrm{n}} - \mathrm{BE}_j/c^2 \tag{2.6}$$

Using this representation of mass in the expression defining Q-value from Eq.(2.5) and conserving nucleons, we readily arrive at

$$Q = \mathrm{BE_Y} + \mathrm{BE}_y - \mathrm{BE}_x - \mathrm{BE_X}$$

The nuclear reaction Q-value can be expressed entirely as a function of the BE of constituent reactants and products. This should not be too surprising since we can consider the nuclear reaction as another form of nucleosynthesis yielding final products Y and y. A positive Q-value then implies $\mathrm{BE}_y + \mathrm{BE}_\mathrm{Y} > \mathrm{BE}_\mathrm{X} + \mathrm{BE}_x$, in agreement with our intuition. The reactants in this instance are of larger mass and, thereby, represent nuclei that are less tightly bound than the product nuclei, *i.e.* in general, exoergic nuclear reactions tend to move the products toward the peak of the BE/A curve.

2.4.1.4 Binding Energy as the Minimum Energy Synthesis

In the first chapter, it was stated that the individual nucleon assembly process is the minimum energy route to nucleosynthesis. Consider, for example, constructing the ^{4}He atom in the following three steps starting from ^{1}H.

1. $\mathrm{n} + {}^1\mathrm{H} \longrightarrow {}^2\mathrm{H} + \gamma_1, \quad Q_1$
2. ${}^1\mathrm{H} + {}^2\mathrm{H} \longrightarrow {}^3\mathrm{He} + \gamma_2, \quad Q_2$
3. $\mathrm{n} + {}^3\mathrm{He} \longrightarrow {}^4\mathrm{He} + \gamma_3, \quad Q_3$

The net reaction is obtained by summing these such that

$$2(\mathrm{n}) + 2({}^1\mathrm{H}) \longrightarrow {}^4\mathrm{He} + \gamma$$

where γ represents the sum of excitation photons released in individual synthesis steps, and $Q = Q_1 + Q_2 + Q_3 = c^2[2m_\mathrm{H} + 2m_\mathrm{n} - m_{^4\mathrm{He}}] = \mathrm{BE}_{^4\mathrm{He}}$. Assembling an arbitrary atom, X, requires $A-1$ processes, but again results in our familiar BE expression

$$Q = \sum_{l=1}^{A-1} = c^2[Zm_\mathrm{H} + Nm_\mathrm{n} - m_\mathrm{X}] = \mathrm{BE}_\mathrm{X}$$

To prove that this is the minimum energy path to synthesis, we would need to identify all other such paths and show that they all lead to a release of greater energy.

As an illustration, let's consider two of the almost limitless possibilities. First, consider assembling ${}^A_Z\mathrm{X}^N$ in $A-1$ steps, this time involving $Z-1$ protons and $N+1$ neutrons to form ${}_{Z-1}^{\ \ A}\mathrm{Y}^{N+1}$. Then, as the final step in synthesis, this will be followed by nuclear beta minus decay process (chpt. 4) that, in essence, converts a nuclear neutron to a proton

$${}_{Z-1}^{\ \ A}\mathrm{Y}^{N+1} \xrightarrow{\beta^-} {}^A_Z\mathrm{X}^N$$

The first $A-1$ steps have the composite $Q_1 = \mathrm{BE}_\mathrm{Y}$

$$Q_1 = c^2[(Z-1)m_\mathrm{H} + (N+1)m_\mathrm{n} - m_\mathrm{Y}]$$

and in the final decay step

$$Q_2 = c^2[m_{\mathrm{Y}} - m_{\mathrm{X}}]$$

The result being

$$Q = Q_1 + Q_2 = \mathrm{BE}_{\mathrm{X}} + c^2[m_{\mathrm{n}} - m_{\mathrm{H}}]$$

Since $m_{\mathrm{n}} > m_{\mathrm{H}}$, then $Q > \mathrm{BE}_{\mathrm{X}}$, and this synthesis route evolves more nuclear energy than the BE route.

Similarly, we could consider assembling $N+2$ neutron and $Z+2$ protons to form ${}^{A+4}_{Z+2}\mathrm{Y}^{N+2}$ followed by the release of ${}^4\mathrm{He}$ to produce ${}^A_Z\mathrm{X}^N$. (This is the α decay process discussed in more detail in chpt. 4.) Again, the first set of synthesis steps yields $Q_1 = \mathrm{BE}_{\mathrm{Y}}$, where

$$Q_1 = c^2[(Z+2)m_{\mathrm{H}} + (N+2)m_{\mathrm{n}} - m_{\mathrm{Y}}]$$

In the α decay step,

$$Q_2 = c^2[m_{\mathrm{Y}} - m_{\mathrm{X}} - m_{^4\mathrm{He}}]$$

yielding the composite

$$Q = Q_1 + Q_2 = \mathrm{BE}_{\mathrm{X}} + \mathrm{BE}_{^4\mathrm{He}}$$

which is again greater than the BE synthesis route.

2.4.2 Momentum Conservation

By considering the simultaneous conservation of linear momentum and mass/energy, we can complete the kinematic description of binary two-product reactions. The simplest case to examine is one in which all reactants are stationary before the reaction, $\mathrm{KE}_x = \mathrm{KE}_{\mathrm{X}} = 0$, so that

$$Q = \mathrm{KE}_{\mathrm{Y}} + \mathrm{KE}_y$$

This case is exoergic by definition, but also appears pathological at first glance since, by assuming this condition, one might wonder how reactions like this can come to be. In practice, however, there is no difficulty. Decay reactions, for example, require no second reactant and, thus, identically satisfy this condition. Furthermore, many exoergic reactions (like fusion) proceed with reactant initial energies much, much less than those of the products and, thereby, very closely approximate this condition.

Conserving linear momentum for this system reveals that the products must be co-linear

$$\mathbf{p}_y = -\mathbf{p}_{\mathrm{Y}}$$

Since this system possesses no (or little) net momentum before the interaction, there is no preferred directionality with respect to an arbitrarily chosen coordinate system for the reaction products. The reaction is then said to yield *isotropic* emission of reaction products.

The product kinetic energies are expressed as functions of their scalar momenta in the non-relativistic limit

$$\mathrm{KE_Y} = \frac{1}{2} m_\mathrm{Y} v_\mathrm{Y}^2 = \frac{1}{2} \frac{p_\mathrm{Y}^2}{m_\mathrm{Y}}$$

and

$$\mathrm{KE}_y = \frac{1}{2} m_y v_y^2 = \frac{1}{2} \frac{p_y^2}{m_y} \tag{2.7}$$

The non-relativistic limit is appropriate for most all of our analyses, at least for the time being. Reaction Qs and participant kinetic energies, KE_j, are always on the order of several MeV at best. This is always much less than the participant rest energies (see example 4 in chpt. 1) except for electrons, which we exclude for the moment. Since the reaction product scalar momenta must be equal ($p_\mathrm{Y}^2 = p_y^2$), the lighter product possesses greater kinetic energy and vise versa, $\mathrm{KE}_y/\mathrm{KE_Y} = m_\mathrm{Y}/m_y$. The momenta and energies are still unknowns, but by incorporating energy balance, we can find unique expressions

$$\mathrm{KE_Y} = Q\left(\frac{m_y}{m_\mathrm{Y} + m_y}\right)$$

and

$$\mathrm{KE}_y = Q\left(\frac{m_\mathrm{Y}}{m_\mathrm{Y} + m_y}\right) \tag{2.8}$$

Now, let's consider the more general problem in which $\mathrm{KE}_x \neq 0$. Again, we can let $\mathrm{KE_X} = 0$ without loss of generality since we can always consider this an inertial transformation from a system where $v_\mathrm{X} \neq 0$. The state of the system before and after the interaction is illustrated in Fig. 2.3. As before, we need not consider the interaction details. The insert here shows momentum conservation vectorially

$$\mathbf{p}_x = \mathbf{p}_\mathrm{Y} + \mathbf{p}_y \tag{2.9}$$

for $\mathbf{p}_\mathrm{X} = 0$. The incident projectile vector momentum and that of either product defines the interaction plane. Since there is no component of linear momentum perpendicular to this plane before the interaction, momentum conservation shows that the other product must also lie in this plane. The reaction products and projectile are all co-planar. The problem, then, reduces to a two-dimensional one where the azimuth is ignorable. The limits on polar angles θ and ϕ are, $0 \leq \theta \leq \pi$ and $0 \leq \phi \leq \pi$, respectively.

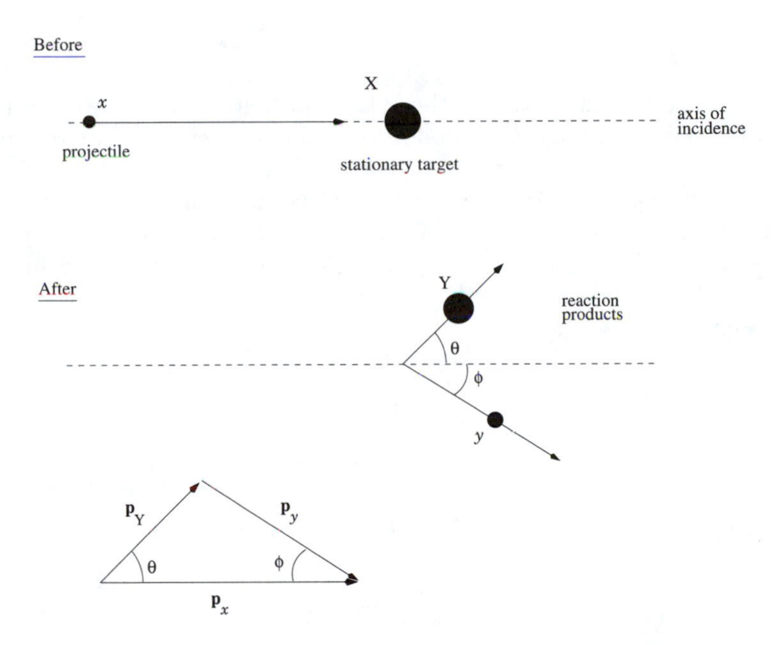

Figure 2.3: Depiction before and after a binary, two-product reaction with a stationary target. The insert at lower left shows the vector momentum balance of products with reactants.

Momentum conservation can be rewritten

$$\mathbf{p}_{\mathrm{Y}} = \mathbf{p}_x - \mathbf{p}_y$$

so that by taking the scalar product with $\mathbf{p}_{\mathrm{Y}}$, we can find

$$p_{\mathrm{Y}}^2 = p_x^2 + p_y^2 - 2p_x p_y \cos\phi \tag{2.10}$$

If we exclude photons and electrons as reaction participants for the time being, and restrict our discussion to non-relativistic particles, then

$$p_j^2 = 2m_j E_j$$

For conciseness we have relaxed the KE_j notation to E_j. Since we are considering only non-relativistic conditions, this convention should cause no confusion, *i.e.* $E_i = \frac{1}{2}m_i v_j^2$. It is reasonable that we need not include relativistic effects. We have already limited considering nuclear reaction projectile energies below about 50 MeV. As well, we have seen that typical Q-values are on the order of a few MeV. When electrons are not involved, the particle kinetic energies can then be only a tiny fraction of their respective rest energy, $E_j \ll m_j c^2$.

Expressing the momentum conservation expression (Eq.(2.9)) as a function of kinetic energy

$$E_{\mathrm{Y}} = \frac{m_x}{m_{\mathrm{Y}}} E_x + \frac{m_y}{m_{\mathrm{Y}}} E_y - 2\frac{\sqrt{m_x m_y}}{m_{\mathrm{Y}}} \sqrt{E_x E_y} \cos\phi$$

Eliminating E_{Y} in favor of $Q = E_{\mathrm{Y}} + E_y - E_x$, we find

$$\begin{aligned} E_y \left(\frac{m_{\mathrm{Y}} + m_y}{m_{\mathrm{Y}}}\right) &- \frac{2}{m_{\mathrm{Y}}} (m_x m_y E_x E_y)^{1/2} \cos\phi \\ &- \left[\left(\frac{m_{\mathrm{Y}} - m_x}{m_{\mathrm{Y}}}\right) E_x + Q\right] = 0 \end{aligned} \tag{2.11}$$

This expression is quadratic in $E_{\mathrm{y}}^{1/2}$ and is solved directly to give

$$\boxed{\begin{aligned} E_y^{1/2} = \frac{(m_x m_y E_x)^{1/2}\cos\phi}{m_{\mathrm{Y}}+m_y} &\pm \left[\frac{m_x m_y E_x \cos^2\phi}{(m_{\mathrm{Y}}+m_y)^2}\right. \\ &+ \left.\frac{m_{\mathrm{Y}}}{m_{\mathrm{Y}}+m_y}\left(\frac{m_{\mathrm{Y}}-m_x}{m_{\mathrm{Y}}}E_x + Q\right)\right]^{1/2} \end{aligned}} \tag{2.12}$$

When E_x is known, as it usually is, we now have the tools to predict the product energies. The product emission angle, ϕ, is also presumed known. We have no means to find this quantity independently. Indeed, we have only two equations (energy and momentum conservation) and three unknowns: E_y, E_Y, and ϕ. It is, perhaps, best to consider the product energies, E_y and E_Y, as functions of ϕ. We will do this later when we describe the details of the ^{7}Li(p,n)^{7}Be reaction.

As a check on our analysis, we can revisit the special case, $E_x = 0$. In Eq.(2.12), this yields

$$\lim_{E_x \to 0} E_y = Q\left(\frac{m_Y}{m_Y + m_y}\right)$$

as we saw earlier, Eq.(2.8).

2.5 The Coulomb Barrier

The analysis of the previous section describes reaction product energies after an interaction. We have not, however, answered the fundamental question, "Given reactant initial conditions, will the reaction occur?" It has been presumed that compound nucleus formation will occur and the reaction will proceed. This, though, pre-supposes that (1) the reactants come into sufficient proximity to allow compound nucleus formation, and (2) the reaction projectile possesses enough energy (in the rest frame of the target) to provide energy and momentum conservation requirements for $Q < 0$ reactions. In other words, there must be sufficient energy in the rest frame of the compound nucleus to supply $-Q$. The latter is the subject of the next section.

Regardless of the magnitude or sign of the Q-value, the reactants must come into close proximity of each other to allow nuclear forces to bring them into coalescence as a compound nucleus. This usually requires approach of the reactants on the order of the nuclear radius, Eq.(1.13). Impeding the approach is the nuclear Coulomb force of repulsion. When targets and projectiles possess $Z \neq 0$ (*i.e.* all except n and γ), Coulomb repulsion may prevent the reactant pair from participating in a nuclear reaction.

Motion against the Coulomb force requires work on the part of the projectile. Penetrating the nucleus implies overcoming the electrostatic barrier provided by this force. Recall that the Coulomb central force law is

$$\mathbf{F}_c = \frac{Z_x Z_X e^2}{4\pi\epsilon_o r^2}\hat{r} \tag{2.13}$$

The work done by the Coulomb force on the projectile (assuming target and projectile are on a direct path) is

$$W_c = \int_{r_i}^{r_f} \mathbf{F}_c \cdot d\mathbf{r} = \frac{Z_x Z_X e^2}{4\pi\epsilon_o} \int_{r_i}^{r_f} \frac{dr}{r^2} \tag{2.14}$$

Let us further assume that our projectile begins its trajectory a distance far from the target where the Coulomb force has negligible influence. Then, we can let $r_i = \infty$. We'll further assign the final position, r_f, to be the distance of closest approach, b. If this distance is close enough to allow nuclear forces to dominate over the Coulomb force, then the reaction can proceed. Otherwise, the reactants will be repelled, reverse their relative motion, and will not undergo a nuclear reaction. This situation is referred to as Coulomb scattering (sec. 6.1.1).

Evaluating Eq.(2.14) yields

$$W_c = -\frac{Z_x Z_X e^2}{4\pi\epsilon_o b}$$

so that, by the work-energy theorem, this represents a reduction in the projectile kinetic energy by the amount W_c. The projectile must initially possess kinetic energy, E_i, equal to $-W_c$ to approach the target to a distance b. The projectile must then satisfy a threshold condition

$$B = E_i = \frac{Z_x Z_X e^2}{4\pi\epsilon_o b} \tag{2.15}$$

the *Coulomb barrier*.

To a reasonable degree of approximation, we can estimate

$$b \sim R_x + R_X = R_o(A_x^{1/3} + A_X^{1/3})$$

(This is only an approximation because the Coulomb and nuclear forces overlap for some finite distance rather than having a sharp cutoff at b as assumed above. As well, quantum mechanical effects allow non-zero probability for particles possessing less than B to penetrate the nucleus. This is called *barrier penetration* or *barrier tunneling*[Eisberg85].) By evaluating the constants in the barrier expression, we have

$$\boxed{B \sim 1.2\frac{Z_x Z_X}{A_x^{1/3} + A_X^{1/3}}, \quad \text{MeV}} \tag{2.16}$$

Of course, when $x = \text{n}$ or γ, $Z_x = 0$, and there is no barrier, $B = 0$. When Z_x and Z_X are non-zero, energy of at least B must be provided as projectile kinetic energy to allow the reaction (in this classical limit). This energy is not wasted, though. In the intermediate state, momentum conservation requires that part of B temporarily provide kinetic energy to the compound nucleus, $E^* = \frac{m_x}{m^*}B$ (if $E_x = B$ and $E_X = 0$) where $*$ properties are those of the compound nucleus. The remainder of B contributes to excitation. In the final state of the system, the entirety of B shows up in reaction product kinetic energy.

2.6 The Threshold Energy for a Nuclear Reaction

To complete our answer to the reaction threshold question, we must also consider energy and momentum conservation requirements. When $Q \geq 0$, a nuclear reaction may proceed at any reactant energies, providing the Coulomb barrier condition is satisfied. For $Q < 0$, however, energy must be supplied to the system. Energy conservation alone dictates that $E_x > -Q$. This is a necessary, but not a sufficient condition for the reaction to proceed. To simultaneously conserve momentum, Eq.(2.12) must be satisfied. The minimum E_x that satisfies this expression is the *threshold energy*, E_x^{t}, for the reaction. Below this energy, the reaction is not possible. The threshold projectile energy occurs when both ϕ and the argument of the radical in Eq.(2.12) are zero,

$$E_x^{\mathrm{t}} = -Q \frac{m_{\mathrm{Y}} + m_y}{m_{\mathrm{Y}} + m_y - m_x} = \frac{-Q}{1 - \frac{m_x}{m_{\mathrm{Y}} + m_y}} \tag{2.17}$$

The threshold energy is always greater than $-Q$, although it may be only slightly greater since usually $m_x \ll m_{\mathrm{Y}} + m_y$. The threshold condition should be recognized as the necessity of providing $-Q$ for the reaction in the rest frame of the compound nucleus. The difference, $E_x^t + Q$, is temporarily consumed in providing momentum to the compound nucleus and eventually contributes to kinetic energy of the reaction products. Since this is a more strict condition, satisfying $E_x \geq E_x^{\mathrm{t}}$ is a sufficient condition for reaction (assuming $E_x \geq B$ also). Note that the minimization condition only occurs at $\phi = 0$. Here $\theta = 0$ also, and all reaction participants are co-linear. This is an intuitive result since momentum must be independently conserved in each direction.

To summarize the sufficient conditions for nuclear reactions, we may condense our discussion to the following. When $Q > 0$, there is no threshold ($E_x^{\mathrm{t}} = 0$). The Coulomb barrier is then the only condition to satisfy, *i.e.* $E_x \geq B$. For $Q < 0$ reactions, the initial projectile energy must at least match the larger of B or E_x^{t}, *i.e.* $E_x \geq \max(B, E_x^{\mathrm{t}})$.

There is another important energy in this problem when $Q < 0$. Closer inspection of Eq.(2.12) reveals that double valued behavior of E_y appears (when the second term is less than the first term in Eq.(2.12)) at energies above E_x^{t}, but below

$$E_x' = -Q \frac{m_{\mathrm{Y}}}{m_{\mathrm{Y}} - m_x} \tag{2.18}$$

i.e. for $E_x^{\mathrm{t}} \leq E_x \leq E_x'$, there are exactly two values of E_y for each E_x that satisfy energy and momentum conservation, with kinetic energy sharing

Example 11: Threshold Conditions

The compound nucleus $^{15}N^*$ can be formed in at least 14 different known ways. A few of these are listed below with their respective end products. Examine the threshold condition for each of the seven paths listed.

$$^4He + {}^{11}B \longrightarrow {}^{15}N^* \longrightarrow \left\{ \begin{array}{ccc} ^{14}N & + & n \\ ^{14}C & + & H \end{array} \right.$$

$$^3H + {}^{12}C \longrightarrow {}^{15}_{7}N^* \longrightarrow {}^{14}C + H$$

$$^2H + {}^{13}C \longrightarrow {}^{15}N^* \longrightarrow \left\{ \begin{array}{ccc} ^{11}B & + & ^4He \\ ^{12}C & + & ^3H \end{array} \right.$$

$$p + {}^{14}C \longrightarrow {}^{15}_{7}N^* \longrightarrow {}^{14}N + n$$

$$n + {}^{14}N \longrightarrow {}^{15}_{7}N^* \longrightarrow {}^{11}B + {}^4He$$

Solution:

path	B (MeV)	Q (MeV)	E_x^t (MeV)	E_x Requirement	$\min(KE_Y + KE_y) = \min(E_x) + Q$ (MeV)
$^{11}B(\alpha, n)^{14}N$	3.15	0.157	0	$E_x \geq B$	3.307
$^{11}B(\alpha, p)^{14}C$	3.15	0.783	0	$E_x \geq B$	3.933
$^{12}C(T, p)^{14}C$	1.93	4.64	0	$E_x \geq B$	6.57
$^{13}C(D, \alpha)^{11}B$	1.99	5.17	0	$E_x \geq B$	7.16
$^{13}C(D, T)^{12}C$	1.99	1.31	0	$E_x \geq B$	3.30
$^{14}C(p, n)^{14}N$	2.11	-0.626	0.671	$E_x \geq B$	1.484
$^{14}N(n, \alpha)^{11}B$	0	-0.157	0.168	$E_x \geq E_x^t$	0.011

among the reaction products being the only difference. There is also a corresponding angle below which double-valued product energies occur. By again examining the conditions under which the argument of the radical in Eq.(2.12) vanishes, we can identify this maximum angle for double-valued behavior as

$$\cos^2 \phi_o = -\frac{m_Y(m_Y + m_y)}{m_y m_x}\left(\frac{Q}{E_x} + \frac{m_Y - m_x}{m_Y}\right) \tag{2.19}$$

At this angle, there is only one product energy for a given E_x. Below ϕ_o there are two allowed product energies at each E_x. The critical angle ϕ_o can be considered a threshold angle as a function of E_x since at $\phi_o = 0, E_x = E_x^t$ and at $\phi_o = \pi/2, E_x = E_x'$. Double-valued behavior then only occurs for $0 \leq \phi \leq \pi/2$; there is no backscatter in the double-valued region. When $Q \geq 0$, only single-valued behavior occurs. These features are illustrated with the following examples in the next two sections.

2.7 The Neutron Detection Reaction: ^{3}He(n,p)^{3}H

The (n,p) reaction in ^{3}He gas is often exploited as a neutron detection reaction[Knoll89]. Since this is an exoergic reaction with $Q = 0.764$ MeV and the projectile is the uncharged neutron, there is neither a Coulomb barrier nor a reaction threshold. Thus, we could use this reaction to detect neutrons of even very low energy. (The reverse reaction ^{3}H(p,n)^{3}He has, of course, $B \sim 0.49$ Mev, and since $Q = -0.764$ MeV, then $E_x^{\rm t} = 1.02$ MeV.)

In the low energy limit (n and ^{3}He having negligible kinetic energy before the reaction), the forward reaction yields products with energies (Eq.(2.8)) $E_{\rm p} = 0.573$ MeV and $E_{^3{\rm He}} = 0.191$ MeV. When $E_{\rm n}$ is not small with respect to Q, we can find product energies by solving the complete problem, Eq.(2.12). Figure 2.4 shows the results of such a calculation as a plot of $E_{\rm p}$ vs. $E_{\rm n}$ for several values of ϕ. There is no double-valued behavior for this case. One can infer, then, a unique neutron projectile energy for a measured proton product energy at a given emission angle. In the limit $E_x \gg Q$, $E_y \propto E_x$, as we expect from Eq.(2.12) and Fig. 2.4.

2.8 The Neutron Production Reaction: ^{7}Li(p,n)^{7}Be

Often times, it is desired to produce neutrons. As mentioned earlier, neutron sources serve many functions from basic quantum physics experiments

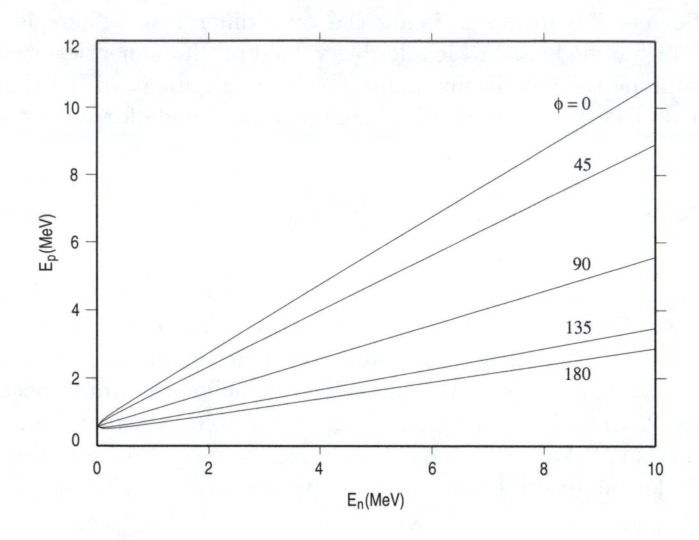

Figure 2.4: Neutron detection reaction ^{3}He(n,p)^{3}H product energy $E_{\rm p}$ vs. $E_{\rm n}$ for several values of ϕ = 0, 45, 90, 153, and 180 degrees.

to determining neutron-induced reaction probabilities to designing neutron detectors. Almost as often, producing neutrons of a fixed energy (monoenergetic) is desired as neutron interaction phenomena are often a sensitive function of energy.

Certain nuclear reactions provide a means for producing very nearly monoenergetic neutrons with neutron energy as a function of emission angle, an excellent source for many experimental needs. Charged particle induced reactions like (p,n), (d,n), (α,n), ... are particularly useful since the projectile energy can be controlled in a particle accelerator, and thereby, produce the desired product neutron energy. A classic example is the (p,n) reaction in ^{7}Li (Fig. 2.5). Since this is an endoergic reaction with $Q = -1.644$ MeV, the kinematics are slightly more involved than in the prior exoergic neutron detection example. We must include the double-valued nature of the product energies for projectile energy in the range $E_{\rm p}^{\rm t} \leq E_{\rm p} \leq E_{\rm p}'$. ($B \sim 1.236$ MeV for this case, so the threshold energy is the limiting condition.) Equations (2.17) and (2.18) yield $E_{\rm p}^{\rm t}$ and $E_{\rm p}'$ of 1.88 and 1.92 MeV, respectively. Figure 2.6 shows the critical angle (Eq.(2.19)) as a function of projectile energy in this range.

Product neutron energy is shown vs. projectile energy for the double-valued range in Fig. 2.7. In this energy range, neutrons carry one of two possible energies on any one interaction for each ϕ below ϕ_o. The maximum

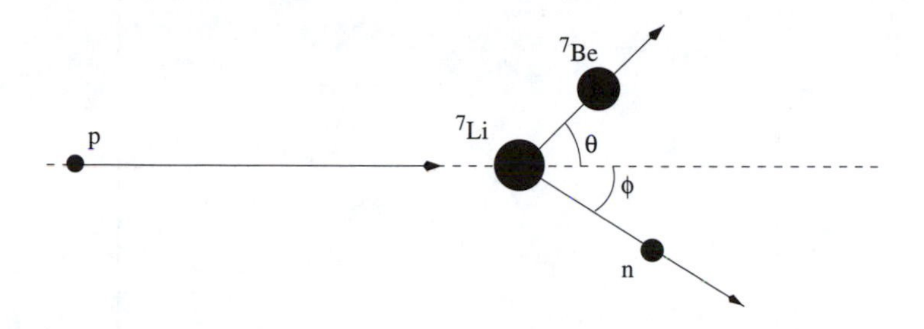

Figure 2.5: The ^{7}Li(p,n)^{7}Be neutron production reaction, before and after interaction.

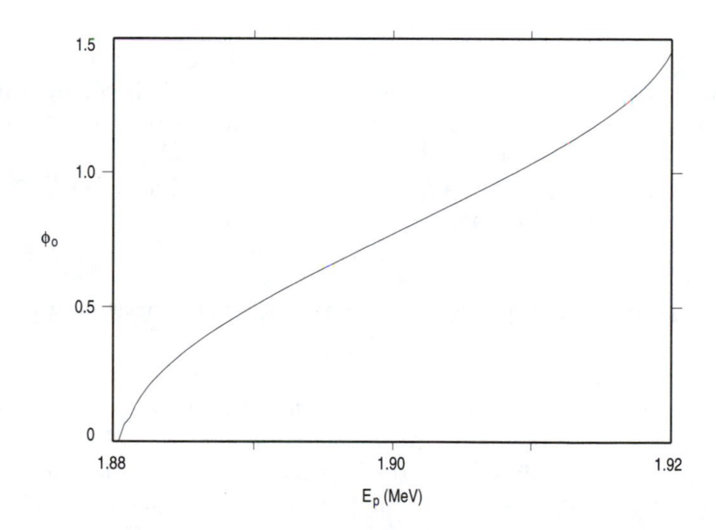

Figure 2.6: Critical angle (radians) for the ^{7}Li(p,n)^{7}Be reaction.

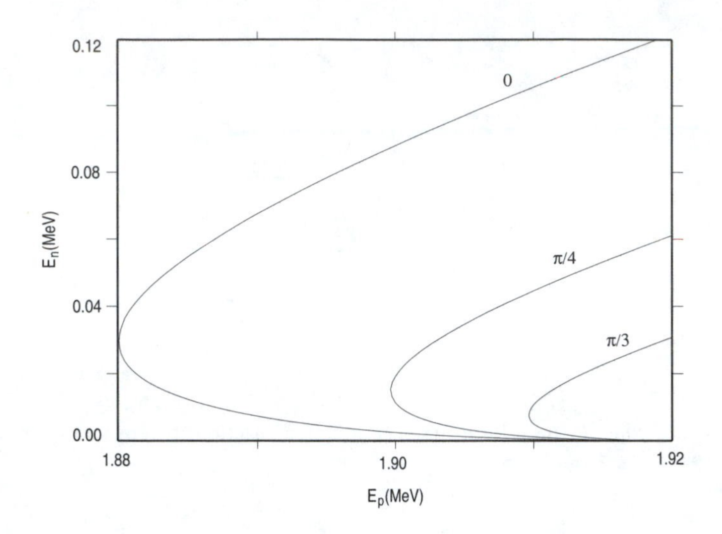

Figure 2.7: Neutron energy vs. proton projectile energy for the ^{7}Li(p,n)^{7}Be reaction in the double-valued energy range for $\phi = 0, \pi/4, \pi/3$.

possible ϕ_o is $\pi/2$, *i.e.* there are no neutrons directed backward in the double-valued energy range. For energies above $E'_{\rm p}$, $E_{\rm n}$ is a single-valued function of $E_{\rm p}$, as shown in Fig. 2.8. Single-valued neutrons at all emission angles $0 \leq \phi \leq \pi$ are observed. We may then produce monoenergetic neutrons for a given ϕ so long as $E_{\rm p} \geq E'_{\rm p}$. Note again that when $E_p \gg\mid Q \mid$, $E_n \propto E_p$, a linear relationship.

2.9 Center of Mass Coordinate System

Until now, we have employed a view of nuclear reactions in which the reaction participant energies and momenta are described in the context of a stationary frame of reference. Most often, we would associate this with the fixed frame of the observer in the laboratory and, therefore, call it the *lab frame*. To simplify matters, we had allowed our target particle to take on zero kinetic energy (and momentum) in this frame. These are natural choices since they are comfortable within our usual human experience. Yet, there may be circumstances for which an alternative approach is desired. A frame of reference which provides some convenience in certain interaction problems is the *center of mass* frame. (For example, the description of interaction probability as a function of product emission angle is often

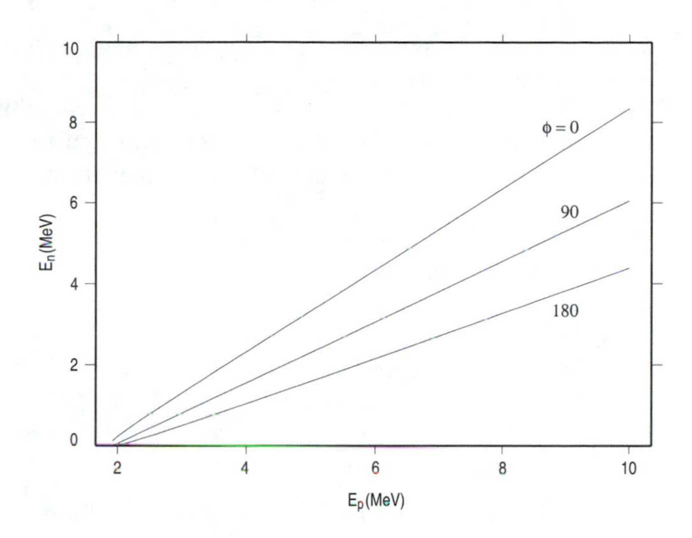

Figure 2.8: Neutron energy vs. proton projectile energy for the ^{7}Li(p,n)^{7}Be reaction for energies above $E'_{\rm p}$.

facilitated by reaction description in the center of mass system[Lamarsh66] since reaction products are often more isotropic in this frame. Though it is well beyond the scope of our endeavors here, it is worth noting this point for subsequent work in reactor and nuclear physics.) As well, certain reaction properties appeal to our sense of physical intuition when expressed in the center of mass. The threshold condition, for example, appears naturally in the kinematics expressions developed in the center of mass.

Since the laws of physics are invariant under all inertial transformations, we are free to choose any inertial frame. The choice that has the most widespread use is that in which the net momentum of the system is identically zero before and after the interaction. This system is identical to one moving with the compound nucleus. Such a reference frame is called the *center of momentum* frame. For the low energy, non-relativistic interactions that we are considering, there is little difference between this frame and the *center of mass* frame. In keeping with tradition, we'll retain the latter terminology. Our lab frame, however, conforms exactly to the center of mass only in the trivial case where $E_x = E_{\rm X} = 0 \;\; (Q > 0)$. The center of mass system is used almost exclusively in nuclear physics and neutron slowing down. Some important nomenclature for this discussion is listed below:

L = The lab coordinate system. The observer is fixed in a frame stationary in the laboratory.
C = The center of mass coordinate system. The observer is in a frame of reference moving with respect to the laboratory frame such that the interaction system has identically zero momentum before and after the interaction.
m_j = The rest mass of species $j(= x, X, Y, y)$, projectile, target, heavy product, and light product, respectively.
$\mathbf{v}_j$ = The velocity of m_j in L.
$\mathbf{v}_j^c$ = The velocity of m_j in C.
$\mathbf{V}_c$ = L velocity of C before the interaction.
$\mathbf{V}_c'$ = L velocity of C after the interaction.
E_j = The kinetic energy of m_j in L.
E_j^c = The kinetic energy of m_j in C.

As usual, our general binary two-product reaction takes the form

$$x + \mathrm{X} \longrightarrow \mathrm{Y} + y$$

Recall that non-relativistic momentum conservation requires

$$m_x \mathbf{v}_x + m_\mathrm{X} \mathbf{v}_\mathrm{X} = m_\mathrm{Y} \mathbf{v}_\mathrm{Y} + m_y \mathbf{v}_y$$

Let's now transform to a frame moving with velocity $\mathbf{V}_c$ with respect to the lab frame so that before the interaction

$$m_x(\mathbf{v}_x - \mathbf{V}_c) + m_\mathrm{X}(\mathbf{v}_\mathrm{X} - \mathbf{V}_c) + (m_x + m_\mathrm{X})\mathbf{V}_c = m_x \mathbf{v}_x + m_\mathrm{X} \mathbf{v}_\mathrm{X} \quad (2.20)$$

and after the interaction, we consider a moving frame at a different lab velocity, $\mathbf{V}_c'$, for generality

$$m_\mathrm{Y} \mathbf{v}_\mathrm{Y} + m_y \mathbf{v}_y = m_\mathrm{Y}(\mathbf{v}_\mathrm{Y} - \mathbf{V}_c') + m_y(\mathbf{v}_y - \mathbf{V}_c') + (m_\mathrm{Y} + m_y)\mathbf{V}_c' \quad (2.21)$$

Since any choice of $\mathbf{V}_c$ will satisfy Eq.(2.20), we'll select $\mathbf{V}_c$ to make the net momentum vanish in the transformed system before the interaction. This implies

$$m_x(\mathbf{v}_x - \mathbf{V}_c) + m_\mathrm{X}(\mathbf{v}_\mathrm{X} - \mathbf{V}_c) \equiv 0$$

so that

$$\boxed{\mathbf{V}_c \equiv \frac{m_x \mathbf{v}_x + m_\mathrm{X} \mathbf{v}_\mathrm{X}}{m_x + m_\mathrm{X}}} \quad (2.22)$$

Similarly, we wish to have the net momentum again vanish after the interaction

$$m_\mathrm{Y}(\mathbf{v}_\mathrm{Y} - \mathbf{V}_c') + m_y(\mathbf{v}_y - \mathbf{V}_c') \equiv 0$$

such that

$$\mathbf{V}'_c \equiv \frac{m_Y \mathbf{v}_Y + m_y \mathbf{v}_y}{m_Y + m_y} = \frac{m_x + m_X}{m_Y + m_y}\mathbf{V}_c \tag{2.23}$$

The lab velocities of C differ before and after the interaction by the factor $(m_x + m_X)/(m_Y + m_y)$, which is typically within 1 part in 10^3 to 10^5 of unity, so there is little practical importance to the distinction between $\mathbf{V}'_c$ and $\mathbf{V}_c$. It is interesting to encounter this result, however, since it is not borne out in our classical interpretation. Classically, we expect these two velocities to be identical for momentum conservation. Reality differs from our classical view because mass is not conserved when $Q \neq 0$. When mass is not conserved, the center of mass velocity must change after the interaction to conserve momentum. When there is no mass change (as in scattering interactions where $Q = 0$), we recover the classical result. We'll see this in the next chapter when we discuss scattering.

Let's define all C velocities as $\mathbf{v}^c_j = \mathbf{v}_j - \mathbf{V}_c$ before the interaction, and $\mathbf{v}^c_j = \mathbf{v}_j - \mathbf{V}'_c$ after, so that

$$m_x \mathbf{v}^c_x + m_X \mathbf{v}^c_X = 0$$

and

$$m_Y \mathbf{v}^c_Y + m_y \mathbf{v}^c_y = 0 \tag{2.24}$$

From this, we can immediately see that the binary reaction participants are co-linear in the C system both before and after the reaction (Fig. 2.9). This description injects symmetry into the formalism, so that our final kinematics expressions will become simplified. The C system emission angle, θ_c, ranges from 0 to π and $\neq f(m_j, E_j)$ (*i.e.* not a function of mass or energy), as with θ in the L system.

Let's now express particle kinetic energies in the new system. Again, we are considering only non-relativistic interactions so that, $E_j = \frac{1}{2} m_j \mathbf{v}^2_j$. In C, the kinetic energy is

$$E^c_j = \frac{1}{2} m_j (\mathbf{v}^c_j)^2 = \frac{1}{2} m_j (\mathbf{v}_j - \mathbf{V}_c)^2 \tag{2.25}$$

Expanding the transformed velocity with the aid of Eq.(2.22), we can express this as a function of particle lab velocities only

$$\mathbf{v}_x - \mathbf{V}_c = \mathbf{v}_x - \frac{m_x \mathbf{v}_x + m_X \mathbf{v}_X}{m_x + m_X} = \frac{m_X}{m_x + m_X}(\mathbf{v}_x - \mathbf{v}_X) \tag{2.26}$$

so that

$$E^c_x = \frac{1}{2} \frac{m_x m^2_X}{(m_x + m_X)^2} (\mathbf{v}_x - \mathbf{v}_X)^2$$

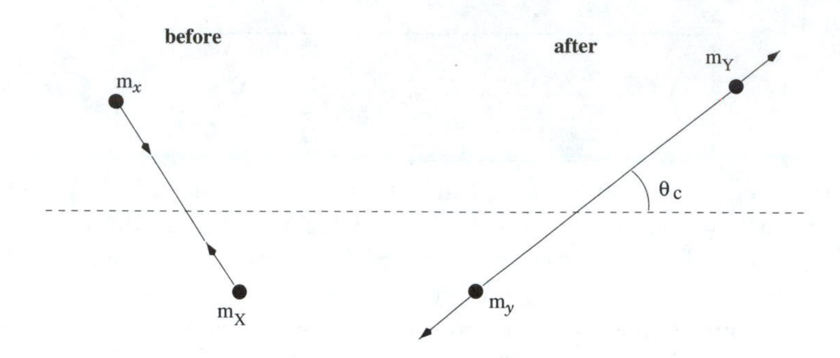

Figure 2.9: Binary, two-product interaction in the center of mass frame.

and by symmetry,

$$E_X^c = \frac{1}{2}\frac{m_X m_x^2}{(m_x + m_X)^2}(\mathbf{v}_X - \mathbf{v}_x)^2$$

Both reactants are in motion in C regardless of their state of motion in L. The labels "projectile" and "target" are then meaningless in C. The sum $E_x^c + E_X^c$ represents the net kinetic energy of the reactants in C. Let's reduce this in favor of lab frame energies

$$\begin{aligned} E_x^c + E_X^c &= \frac{1}{2}\frac{m_x m_X}{m_x + m_X}(\mathbf{v}_x - \mathbf{v}_X)^2 \qquad (2.27) \\ &= \frac{1}{2}\frac{m_x m_X}{m_x + m_X}(v_x^2 + v_X^2 - 2\mathbf{v}_x \cdot \mathbf{v}_X) \end{aligned}$$

Following some rearrangement of terms, this becomes

$$E_x^c + E_X^c = E_x + E_X - \frac{1}{2}(m_x + m_X)\left(\frac{m_x\mathbf{v}_x + m_X\mathbf{v}_X}{m_x + m_X}\right)^2$$

where the last term on the right hand side should be recognized as just $\frac{1}{2}(m_x + m_X)\mathbf{V}_c^2 = \mathrm{KE}_c$, the lab energy of the center of mass before the interaction. Therefore, the C reactant kinetic energies are related to the L reactant kinetic energies through

$$\boxed{E_x^c + E_X^c + \mathrm{KE}_c = E_x + E_X} \qquad (2.28)$$

Similarly, we can show

$$E_Y^c + E_y^c + \mathrm{KE}_c' = E_Y + E_y \tag{2.29}$$

where $\mathrm{KE}_c' = \frac{1}{2}(m_Y + m_y)(\mathbf{V}_c')^2$. From the definition of the Q-value, (Eq.(2.5))

$$\begin{aligned} Q &= E_Y + E_y - E_x - E_X \\ &= E_Y^c + E_y^c - E_x^c - E_X^c + (\mathrm{KE}_c' - \mathrm{KE}_c) \end{aligned} \tag{2.30}$$

where the correction term $\mathrm{KE}_c' - \mathrm{KE}_c = \frac{1}{2}Q\frac{m_x+m_X}{m_Y+m_y}\frac{V_c^2}{c^2}$ is usually tiny compared to unity. Note that when $Q > 0$, the correction term is also > 0. Since the products possess less mass than the reactants in this case, the center of their masses must attain a higher velocity to conserve momentum. We can then write mass/energy balance in the transformed system

$$Q = \frac{E_Y^c + E_y^c - E_x^c - E_X^c}{1 - \frac{\mathrm{KE}_c}{(m_Y+m_y)c^2}} \tag{2.31}$$

Because of co-linearity, Eq.(2.24), the reaction participant momenta and energies are simply related, and independent of emission angle

$$\begin{aligned} m_x v_x^c &= m_X v_X^c \\ m_x E_x^c &= m_X E_X^c \\ m_Y v_Y^c &= m_y v_y^c \\ m_Y E_Y^c &= m_y E_y^c \end{aligned} \tag{2.32}$$

For given values of E_x, E_X, Q, then E_Y^c, E_y^c, v_Y^c, and v_y^c are fixed and $\neq f(\theta_c)$.

An arbitrary binary reaction is depicted in Fig. 2.10 in both the L and C systems. Here, the open circles represent the center of mass, moving with $\mathbf{V}_c$ and $\mathbf{V}_c'$ in the lab frame, before and after the interaction, respectively. An observer in C traveling with the center of mass would perceive the interaction on the dotted trajectories, initially toward (before the interaction) and then away (after). The trajectory lengths s_j and x_j represent the L and C distance traveled by m_j, respectively. Since $v_Y^c = \frac{m_y}{m_Y}v_y^c$, we have

$$\frac{x_Y}{x_y} = \frac{v_Y^c}{v_y^c} = \frac{m_y}{m_Y}$$

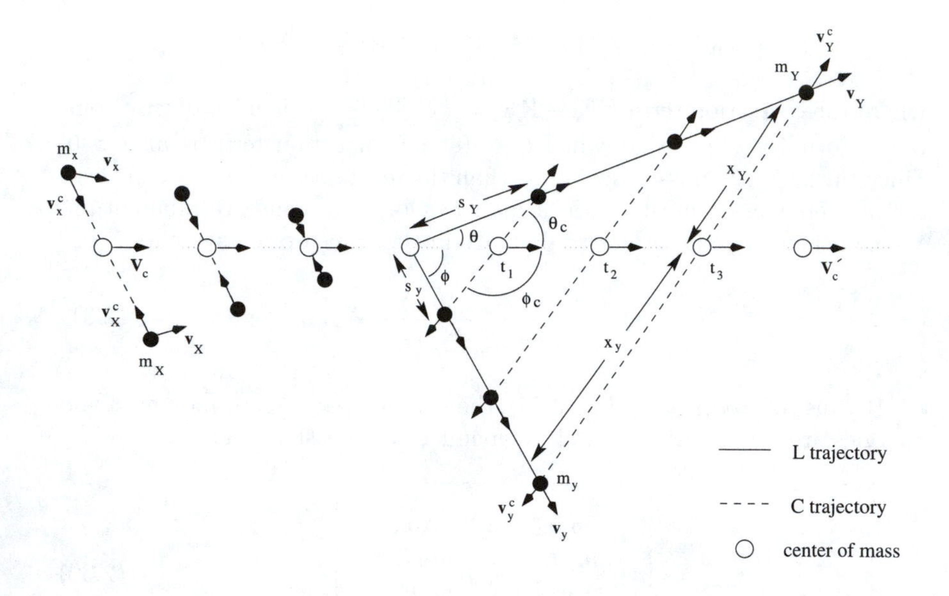

Figure 2.10: Binary, two-product interaction in the L and C frames for comparison. The C frame is in translation to the right at V_c before and V_c' after the interaction in the L frame. In the C system, reaction participants are co-linear both before and after the interaction.

and

$$\frac{s_Y}{s_y} = \frac{v_Y}{v_y} = \frac{|\mathbf{v}_Y^c + \mathbf{V}_c'|}{|\mathbf{v}_y^c + \mathbf{V}_c'|}$$

As special cases, when we have $\mathbf{V}_c = 0$ (*i.e.* $\mathbf{v}_x = -\frac{m_X}{m_x}\mathbf{v}_X$) the interaction is identical in both L and C frames. If $m_X \gg m_x$ (the target is much more massive than the projectile), then

$$\mathbf{V}_c \sim \frac{m_x}{m_X}\mathbf{v}_x + \mathbf{v}_X \sim \mathbf{v}_X \tag{2.33}$$

as long as $\mathbf{v}_x \ll \frac{m_X}{m_x}\mathbf{v}_X$, and the center of mass resides with the target particle.

2.9.1 Energy Relationships for a Stationary Target

Since the situation with the most practical importance is still the one in which $\mathbf{v}_X = 0$ (stationary target in the lab frame), let's describe our L–C transformation in more detail under these conditions. For this case

$$\mathbf{V}_c = \frac{m_x \mathbf{v}_x}{m_x + m_X} \tag{2.34}$$

so

$$\begin{aligned} \mathbf{v}_x^c &= \mathbf{v}_x - \mathbf{V}_c = \frac{m_X}{m_x + m_X}\mathbf{v}_x \\ \mathbf{v}_X^c &= -\mathbf{V}_c = -\frac{m_x}{m_x + m_X}\mathbf{v}_x \end{aligned} \tag{2.35}$$

The respective reactant kinetic energies in C can be expressed as functions of the lab energy, E_x

$$\boxed{\begin{aligned} E_x^c &= \left(\frac{m_X}{m_x+m_X}\right)^2 E_x \quad , \le E_x \\ E_X^c &= \frac{m_x m_X}{(m_x+m_X)^2} E_x \quad , \le E_x \\ \mathrm{KE}_c &= \frac{m_x}{m_x+m_X} E_x \quad , \le E_x \end{aligned}} \tag{2.36}$$

To determine the product energies, we must again look to the reaction mass/energy balance, Eq.(2.31). With $E_Y^c = \frac{m_y}{m_Y} E_y^c$ and $E_X^c = \frac{m_x}{m_X} E_x^c$, and Eq.(2.36), we find directly

$$\boxed{E_y^c = \frac{m_Y}{m_Y + m_y}\left[Q + \left(1 - \frac{m_x}{m_Y + m_y}\right) E_x\right]} \tag{2.37}$$

and

$$E_{\mathrm{Y}}^c = \frac{m_y}{m_{\mathrm{Y}} + m_y}\left[Q + \left(1 - \frac{m_x}{m_{\mathrm{Y}} + m_y}\right)E_x\right] \tag{2.38}$$

An obvious advantage of this formalism is the direct recognition of the threshold condition from the reaction product energy relations in C. At threshold $E_y^c = E_{\mathrm{Y}}^c = 0$, the products appear as separate entities but move together with the center of mass. Then, from Eq.(2.37) or (2.38), we find directly, $E_x^t = -Q/(1 - \frac{m_x}{m_{\mathrm{Y}}+m_y})$ in agreement with Eq.(2.17) through a much more involved analysis. The C reactant energies at threshold from Eq.(2.31) are, $E_x^c|_t + E_{\mathrm{X}}^c|_t = -Q$ ignoring the small correction term, consistent with our interpretation of threshold. Finally, these expressions readily reduce to the simple case (Eq.(2.8)) of $E_x = E_{\mathrm{X}} = 0$, where for this particular situation $E_y^c = E_y$, the L and C systems are coincident.

2.9.2 Neutron Detection and Production Reactions in the Center of Mass

The C system solutions appear algebraically simpler and lack angular dependence in the product energy solutions, Eqs.(2.37) and (2.38). Relating the C energies back to the L system does, however, require incorporation of the emission angle and reintroduces complexity. To see this, we can expand

$$E_y = \frac{1}{2}m_y\mathbf{v}_y^2 = \frac{1}{2}m_y(\mathbf{v}_y^c + \mathbf{V}_c')^2$$

to show

$$E_y = E_y^c + \frac{m_x m_y}{(m_{\mathrm{Y}} + m_y)^2}E_x + \frac{2\cos\phi_c}{m_{\mathrm{Y}} + m_y}\sqrt{m_x m_y E_x E_y^c} \tag{2.39}$$

Notice that E_y is maximized for forward emission ($\phi_c = 0$) and minimized for backward emission ($\phi_c = \pi$). Also, when $E_x \to 0$, the product energy $E_y = E_y^c = m_{\mathrm{Y}}Q/(m_{\mathrm{Y}} + m_y)$, as we saw earlier (Eq.(2.8)) for this limit. We can use expressions (2.37) and (2.39) to straightforwardly show that, in general, $E_y > E_y^c$ when

$$\cos\phi_c > -\frac{1}{2}\sqrt{\frac{m_x m_y}{m_{\mathrm{Y}}(m_{\mathrm{Y}} + m_y)[\frac{Q}{E_x} + 1] - m_x m_{\mathrm{Y}}}}$$

In the scattering limit, this reduces to $\cos\phi_c = -\frac{1}{2}\frac{m_x}{m_{\mathrm{X}}}$, which approaches 0 for heavy targets, so that $E_y > E_y^c$ for $\phi_c < \pi/2$ as we would expect.

We can also express the lab emission angle as a function of that in C by utilizing the law of sines. With the aid of Fig. 2.10

$$v_y \sin\phi = v_y^c \sin\phi_c$$

so that

$$\left(\frac{v_y}{v_y^c}\right)^2 = \frac{E_y}{E_y^c} = \frac{1-\cos^2\phi_c}{1-\cos^2\phi}$$

then

$$\cos^2\phi = 1 - \frac{E_y^c}{E_y}\left(1-\cos^2\phi_c\right) \tag{2.40}$$

As a check, let's use these expressions to determine product energies for our neutron detection and production reactions investigated earlier. We had found for the neutron detection reaction, ^{3}He(n,p)^{3}H, that $Q = 0.764$ MeV. As a particular point of comparison, let's choose $E_x = E_{\rm n} = 4$ MeV. Then from Eq.(2.37), $E_y^c = 2.819$ MeV. With $\phi_c = \pi/2$, this yields $E_y = 3.07$ MeV and $\phi = 73.4^o$ from Eqs.(2.39) and (2.40), respectively, confirmed by the predictions in the L analysis, Eq.(2.12).

In the neutron production reaction, ^{7}Li(p,n)^{7}Be with $Q = -1.644$ MeV, we saw double-valued product energy behavior in the energy range $E_{\rm p}^t =$ 1.88 MeV to $E_{\rm p}' = 1.92$ MeV. Selecting $E_{\rm p} = 1.9$ MeV, we find $E_y^c = 0.0152$ MeV. Choosing, for example, $\phi = 0$, we can satisfy Eq.(2.40) with either $\phi_c = 0$ or π, yielding $E_y = 0.00248$ or 0.0.0879 MeV, identical to our predictions of sec. 2.8.

References

[Eisberg85] Eisberg, R., and Resnick, R., *Quantum Physics of Atoms, Molecules, Solids, Nuclei, and Particles*, John Wiley & Sons, NY, 1985.

[Kaplan63] Kaplan, I., *Nuclear Physics*, Addison-Wesley, Inc., Reading, 1963.

[Knoll89] Knoll, G. F., *Radiation Detection and Measurement*, 2^{ed}, John Wiley & Sons, NY, 1989.

[Krane88] Krane, K. S., *Introductory Nuclear Physics*, John Wiley & Sons, NY, 1988.

[Lamarsh66] Lamarsh, J. R., *Introduction to Nuclear Reactor Theory*, Addison-Wesley, Reading, MA, 1966.

Problems and Questions

1. Determine the de Broglie wavelength for a 1, 5, 10, 50, and 100 MeV neutron and compare this with the nuclear dimension of a single nucleon. Discuss this result in the context of direct and compound nuclear interaction mechanisms. Repeat for an alpha particle.

2. Complete the following interactions and show what the compound nucleus would be in each case:

   (a) ^{7}Li(α,n)

   (b) ^{60}Ni(α,n)

   (c) ^{63}Cu(p,2n)

   (d) ^{115}In(d,p)

   (e) ^{235}U(n,γ)

3. Repeat the exercise of example 10 for the compound nucleus ^{64}Zn* formed from α bombardment of ^{60}Ni using all exit channels from page 68. Discuss the product possibilities for incident α energies of 1, 5, 10, 20, and 40 MeV.

4. For an arbitrary nuclear reaction, list:

   (a) three things that are conserved and

   (b) two things that are not conserved.

5. Consider the α decay reaction

$$^{210}_{84}\mathrm{Po} \rightarrow {}^{206}_{82}\mathrm{Pb} + {}^{4}_{2}\mathrm{He}$$

   (a) Determine the binding energy for the parent nucleus and each of the product nuclei.

   (b) Express Q_α for this reaction in terms of the binding energies only, and determine Q_α in MeV.

   (c) Calculate the product kinetic energies for this decay assuming that $^{210}_{84}$Po is initially at rest.

6. Demonstrate that the relationship between Q for a binary two-product reaction and the nuclear binding energies of the reactants and products is

$$Q = \mathrm{BE}_\mathrm{Y} + \mathrm{BE}_y - \mathrm{BE}_x - \mathrm{BE}_\mathrm{X}$$

   (Hint: Begin with the expression for the Q value as a function of atomic masses.)

7. Use the above expression in problem 6 in the three nuclear synthesis reactions for ^{4}He to show that

$$Q = \mathrm{BE}_{^4\mathrm{He}} = Q_1 + Q_2 + Q_3$$

8. Determine the Q-values for the following reactions

$$^4\mathrm{He} + {}^9\mathrm{Be} \longrightarrow {}^{12}\mathrm{C} + \mathrm{n}$$

and

$$\gamma + {}^9\mathrm{Be} \longrightarrow {}^8\mathrm{Be} + \mathrm{n}$$

using both expressions for Q (*i.e.* as a function of mass and of the binding energy, BE.)

9. In each of the interactions in question 2, approximate how much work is done by the projectile to overcome the Coulomb force, assuming the distance of the closest approach is approximately equal to the nuclear radius.

10. Consider the nuclear reaction

$$\mathrm{p} + {}^9\mathrm{Be} \longrightarrow \begin{cases} ^{10}\mathrm{B} + \gamma \\ ^9\mathrm{B} + \mathrm{n} \\ ^9\mathrm{Be} + \mathrm{p} \\ ^8\mathrm{Be} + \mathrm{d} \\ ^7\mathrm{Be} + {}^3\mathrm{H} \\ ^6\mathrm{Li} + \alpha \end{cases}$$

What is the necessary condition for each exit channel?

11. The nuclear reaction n + ^{6}Li forms the compound nucleus $^7\mathrm{Li}^*$ which may decay by one of these five routes

$$\mathrm{n} + {}^6\mathrm{Li} \longrightarrow \begin{cases} ^7\mathrm{Li} + \gamma \\ ^6\mathrm{Li} + \mathrm{n} \\ ^6\mathrm{He} + \mathrm{p} \\ ^5\mathrm{He} + \mathrm{d} \\ ^3\mathrm{H} + \alpha \end{cases}$$

What is the necessary condition for each exit channel?

12. For parts (a)–(d) in question 2, determine the minimum value of the kinetic energy of the projectile necessary to cause the interaction.

13. Consider the neutron-producing nuclear reaction

$$\mathrm{p} + {}^{7}\mathrm{Li} \longrightarrow {}^{7}\mathrm{Be} + \mathrm{n}$$

Determine the neutron energy as a function of the incident proton energy and the neutron emission angle (ϕ). Show your results as a family of curves for E_n vs. E_p for a given ϕ. Your results should show clearly the turning conditions. Also, clearly label the critical angle.

14. Show the intermediate steps required in obtaining Eq.(2.12) from Eq.(2.9).

15. Plot the neutron product energy vs. projectile energy for the $^{14}\mathrm{C(p,n)}^{14}\mathrm{N}$ reaction as a family of curves for different emission angles. Plot this on two separate curves so that you may clearly show the double-valued behavior for the energy range, $E_\mathrm{p}^\mathrm{t} \leq E_\mathrm{p} \leq E_\mathrm{p}'$. (Ignore the fact that $B > E_\mathrm{p}'$ for this reaction. Consider the test projectile proton to quantum mechanically tunnel through the Coulomb barrier, *barrier tunneling* (a process which has small, but non-zero probability).

16. An experimental nuclear physicist becomes the proud recipient of a 5 MeV alpha particle accelerator and wishes to use it to induce nuclear reactions. Determine which of the following reactions are possible with this instrument.

    (a) $^{4}\mathrm{He} + {}^{9}\mathrm{Be} \longrightarrow {}^{12}\mathrm{C} + \mathrm{n}$
    (b) $^{4}\mathrm{He} + {}^{7}\mathrm{Li} \longrightarrow {}^{10}\mathrm{B} + \mathrm{n}$
    (c) $^{4}\mathrm{He} + {}^{3}\mathrm{He} \longrightarrow {}^{6}\mathrm{Li} + \mathrm{p}$
    (d) $^{4}\mathrm{He} + {}^{3}\mathrm{H} \longrightarrow {}^{6}\mathrm{Li} + \mathrm{n}$

17. For the general n,γ reaction

$$\mathrm{n} + \mathrm{X} \rightarrow \mathrm{Y} + \gamma$$

    (a) Derive an expression for the photon energy as a function of m_Y and Q in the special case where $E_\mathrm{n} = E_\mathrm{X} = 0$.
    (b) From the above expression, find the photon energy in the limit $m_\mathrm{Y}c^2 \gg Q$.
    (c) What is the threshold energy for this reaction?

18. In a nuclear reactor, (n,γ) reactions with the fuel, *i.e.*

$$\mathrm{n} + {}^{235}_{92}\mathrm{U} \rightarrow {}^{236}_{92}\mathrm{U} + \gamma$$

may rob neutrons from participating in fission reactions. Consider the reaction kinematics of these reactions by (a) writing down mass/energy and momentum balance equations, (b) expressing the Q-value for this reaction both as a function of participant masses and kinetic energies, (c) determining Q in MeV, and (d) determining the threshold energy and Coulomb barrier. If the reactant energies are zero, what energy will the ^{236}U atom attain after the reaction. (Express your final answer as a numerical value in eV. You may assume that the photon carries most of the reaction energy, *i.e.* $E_\gamma \gg E_{236}$.)

19. Consider the following two reactions with 5.5 MeV alpha particles

$$\begin{aligned} \alpha + {}^{7}\text{Li} &\longrightarrow {}^{10}\text{B} + \text{n} \\ \alpha + {}^{9}\text{Be} &\longrightarrow {}^{12}\text{C} + \text{n} \end{aligned}$$

    (a) Determine the Q-value for each interaction.

    (b) What are the kinetic energies of the neutrons in each reaction if they are emitted at an angle of 0° from the projectile direction?

    (c) What are the energies of the neutrons if they are emitted at angles of 30°, 45°, 90°, and 180°?

20. Calculate the neutron kinetic energy in the first reaction of problem 8 above when the initial ^{4}He kinetic energy is 5 MeV and the neutron is emitted at an angle of 0° from the axis if incidence. Repeat for neutron emission angles of 5°, 20°, 45°, 80°, and 90°.

21. In both the L and C coordinate systems, sketch the positions of the reactants and products of a binary, two-product interaction at several times before and after the interaction when the target is stationary in L before the interaction. Show L and C velocities and the velocity of the center of mass.

22. For the center of mass coordinate system:

    (a) Show that the center of mass velocity can be written

$$\mathbf{V}_c = \frac{m_x \mathbf{v}_x + m_{\text{X}} \mathbf{v}_{\text{X}}}{m_x + m_{\text{X}}}$$

    by requiring that net momentum be identically zero in C.

    (b) For a stationary target in L ($\mathbf{v}_{\text{X}} = 0$), obtain the expressions for the reactant kinetic energies before the interaction (E_x^c and E_{X}^c) as functions of E_x only.

    (c) Find E_{Y}^c and E_y^c as functions of E_x only.

23. For the reaction $^9\text{Be}(\alpha,\text{n})^{12}\text{C}$, determine the minimum projectile energy required. With a projectile kinetic energy of 3 MeV, determine the neutron lab and center of momentum energies, and the lab emission angle for $\phi_c = 0^o, 15^o, 30^o, 45^o, 60^o$, and 90°.

24. Consider the inverse fusion reaction

$$\text{n} + \ ^3\text{He} \longrightarrow \ \text{D} + \text{D}$$

   (a) Determine the minimum neutron kinetic energy required for this reaction.

   (b) Determine the C energies of the products for an initial neutron kinetic energy of 10 MeV.

   (c) For the case where the D-D products have C emission angles of $-\pi/2$ and $\pi/2$, this situation is symmetric in the lab system. Determine the product energies and emission angles in the lab system for an initial neutron energy of 10 MeV.

25. Some proponents for cold fusion argue that experimenters see no fusion products because the fusion reaction

$$\text{D} + \text{D} \longrightarrow \ ^4\text{He}$$

takes place instead of that in problem 24. If this were the case, there would be no product to detect since it is likely that the He product would be trapped in the lattice of the cold fusion cell. From what you know about reaction conservation principles, quantitatively discuss the physical possibility of the above reaction occurring in free space (outside a material lattice). Consider simultaneous mass/energy and momentum conservation in the limit where both reactants have negligible kinetic energy before the interaction.

Chapter 3

Neutron Interactions

Neutrons are by far the most important of the sub-atomic particles for nuclear engineers and the producers and users of radioisotopes. The majority of radioisotopes for medical and other applications, as discussed in chapter 1, are generated in neutron-induced nuclear interactions. The fission reaction can likewise be induced by neutrons which, in turn, produce even more neutrons (chpt. 5), making the fission chain reaction and nuclear reactors possible. Neutron-induced fission in many isotopes, including ^{235}U, is most favorable when the projectile neutron is of low kinetic energy, a slow neutron. Yet, those produced in the fission reaction are of relatively high kinetic energy, $\sim$ 2 MeV on average. To sustain a chain reaction, there must be some means by which neutrons can dissipate much of their post-fission energy. As we'll see shortly, elastic scattering interactions in certain materials, like ordinary water, are excellent mechanisms for doing the required neutron slowing or moderating (Fig. 3.1).

For these and other applications, we often require quantitative knowledge of the rate at which these interactions occur in a macroscopic system. To this point in our discussions, we have considered only the microscopic world of individual, isolated nuclear interactions. In this view, at most two entities exist simultaneously (either two reactants or two products), never more. The conservation of energy and momentum for these two completely describes the system. As we back away from this magnified view, we immediately encounter systems of enormous numbers, some 10^{23} particles per cm^3 for most solids, and perhaps some 10^5–10^9 or more neutrons in the same volume. Although the vast majority of material media target atoms do not interact with even a single neutron, the number of interactions is still so enormous that one could not hope to track individual neutrons and their energy exchanges in such a system. A new set of fundamental questions now appears: How do we describe the rate of interactions among an

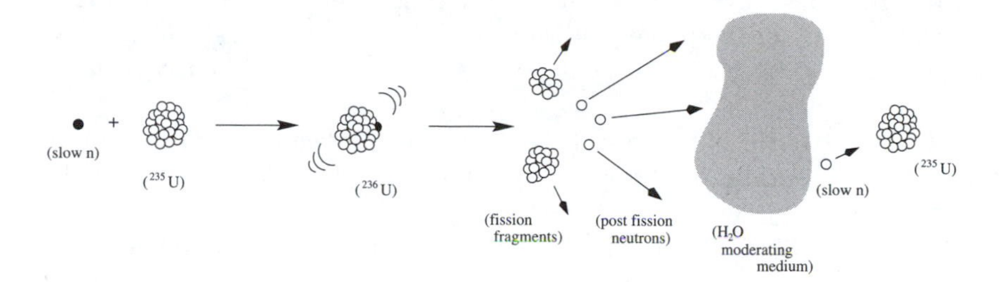

Figure 3.1: Slow neutron-induced fission chain reaction in ^{235}U. Low energy neutrons are readily absorbed in the ^{235}U nucleus, forming the unstable compound nucleus ^{236}U which may decay by the fission exit channel, thus, producing fission fragments and fast neutrons. These fast neutrons from fission must be slowed (moderated in energy) before inducing further fission reactions and completing the chain.

ensemble of many, many reaction projectiles with an even larger number of potential target atoms in a material object? How are the numbers and energies of such a projectile ensemble altered in interactions with the material object? How is the material altered by the projectile ensemble? The answer to these and many similar questions requires developing the concepts of multiple scattering, cross section, and flux. Introducing these will be our undertaking in this chapter.

3.1 Neutron Scattering

Because it is responsible for the majority of neutron moderation (slowing down) in nuclear reactors, elastic scattering is the most important scattering interaction for neutrons.

$$\mathrm{n} + \mathrm{X} \longrightarrow \mathrm{X}' + \mathrm{n}'$$

Since the identity of reactants and products has not changed, the prime notation will be used to indicate that there is an energy exchange between projectile and target. Since there is no mass loss (or gain) in scattering, Q must be identically zero, implying kinetic energy conservation and revealing the elastic nature of the interaction.

$$E_{\mathrm{n}} + E_{\mathrm{X}} = E_{\mathrm{X}'} + E_{\mathrm{n}'}$$

Though the total kinetic energy must be conserved, energy is exchanged among reactants so that $E_{\mathrm{n}'} \neq E_{\mathrm{n}}$ and $E_{\mathrm{X}'} \neq E_{\mathrm{X}}$. Since the projectile neutron is almost always of much greater kinetic energy than the target (which is usually in thermal equilibrium with its surroundings), then $E_{n'} \leq E_n$ and $E_{X'} \geq E_X$, and the neutron is scattered down in energy, *down scattered.*

Elastic scattering of neutrons (Fig. 3.2) is the perfect application for our center of mass formalism developed in chapter 2. We can solve explicitly for the C and L system product energies by directly employing expressions (2.36) through (2.39). Since these expression are functions of mass ratios only, we are free to choose any convenient mass unit. In atomic mass units, we can write $m_x = m_y = m_n \sim 1$ and $m_X = m_Y \sim A$ so that

$$\begin{aligned} E_x^c &= \left(\frac{m_X}{m_x + m_X}\right)^2 E_x = \frac{A^2}{(1+A)^2} E_x \\ E_X^c &= \frac{m_x m_X}{(m_x + m_X)^2} E_x = \frac{A}{(1+A)^2} E_x \\ \mathrm{KE}_c &= \mathrm{KE}_c' = \frac{m_x}{m_x + m_X} E_x = \frac{1}{1+A} E_x \end{aligned} \tag{3.1}$$

For the products in the C system with $Q = 0$

$$E_y^c = \frac{m_Y}{m_Y + m_y}\left[Q + \left(1 - \frac{m_x}{m_Y + m_y}\right) E_x\right] = \frac{A^2}{(1+A)^2} E_x \tag{3.2}$$

and

$$E_Y^c = \frac{m_y}{m_Y + m_y}\left[Q + \left(1 - \frac{m_x}{m_Y + m_y}\right) E_x\right] = \frac{A}{(1+A)^2} E_x \tag{3.3}$$

From the above expressions, it can be immediately recognized that $E_y^c = E_x^c$ and $E_Y^c = E_X^c$, so that $v_y^c = v_x^c$ and $v_Y^c = v_X^c$. The C systems speeds are unchanged in the interaction, a consequence unique to elastic scattering.

To find the L energy of the products, Eq.(2.39) can be straightforwardly reduced to

$$\boxed{\frac{E_y}{E_x} = \frac{A^2 + 1 + 2A\cos\phi_c}{(1+A)^2} = \frac{E_{n'}}{E_n}} \tag{3.4}$$

and

$$\boxed{\frac{E_Y}{E_x} = \frac{2A(1 - \cos\phi_c)}{(1+A)^2}} \tag{3.5}$$

We now have our desired result, $E_y, E_Y = f(E_x, \phi_c)$. As usual, both the projectile energy and light product scattering angle need to be specified.

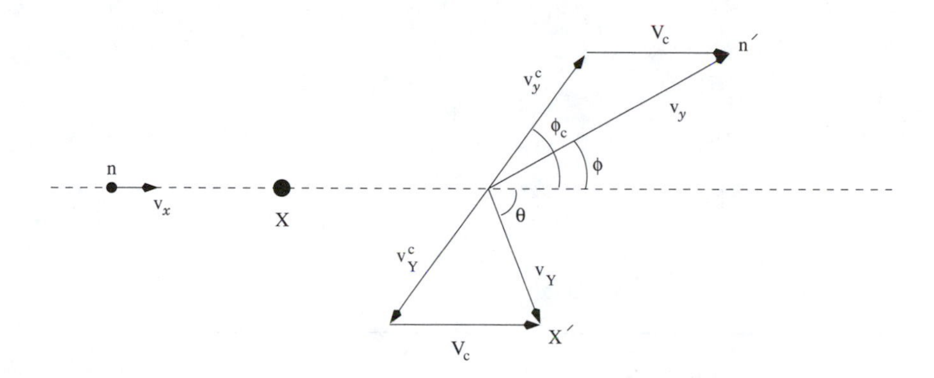

Figure 3.2: Elastic neutron scattering in the L and C frames. The notation is identical with that of Fig. 2.10. Here, $V_c' = V_c$ since there is no mass change in scattering.

Although the C system particle speeds are unaffected by the interaction, the above expressions show that $E_y \leq E_x$ (the neutron loses energy in the lab system) and $E_Y \geq E_X$ (the target gains energy in L). Neutron energy loss is consistent through multiple interactions, so long as E_n remains well above the thermal energy of the system. Energy gained by the target atom is quickly dissipated as thermal energy in the solid or fluid target material. Should we desire the C velocity, it can be found in the same manner, $\mathbf{V}_c = \frac{1}{1+A}\mathbf{v}_x$.

As we did in the development of Eq.(2.40), we may again employ the law of sines to express the L scattering angle ϕ as a function of that in the C system, ϕ_c

$$\cos\phi = \frac{1 + A\cos\phi_c}{(A^2 + 1 + 2A\cos\phi_c)^{1/2}} \tag{3.6}$$

where it should be remembered that this expression applies only to the special case $Q = 0$.

The range of scattering angles in the C system is by definition $0 \leq \phi_c \leq \pi$. As we saw for the more general case ($Q \neq 0$), $0 \leq \phi \leq \pi$ for elastic scattering also. (The equality applies in the limits at 0 and π, whereas $\phi < \phi_c$ at intermediate angles.) The exception to this rule is for $A = 1$, a special case deserving further attention. When $A = 1$ and $\phi_c = 0$, Eq.(3.6) yields $\phi = 0$ without difficulty. However, for $\phi_c = \pi$ Eq.(3.6) is

Example 12: Elastic Scattering

A 2 MeV neutron elastically scatters on a ^{12}C nucleus at $\phi_c = \frac{\pi}{2}$. Determine the product energies and lab scattering angle.

Solution:

Employing Eqs.(3.2) through (3.6), we can find

$$
\begin{aligned}
E_y^c &= \frac{12^2(2\text{ MeV})}{13^2} = 1.7\text{ MeV} \\
E_Y^c &= \frac{12(2\text{ MeV})}{13^2} = 0.142\text{ MeV} \\
E_Y^c + E_y^c &= \frac{12(2\text{ MeV})}{13} = 1.85\text{ MeV}
\end{aligned}
$$

so that

$$
\begin{aligned}
\cos\phi &= \frac{1}{145^{1/2}} \quad \text{so} \quad \phi = 85.2^{\text{o}} < \phi_c \\
E_y &= \frac{145}{169}(2\text{ MeV}) = 1.716\text{ MeV} \\
E_Y &= E_x - E_y = 0.284\text{ MeV}
\end{aligned}
$$

indeterminant. The limit via L'Hospital's rule yields $\phi = \pi/2$, so for $A = 1$, $0 \leq \phi \leq \pi/2$. Neutrons cannot be backscattered on collisions with protons.

Another interesting limit appears at large A. In the limit $A \to \infty$ (very massive target), then $\cos\phi \sim \cos\phi_c$. Neutron scattering appears identical in both systems for this special case, a result we saw earlier (sec. 2.9) for the more general interaction formulation.

Returning to the scattering angle limits, $\phi_c = [0, \pi]$, we can assign some significance to each. At small scattering angles, $\phi_c \sim 0$, the neutron suffers a *glancing* interaction with the target nucleus. In the limit $\phi_c = 0$, $E_y = E_x$ and there is no evidence of an interaction.

At the other extreme, $\phi_c = \pi$, the greatest energy exchange occurs for these *head-on* interactions. In this limit,

$$\left.\frac{E_y}{E_x}\right|_{\pi} = \left(\frac{A-1}{A+1}\right)^2 = \alpha \tag{3.7}$$

where α is called the collision parameter, so that at maximum energy transfer, $E_y = \alpha E_x$. When $A = 1$, $\alpha = 0$ and the target proton receives all the incident neutron projectile energy ($E_{\mathrm{Y}} = E_x$) in a single head-on elastic interaction. The neutron comes to rest, $E_y = 0$, in place of the target proton. In the other extreme, $A = \infty$, $\alpha = 1$ and $E_y = E_x$ so the neutron loses no energy in the interaction. It does, however, reverse direction.

To be complete, a few comments should be made about inelastic scattering

$$\mathrm{n} + \mathrm{X} \longrightarrow (\mathrm{X} + \mathrm{n})^* \longrightarrow \mathrm{X}^* + \mathrm{n}' \longrightarrow \mathrm{X}' + \mathrm{n}' + \gamma$$

The decay of nuclear excitation by photon emission implies that massive particle kinetic energy is not conserved here, $E_{\mathrm{n}} + E_{\mathrm{X}} > E_{\mathrm{X}'} + E_{\mathrm{n}'}$. The difference appears as E_γ. Although inelastic scattering is a rather unimportant reaction in neutron slowing down, it may play an important role in radiation shielding and other applications.

By considering the intermediate process $\mathrm{n} + \mathrm{X} \longrightarrow \mathrm{X}^* + \mathrm{n}'$, we can again employ the binary, two-product formalism developed in chapter 2. Mass/energy balance yields

$$E^* = E_{\mathrm{n}} + E_{\mathrm{X}} - E_{\mathrm{X}'} - E_{\mathrm{n}'}$$

which takes the place of $-Q$ in Eq.(2.37). (Of course, by our strict definition of the *Q-value* in Eq.(2.5), $Q = 0$ for the overall process since $\Delta m = 0$.) Following a bit of rearrangement, we find

$$E_{\mathrm{n}'} = \frac{1}{2}(1+\alpha)E_{\mathrm{n}} + \frac{1}{2}(1-\alpha)(1-\epsilon)^{1/2}E_{\mathrm{n}}\cos\phi_c - \frac{A}{A+1}E^* \tag{3.8}$$

where $\epsilon = \frac{A+1}{A}\frac{E^*}{E_n}$. Of course, the condition $\epsilon < 1$ must be satisfied, which requires $E_n > \frac{A+1}{A}E^*$. This expression reduces to the elastic scattering result, Eq.(3.4), when $\epsilon \to 0$.

As with elastic scattering, $E_{n'}$ is maximized and minimized at $\phi = 0, \pi$, respectively, for which Eq.(3.8) reduces to

$$\left.\frac{E_{n'}}{E_n}\right|_{\substack{max\\min}} = \left[\frac{A(1-\epsilon)^{1/2} \pm 1}{A+1}\right]^2 \tag{3.9}$$

where the upper (lower) limit is evaluated with the + (−) root. This too reduces to the elastic limit as $\epsilon \to 0$. And finally, with the aid of Eq.(2.40), we may relate the L and C system scattering angles

$$\cos\phi = \frac{1 + g\cos\phi_c}{(g^2 + 1 + 2g\cos\phi_c)^{1/2}} \tag{3.10}$$

where $g^2 = A^2 - A(A+1)E^*/E_n$. Since $g \to A$ as $E^* \to 0$, the above expression reduces to Eq.(3.6) in the elastic limit.

3.2 Lethargy and Multiple Elastic Collisions

The expressions developed in the preceding section apply only to single neutron scattering events. In one particular single, and rather fortunate elastic scattering event, a post-fission (or otherwise energetic) neutron may scatter on a proton from some relatively high energy of, perhaps, a few MeV to near thermal ~ 0.025 eV where it is much more likely to initiate another fission event. The foregoing situation requires scattering preferentially at nearly $\phi_c \sim \pi$, and is, therefore, quite rare. Elastic scattering interactions sample all of emission space nearly equally since scattering is often isotropic in the C system. It, therefore, takes several to many such events on average to reduce a single neutron kinetic energy to the thermal region.

Recall that on a single elastic collision an incident neutron leaves the interaction with an energy dependent on the C scattering angle

$$\frac{E_{n'}}{E_n} = \frac{E_f}{E_i} = \frac{1 + A(A + 2\cos\phi_c)}{(1+A)^2} \tag{3.11}$$

where we'll temporarily adopt a slightly new notation for the neutron final energy, E_f, and E_i for the initial energy. To facilitate the analysis of multiple such events, let's introduce a new quantity u such that

$$u = \ln\frac{E_o}{E}$$

The quantity u is a function of the state of the system since it is evaluated at the instantaneous energy, E (a state variable), relative to some reference energy, E_o. The reference energy might naturally take on the identity of the source energy for a monoenergetic neutron source, or is usually chosen to be 10MeV for a fission source. Regardless, the choice is arbitrary, but it is convenient to choose E_o as the highest energy in the problem so that u is monotonic. With u as a monotonically increasing function as neutron energy decreases, it calls to mind the notion of sluggishness and is, thereby, called the neutron "lethargy." Neutrons become more "lethargic" as they slow down. It is a strict property of the system (at this moment, a single neutron particle) and is completely independent of any interaction or process details. Historically, the lethargy variable has been used in neutron slowing down calculations because it is found that the neutron slowing down rate is only a weak function of u and, thereby, much easier to approximate in u than in E.

Recognizing that the state of the system changes during an interaction, we exploit changes in the state variable, lethargy, to quantify the process. Before an interaction, a neutron will have an absolute lethargy given by

$$u_i = \ln \frac{E_o}{E_i}$$

After said interaction, the neutron energy will change to E_f so that the absolute lethargy is now

$$u_f = \ln \frac{E_o}{E_f}$$

The change in lethargy during the process is then

$$\Delta u = u_f - u_i = \ln \frac{E_i}{E_f}$$

completely independent of the reference state, E_o. This new quantity, Δu, is only a function of the initial and final states. (Some find it advantageous to draw parallels with thermodynamic state variables like entropy.)

The advantage of expressing energy changes in this way is that the net effect for multiple events (the additivity property) is independent of intermediate processes and states. For example, let's consider three consecutive neutron elastic scattering events involving a single neutron with different targets (all at rest in the lab frame). These are consecutive so that the neutron energy leaving the first event is that entering the second, etc. On the first interaction, we quantify

$$\Delta u_1 = \ln \frac{E_{i_1}}{E_{f_1}}$$

For the second

$$\Delta u_2 = \ln \frac{E_{i_2}}{E_{f_2}}$$

and finally the third

$$\Delta u_3 = \ln \frac{E_{i_3}}{E_{f_3}}$$

However, since $E_{i_2} = E_{f_1}$ and $E_{i_3} = E_{f_2}$, the total lethargy change after all three interactions is

$$\Delta u_{\text{tot}} = \sum_{j=1}^{3} \Delta u_j = \ln \frac{E_{i_1}}{E_{f_3}}$$

again, a function of only the initial and final states of the system. In this way, the concept of lethargy is very similar to that of the thermodynamic variables $\Delta s, \Delta h, \ldots$ depending only on system states and not on process details.

To be useful in our neutron slowing down context, we need to express Δu as a function of the target mass, A, for an individual event. This does, however, require details of the process (*i.e.* $E_i, A, \cos\phi_c, \ldots$). For a single neutron elastic scattering event

$$\Delta u = \ln \frac{E_i}{E_f} = \ln \frac{(1+A)^2}{1 + A(A + 2\cos\phi_c)}$$

Note that our macroscopic physical system (reactor, neutron beam and target, ...) involves an ensemble of many neutrons, each of which may require several to many elastic interactions in slowing to thermal energies. Then, the elastic interaction behavior of the entire ensemble is well represented by the average behavior of one neutron on a single scattering event. (This is not an acceptable argument for a single neutron interacting in a macroscopic medium alone. Such a situation would violate the statistical significance of a large sample.) The average behavior of a neutron on a single elastic event can be quantified by determining the average lethargy change per elastic collision. Since elastic scattering is nearly isotropic in C (*i.e.* scattering occurs with relative probability that is independent of emission angle), the likelihood of finding a scattered neutron in any $\Delta\phi_c$ about ϕ_c is then identical to the probability of scattering into the spherical surface area ΔA subtended by $\Delta\phi_c$, relative to the entire sphere surface area (cf: Fig. 3.3). In the differential limit, this ratio is proportional to the three dimensional (solid) angle, Ω, measured in steradians, *i.e.* $dA = d\Omega\, r^2 = r\sin\phi\, d\Phi\, r\, d\phi$ where $\int d\Omega = 4\pi$, and Φ is the azimuthal angle ($0 \le \Phi \le 2\pi$). Hence, the average of Δu is computed with respect to Ω

$$\xi = \langle \Delta u \rangle_\Omega = \frac{\int d\Omega\, \ln \frac{E_i}{E_f}}{\int d\Omega} \tag{3.12}$$

Then, the average lethargy change in an isotropic, elastic scattering event is

$$\begin{aligned}\xi &= \frac{1}{4\pi}\int_0^{2\pi} d\Phi \int_0^{\pi} \sin\phi_c\, d\phi_c \ln\frac{(1+A)^2}{1+A(A+2\cos\phi_c)} \\ &= 1 - \frac{(A-1)^2}{2A}\ln\frac{A+1}{A-1}\end{aligned} \tag{3.13}$$

This function ξ is well behaved except at $A = 1$. This is such an important

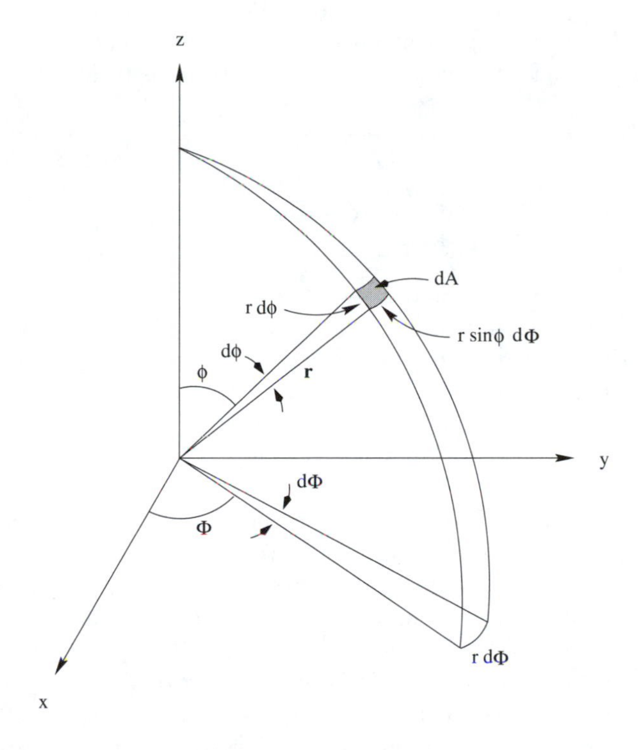

Figure 3.3: Solid angle in spherical coordinates. The azimuthal angle is indicated by Φ. The center of mass scattering angle ϕ_c is coincident with the polar angle ϕ for scattering interactions at the origin.

case that special treatment is warranted. Writing ξ as a function of α expedites the analysis

$$\boxed{\xi = 1 + \frac{\alpha}{1-\alpha}\ln\alpha} \tag{3.14}$$

Using L'Hospital's rule, one can readily show that $\lim_{\alpha\to 0}\xi = 1$. On average, then, neutrons can suffer a fractional energy decrement of e^{-1} on an average single collision with hydrogen. (In the other extreme of large A, $\lim_{\alpha\to 1}\xi = 0$.) For n such events occurring sequentially

$$\Delta u_{\text{tot}} = \sum_{j=1}^{n} \Delta u_j = \ln \frac{E_{i_1}}{E_{f_n}}$$

where E_{i_1} is the initial neutron energy and E_{f_n} is the final energy after n collisions, for an average neutron in the ensemble. For a large sample, $\sum_j \Delta u_j \sim n\xi$ so that the average number of collisions to attain a lethargy gain to $u(E_{f_n})$ is

$$\boxed{n \simeq \frac{1}{\xi} \ln \frac{E_{i_1}}{E_{f_n}}} \tag{3.15}$$

This result is statistically valid even when n is small by virtue of the extremely large number of neutrons in any real sample. The fate of an individual neutron is irrelevant. The effectiveness of a neutron moderator (material for reducing neutron energies) as a thermalizing medium can be gauged, in part, by n. A smaller value of n indicates more effective neutron energy loss to a given target material (*i.e.* greater energy loss per average event).

This is not the complete picture, however. Some materials that possess efficient thermalization properties, like hydrogen (in water), have an affinity for absorbing neutrons. Their high neutron absorption coefficient (cross section) detracts from its usefulness in a neutron multiplication medium like a reactor where neutron inventory control is at a premium. Robbing neutrons from the population in this way is a poor quality of a moderator. A more complete figure of merit that includes the absorbing effect is called the "moderator ratio"$\equiv f\xi$ (here, f is the ratio of relative probabilities a neutron will undergo a scattering interaction to that in which it will undergo an absorption interaction, Σ_s/Σ_a). Larger values of this quantity are obviously desirous. The quantities used to determine the f parameter introduced above will be explicitly defined in a later section (sec. 3.5). Some sample calculations are described in example 13.

While ξ represents the average logarithmic energy change per collision, the average scattered energy in a single elastic and isotropic event can be likewise calculated

$$\frac{\langle E_{\text{n}'}\rangle}{E_{\text{n}}} = \frac{\int d\Omega \, \frac{E_{\text{n}'}}{E_{\text{n}}}}{\int d\Omega} = \frac{1+\alpha}{2} \tag{3.16}$$

for which the minimum occurs at $A = 1$ ($\alpha = 0$) where $\langle E_{n'} \rangle / E_n = 1/2$. The average energy transfer in a single such interaction is then $1 - \langle E_{n'} \rangle / E_n = (1 - \alpha)/2$, which takes on maximum value of 1/2 at $\alpha = 0$.

Example 13: Moderator Effectiveness

For ^{235}U thermal fission neutron moderation in H, with $\xi = 1$, $E_{i_1} = 2$ MeV, $E_{f_n} \sim 0.025$ eV, then

$$n = \ln \frac{2 \times 10^6 \text{ eV}}{0.025 \text{ eV}} = 18.2 \quad \text{collisions on average}$$

Other examples of thermalization of a 2 MeV neutron include:

moderator	A	α	ξ	n	$f\xi$
H	1	0	1	18.2	—
H_2O	1,16	—	0.92	19.8	71
D	2	0.111	0.725	25	—
D_2O	2,16	—	0.51	35.7	5670
He	4	0.36	0.425	42.8	83
Be	9	0.64	0.21	87	143
C	12	0.716	0.16	115	192
^{238}U	238	0.983	0.0084	2172	0.0092

3.3 The Concept of Cross Section

We have discussed in detail the necessary conditions for a nuclear reaction to occur and then the details of the reaction product kinematics. What we have ignored thus far is a consideration for the likelihood (or relative probability) of such interactions occurring in macroscopic systems. This also naturally develops into a model to quantify the rate at which such interactions occur. Constructing these models provides yet another step in our gradual development of a macroscopic description for nuclear processes.

Consider the following hypothetical physical situation in which we are provided, by some means, a well behaved beam of neutrons. By "well behaved," we mean that this beam is nearly completely monoenergetic (all neutrons in the beam have the same energy) and monodirectional (all neutrons in the beam travel in the same direction). This idealistic picture will serve us well to develop a basic understanding from which we may later extrapolate to more general conditions. Furthermore, we are free to place, in the path of this beam, a very thin target of material of our choosing for the neutron beam to interact with. We'll insist for now that the target be made sufficiently thin so that there is reasonable assurance the beam neutrons will interact with at most one atom as they pass through the target. While we run the risk (and in fact the likelihood) that most neutrons will pass directly through such a target without interacting at all, this is a small price to pay for the inordinate complications that would arise were we to allow multiple interactions. In addition, a thin target, having small interaction probability, ensures that the neutron beam intensity remains nearly constant through the target. All target atoms are then "equally accessible" to the incident beam. We are also provided with a "perfect" detector, which we can arrange to "detect" the presence of scattered neutrons or reaction products that emanate from interactions within the target. This detector should be perfect so that it measures all particles that enter its front face without loss and without deference for the particle type. A schematic of the proposed experimental arrangement is shown in Fig. 3.4.

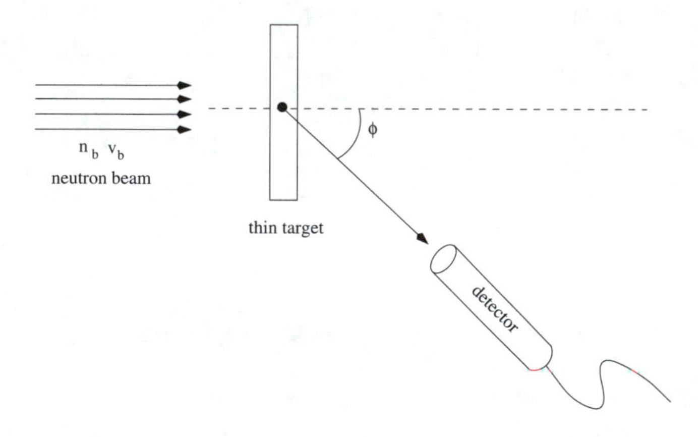

Figure 3.4: Hypothetical experimental setup for measuring cross section. An ideal beam of neutrons (monoenergetic and monodirectional) strikes a thin target to guarantee no change in beam intensity and at best only one interaction. The product of this reaction, emitted at lab angle ϕ, is collected with 100% efficiency by a "perfect" detector.

Our objective is to identify the rate of occurrence for interactions between beam neutrons and target atoms (interactions/s). If we define n_b, v_b to be the beam density (n/cm^3) and beam speed (cm/s), respectively, and $N, A, \Delta x$ to be the target atom density (atoms/cm^3), projected beam area on target (cm^2), and target thickness (cm), respectively, then we can define the reaction rate, R, by inspection.

$$R \propto n_b v_b N A \Delta x$$

Since the proportionality constant required to satisfy the above expression has units of a cross-sectional area, we call this quantity a *cross section* although, in general, it has not a thing to do with any geometric cross-sectional area in the problem. Rather, we should interpret the cross section as related to the relative probability of an interaction, a concept we shall be developing shortly. Most often in the study of neutron physics this quantity is given the symbol σ, so that the expression

$$\sigma \equiv \lim_{\Delta x \to 0} \frac{R}{n_b v_b N A \Delta x} \tag{3.17}$$

defines σ. The limit reminds us of our obligation to keep the target thin and that σ is the "expected value" for Δx made vanishingly small. It often times proves much more convenient, especially for more involved problems including spatial variation, to define a new quantity called the reaction rate density as the number of interactions per second per unit target volume. Using our definitions above the reaction rate density is

$$\mathcal{R} = \frac{R}{A \Delta x} = \sigma n_b v_b N \tag{3.18}$$

Let's now examine the probabilistic or statistical nature of what we have just defined. Firstly, we have insisted that our interaction target be very thin so that we could guarantee at best only a single interaction per neutron in the target and avoid beam attenuation. For example, consider an aluminum sheet target of 10 micron thickness ($\Delta x = 10\ \mu\text{m} = 10^{-5}$ m). This material contains about $6 \times 10^{19} (= N A \Delta x)$ potential target atoms of Al when the beam illumination area on target is $A = 1\ \text{cm}^2$. Supposing our incident neutron beam passes, for example, 10^{10} n/s $(= n_b v_b A)$ through the target for which there result 10^6 interactions/s, then only one in 10^4 beam neutrons interact in any fashion. We, therefore, assign an interaction probability of $p = 10^{-4}$ per beam particle for this situation. For a general beam, thin target problem, we write

$$p = \frac{R}{n_b v_b A} = \sigma N \Delta x \tag{3.19}$$

as the interaction probability. The quantity σN then can be interpreted as the probability per unit path length of interacting. We'll revisit this later (sec. 3.5). It should be obvious that by doubling the target thickness, the interaction probability will double (so long as we don't make it too thick and cause multiple interactions and/or beam attenuation), but the probability per unit length remains the same. The latter is always true, so long as we do not alter the composition of the target. By doubling N, however, we could double both p and $p/\Delta x$, although its not clear how to alter the physical density of Al.

The results justify our thin target assumption. An interaction probability of 10^{-4} is indeed low. The incident neutron intensity is diminished by only this small fraction. Furthermore, assuming an unchanged interaction probability, a reasonable initial estimate for the second interaction fraction (fraction of incident neutrons that interact at least twice) is exceedingly small, $\sim 10^{-8}$. One may question the validity of this assumption since collided neutrons may possess properties (*i.e.* energy) different from the incident, uncollided beam. They will, in general, possess different interaction probability. Nonetheless, our rough approximation captures the appropriate trend that the second interaction probability is negligibly small for all practical purposes.

As a subtle but important point, we should note that we are now beginning to treat, as continuous processes, nuclear interactions which are inherently discrete and statistical in nature. It is the very presence of large numbers of both projectiles and target atoms that allow this description. We say that the statistics are "good" when it is appropriate to apply the concepts of continuum mechanics to large numbers of discrete phenomena. If instead we were dealing with a much smaller sample (perhaps 10 neutrons for the sake of argument), the statistics would obviously be much poorer and we shouldn't expect the results to be close to the average for a large sample. If we were to repeat this small sample experiment 10^9 times, however, our average results would again be very close to the large sample of 10^{10} neutrons. In keeping with our probabilistic description, we should then interpret R as the "expected value" of interactions per second. When the statistics are good, as for a large sample, we will measure very close to this value.

Another implicit assumption in the preceding arguments regarding cross section and interaction rates is that, for our large projectile sample experiment, individual neutrons interact only with target atoms. They do not interact with each other. As well, we must insist that no two neutrons interact with a single target atom simultaneously. Both of these criterion are easily satisfied when $n_b \ll N$.

Completely generalizing our cross section and reaction rate formalism is often an onerous task and the subject of entire volumes (cf: [Duderstadt76], [Lamarsh66] for some examples). Neutrons in nuclear reactors and other real material media move in all different directions and at a continuum of speeds. They can interact with many different materials and suffer several possible types of interactions (sec. 3.5) in a time dependent fashion. The concept of cross section and reaction rate must correctly include all of these dependencies. To be sure, the quantity σ is, in many important cases, a strong function of incident neutron energy and product emission angle, but cannot depend on neutron approach direction or time. To avoid complication in favor of illustrating some basic concepts, energy and emission angle independence will be assumed. The reader is directed to more advanced treatments in the references.

3.4 Neutron Flux

Let's return for the moment to our neutron beam example from above where we have a very idealized stream of monoenergetic and monodirectional beam particles. With the target removed, consider placing our "perfect" particle detector in the path of this beam to intercept neutrons passing through its front surface of area ΔA (Fig. 3.5a). The detector counts only particles passing through the front face, yet it counts every such particle. Then, the number of counts per second recorded on the detector is equal to the number of beam neutrons crossing ΔA per second, $\dot{\mathcal{N}}$. The *neutron flux* is defined to be $\dot{\mathcal{N}}$ per unit detector area, *i.e.*

$$\varphi(\mathbf{r}, t) \equiv \lim_{\Delta A \to 0} \frac{\dot{\mathcal{N}}}{\Delta A} \tag{3.20}$$

which has the dimensions of a particle flux ($\frac{\text{neutrons}}{\text{cm}^2\ \text{s}}$). For the special case of a monoenergetic beam, $\dot{\mathcal{N}} = n_b v_b \Delta A$, so that

$$\varphi(\mathbf{r}, t) = n_b v_b$$

The limit required by Eq.(3.20) serves to define the flux at a particular point in space located with respect to the origin of coordinates by $\mathbf{r}$. Should we desire the value of the flux at another location, the experiment must be repeated at the new position. Likewise, identifying the flux at a new time at any position requires a new measurement. Hence, flux is a local quantity in both time and space.

The spatial and temporal dependencies have been explicitly noted in the quantity $\varphi(\mathbf{r}, t)$. In a more general treatment (cf: [Duderstadt76]), explicit angular and energy dependencies are required in a related quantity called the angular flux. We interpret the angular flux to be a continuous function of all four variables, taking on a unique value at each spatial position, at all angles and neutron energies, and at each instant in time. As with cross section, the more general description provides much more information than is presently required and is the subject of more advanced treatment.

Instead, let's extend the concept of the scalar or total flux, $\varphi(\mathbf{r}, t)$, a bit further by considering neutrons traveling in several different directions

a)

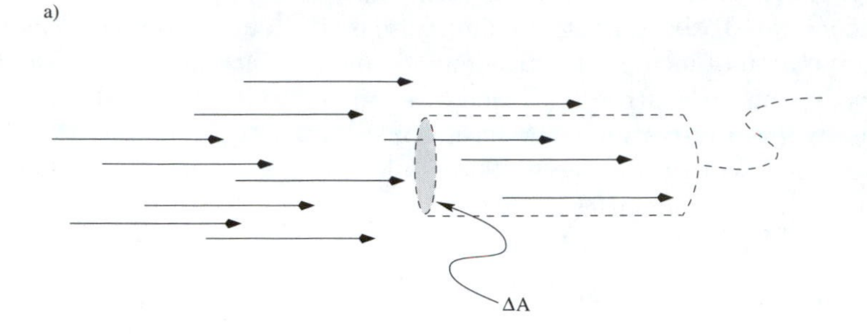

b)

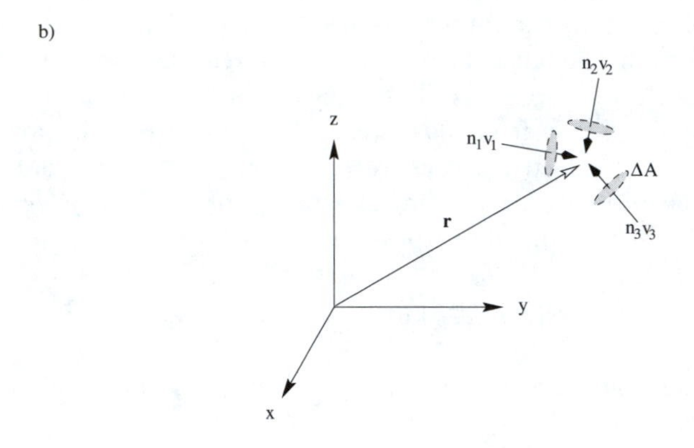

c)

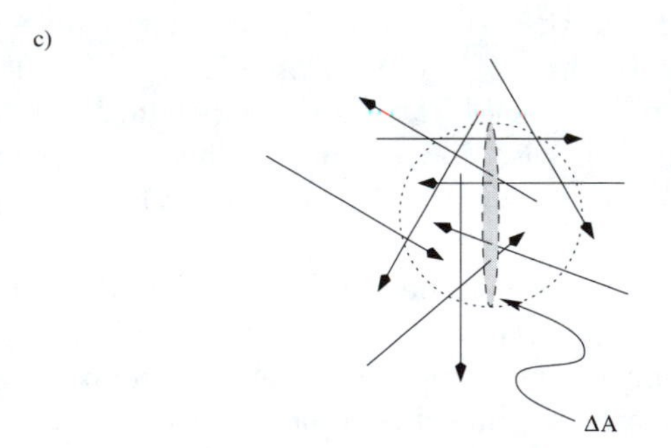

Figure 3.5: Neutron flux characterization: (a) flux from an idealized neutron beam (monoenergetic, monodirectional), (b) contributions to the flux of neutrons at a point **r** originating from several different directions, (c) concept of flux extended to multidirectional neutrons passing through a "perfect" spherical detector.

(Fig. 3.5b). These neutrons no longer emanate from a monodirectional beam. They may be considered to be coming from several different beams at different locations and converging at $\mathbf{r}$. As well, these beams may have different neutron densities, but all produce neutrons of the same speed. If we now imagine these particles all passing through detection areas, ΔA, perpendicular to their respective beams, then the total neutron flux at $(\mathbf{r}, t)$ is the scalar sum of individual beam fluxes

$$\varphi(\mathbf{r}, t) = n_1 v_1 + n_2 v_2 + n_3 v_3 = v(n_1 + n_2 + n_3)$$

since $v_1 = v_2 = v_3 = v$. In the above description, n_1, n_2, n_3 represent partial neutron densities at $\mathbf{r}$ contributed from each of three sources. The total density at the point of intersection is $n = n_1 + n_2 + n_3$, so again

$$\varphi(\mathbf{r}, t) = nv$$

Extending the concept of flux further still, consider many monoenergetic neutrons emanating from many directions simultaneously (Fig. 3.5c). (Such might be an approximation to the situation at the center of a large nuclear reactor surrounded by a large number of fissioning uranium atoms (chpt. 5) as neutron sources.) We may now imagine quantifying the neutron flux in such a situation by centering our neutron detector with detection face ΔA at $\mathbf{r}$ and counting only those neutrons that pass perpendicular to the surface through ΔA into the detector while rotating the detector through π radians. This scenario is equivalent to imagining an idealized spherical detector centered at $\mathbf{r}$ with projected area ΔA. The total number of monoenergetic neutrons entering this detector per second is

$$\dot{\mathcal{N}} = v \sum_j n_j \, \Delta A$$

where n_j is the partial neutron density contributed along direction j. The total neutron density at $(\mathbf{r}, t)$ is then $n(\mathbf{r}, t) = \sum_j n_j$ so that

$$\varphi(\mathbf{r}, t) = vn(\mathbf{r}, t) \tag{3.21}$$

is identical in form to the single beam case. Now, however, we recognize the quantities $\varphi(\mathbf{r}, t)$ and $n(\mathbf{r}, t)$ as being comprised of multidirectional contributions.

Care should be taken in the interpretation of the neutron flux. Indeed, the nomenclature may be a bit misleading. The quantity $\varphi(\mathbf{r}, t)$ does not behave at all like other particle or field fluxes encountered in the physical sciences or engineering, like electric current density or heat flux to name a few. The latter are vector quantities and sum as vectors, whereas $\varphi(\mathbf{r}, t)$ is a scalar, each of its constituents making a scalar contribution to the total. Perhaps a reasonable working interpretation is to regard $\varphi(\mathbf{r}, t)$ as characterizing "...the *total* rate at which neutrons pass through a unit area, regardless of orientation."[Duderstadt76] The utility of the scalar flux is in correctly determining neutron reaction rates as we shall develop shortly.

To be sure, our monoenergetic description of neutrons ($v_j = v$, all j) may only grossly approximate real physical systems. We'll investigate some problems for which this is a reasonable approximation in this and the next chapter. Within the confines of a monoenergetic description, we interpret $\varphi(\mathbf{r}, t)$ as the "one speed" flux. Many times, the one speed of choice is 2200 m/s. Not because this is a convenient beam energy, rather, 2200 m/s is the speed of neutrons with kinetic energy $= kT_o = 0.0253$ eV, where $T_o = 293.15$ K (room temperature). An ensemble of neutrons in thermal equilibrium with their room temperature surroundings (as in a thermal fission reactor operating at low power) has an energy distribution characterized by T_o. The reference speed of 2200 m/s is representative, in some sense, of this distribution. Energy dependent neutron cross sections are often referenced to their value at this speed, for example.

3.5 More on Cross Sections: Microscopic vs. Macroscopic

The quantity, σ, encountered in the reaction rate expression (Eq.(3.17)) is the effective cross section that a single nucleus in the target presents to the projectile for any type of interaction. This quantity only depends on the type of beam and target material, interaction type and, perhaps, the speed of the projectile. It is completely independent of beam or target material density, size of the system, ... It is thereby, referred to as a "microscopic cross section." In practical situations, σ takes on extremely small magnitude. Again, turning to our hypothetical example of a thin aluminum target from above (sec. 3.3), an interaction probability of 10^{-4} for neutrons passing through a 10^{-5} m thick Al target with $N \sim 6 \times 10^{22}$ cm^{-3}, implies $\sigma \sim 1.7 \times 10^{-24}$ cm^2. This is a typical order of magnitude for an interaction cross section. As such, we may adopt a more convenient length scale. For historical reasons, we discuss cross sections in units of the "barn," given the symbol "b," where

$$1 \text{ b} = 10^{-24} \text{ cm}^2$$

For the example above, the effective target cross section is 1.7 b.

We should again be a bit careful about the interpretation here. The interaction cross section, σ, does not imply a cross-sectional area of the target atom or nucleus. We could estimate the projected geometric cross section of the target nucleus as $\sigma_g \sim \pi R^2$, where R is the nuclear radius (Eq.(1.13)) so that $\sigma_g(\text{b}) \sim 0.045A^{2/3}$. For $10 \leq A \leq 200$, this yields $0.2 \text{ b} \lesssim \sigma_g \lesssim 3 \text{ b}$. The true interaction cross section, however, may range from $10^{-19} \text{ b} \lesssim \sigma \lesssim 10^6 \text{ b}$.

The exception to this general rule is "potential scattering." Potential scattering constitutes the mechanistic description of low energy scattering of neutrons in which the neutron elastically scatters off the nuclear potential of the target nucleus. In this way, it can be envisioned much like a hard sphere, billiard ball type collision. Potential scattering is characterized by a very constant cross section over a very wide energy range (thermal to few MeV range), where the scattering cross section can be well represented by the geometrical spherical surface area of the target atom, *i.e.* $\sigma_{\text{p.s.}} \sim 4\pi R^2$. For the example above of Al at low energy, potential scattering is responsible for about 1.62 b of the total cross section of about 1.7 b. Neutron absorption reactions (reactions that absorb neutrons into the nucleus) are responsible for the remainder of the total cross section.

The quantities σ and N always appear together in the reaction rate equation. It has also been established (sec. 3.3) that the interaction probability in our thin target is $\sigma N \Delta x$. We can then interpret

$$\sigma N = \text{ the probability per unit length of a neutron interacting in the target material } = \Sigma$$

where Σ is called the *macroscopic* cross section since it does depend on a macroscopic quantity, the target atom density. Since it is a function of σ, Σ will also depend on interaction type and projectile energy. Recall that the interaction probability in a thin target is $p = \Sigma\Delta x$, so that Σ must obviously take on the dimension of cm^{-1}. Note that the interaction probability per unit length (Σ) is a constant in a thin target. This concept will be generalized in sec. 3.7. The macroscopic cross section may also be interpreted as the effective target cross section for all target atoms per unit volume of target material.

The statistical nature of neutron interactions suggests that reactions of different types are independent. Thereby, their relative probabilities should be additive. The total reaction rate density (Eq.(3.18)) should be a linear sum over reactions of all types j, so that

$$\Sigma = \sum_j \Sigma_j$$

since the neutron flux must be independent of j. The above expression must also hold for the microscopic cross section, $\sigma = \sum_j \sigma_j$, as N is also completely independent of j. For neutrons, all reactions fit into one of two categories, scattering or absorption, so that

$$\sigma = \sigma_s + \sigma_a$$

Scattering is naturally limited to elastic, σ_{s_e}, and inelastic, σ_{s_i}, only. Many absorption reactions are possible

$$\sigma_a = \sigma_\gamma + \sigma_\alpha + \sigma_p + \sigma_f + \ldots$$

where the subscript f is intended to represent fission. Absorption reactions generally show much more energy dependence that does scattering. Obviously, when there is a threshold for neutron energies below $E_{\rm n}^{\rm t}$, $\sigma_j = 0$. Neutron absorption and fission cross section data at 2200 m/s can be found for the nuclides in [GE96]. Neutron scattering cross sections are provided for some elements and compounds in Appendix C.

In keeping with our probability language, we should interpret individual reaction cross sections, σ_j, as being proportional to the interaction probability for a specific reaction type, *e.g.* the probability of observing a fission reaction in ^{235}U among all possible reactions is $\sigma_f^{235}/\sigma^{235}$. Likewise, the probability of observing a fission reaction among all possible absorption reactions in ^{235}U is

$$\frac{\sigma_f^{235}}{\sigma_a^{235}} = \frac{582.2\text{b}}{680.8\text{b}} = 0.855$$

for example.

3.6 The Reaction Rate Equation

Consider now an arbitrary interaction volume of substantial dimensions so as to intentionally violate the thin target conditions. The reaction rate density expression developed earlier (Eq.(3.18)) may be extended to treat this situation by quantifying monoenergetic neutron interactions of type j in infinitesimal volume dV (Fig. 3.6) as

$$\begin{aligned}\mathcal{R}(\mathbf{r},t)\,dV &= \sigma_j n(\mathbf{r},t) v N(\mathbf{r},t)\,dV \\ &= \varphi(\mathbf{r},t)\Sigma_j(\mathbf{r},t)\,dV \\ &= \text{the expected rate of interactions of type } j \text{ in } dV \text{ about } \mathbf{r} \text{ at time } t\end{aligned}$$

so that

$$\mathcal{R}_j(\mathbf{r},t) = \varphi(\mathbf{r},t)\Sigma_j(\mathbf{r},t) \tag{3.22}$$

Example 14: Cross Sections and Reaction Types
Describe some possible reactions for exit channels from the compound nucleus formed via $n + {}^{A}_{Z}X^{N}$ by showing reaction products and cross sections.

Solution:

$$
\begin{aligned}
n + {}^{A}_{Z}X^{N} \xrightarrow{\sigma} \left[{}^{A+1}_{Z}X^{N+1}\right]^{*} & \xrightarrow{\sigma_\gamma} {}^{A+1}_{Z}X^{N+1} + \gamma \\
& \xrightarrow{\sigma_p} {}^{A}_{Z-1}Y^{N+1} + p \\
& \xrightarrow{\sigma_\alpha} {}^{A-3}_{Z-2}Y^{N-1} + \alpha \\
& \xrightarrow{\sigma_f} X_1 + X_2 + (2-5)n \\
& \xrightarrow{\sigma_{s_i}} n' + {}^{A}_{Z}X^{N} + \gamma \\
& \xrightarrow{\sigma_{s_e}} n' + {}^{A}_{Z}X^{N}
\end{aligned}
$$

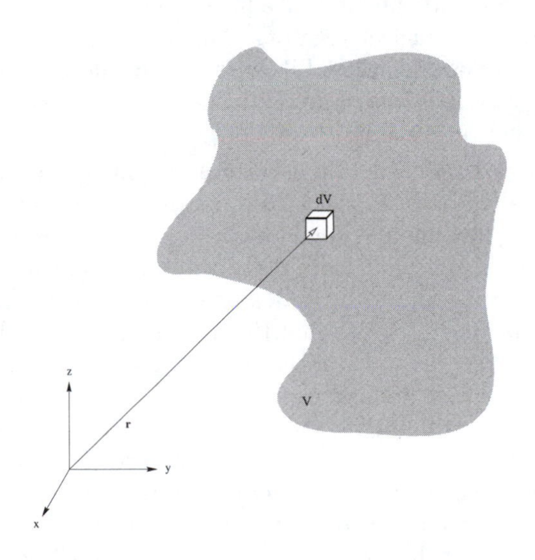

Figure 3.6: Neutron interactions, in general, occur with varying rate density in an arbitrary region of space (V) because of flux or composition profiles. The interaction rate can be determined by summing the incremental contributions in dV over all V (*i.e.* integrating).

The potential space and time dependence in $\Sigma(\mathbf{r}, t)$ emanates from the possibility for target medium concentration gradients and temporal changes in concentration through transmutation (alteration by nuclear reactions) or other physical changes to the system (fluid flow, expansion, ...), and not changes in the microscopic cross section, which is completely independent of space and time. The subscript j in the expression above refers to reactions of type j, *i.e.* we write a new reaction rate density equation for each individual reaction type; $\mathcal{R}_\gamma$ for (n,γ) reactions, $\mathcal{R}_\alpha$ for (n,α) reactions, etc. The microscopic and macroscopic cross sections vary accordingly (σ_γ and Σ_γ for (n,γ), and σ_α and Σ_α for (n,α), respectively), while we realize that the neutron flux cannot be a function of reaction type. Of course, when we are interested in quantifying reactions of any type, the total reaction rate density expression reverts to

$$\mathcal{R}(\mathbf{r}, t) = \varphi(\mathbf{r}, t)\Sigma(\mathbf{r}, t)$$

Finally, we refer to $\mathcal{R}(\mathbf{r}, t)$ as the "expected" value in keeping with the statistical nature of interaction between projectile and target. But again, since sample numbers are large, the statistics are good.

Uses for this expression abound. We have obvious interest in the fission reaction rate density in nuclear reactors. This quantity is directly proportional to the reactor power density. Likewise, we may need the rate at which (n,γ), (n,α), ... reactions are occurring in a radiation barrier (shield) or biological tissue, etc.

Use of the reaction rate equation in this way requires independent knowledge of $\varphi(\mathbf{r}, t)$. Calculations to yield this quantity comprise the subjects of nuclear reactor engineering and shield design. Local neutron flux depends on detailed knowledge of the medium composition, the nature and distribution of neutron sources, and a theoretical description of neutron transport in the medium. We also recognize that the more general treatment requires the quantities in Eq.(3.22) to be speed or energy dependent. We have restricted ourselves in prior discussions to one speed, so we do not explicitly carry this dependence, although it should be realized that it is always there. Handling these subjects constitutes a more advanced treatment and is well beyond the scope of our interests here. Our approach, while restrictive, will enable an appreciation for the concepts and treatment of a reasonable set of problems, ones in which $\varphi(\mathbf{r}, t)$ is determined *a priori*, or the geometry and neutron interaction physics is simple enough (sec. 3.7). The references [Lamarsh83, Lamarsh66, Duderstadt76] at the end of this chapter provide a more complete description for the advanced reader.

By summing $\mathcal{R}(\mathbf{r}, t)$ over all differential volume elements in V (*i.e.* integrating), we arrive at the rate of interactions of type j in the entire volume V

$$R_j(t) = \int_V \mathcal{R}(\mathbf{r}, t)\, dV = \int_V \varphi(\mathbf{r}, t)\Sigma_j(\mathbf{r}, t)\, dV \tag{3.23}$$

In some time interval from t_1 to t_2, the total number of interactions of type j in volume V is

$$\mathbb{R}_j = \int_{t_1}^{t_2} R_j(t)\, dt \tag{3.24}$$

There are many practical instances that can simplify calculations. For example, when the neutron flux is time independent, $\varphi(\mathbf{r}, t) = \varphi(\mathbf{r})$, we may say the flux is *steady*. This is often a reasonable approximation for constant reactor or beam experiment conditions. In certain situations, the neutron flux may be approximated as independent of spatial position (*i.e.* perhaps at the center of a very large reactor, far from any boundaries or other material composition changes), then $\varphi(\mathbf{r}, t) = \varphi(t)$, and the flux is said to be *uniform*. Most often, the interaction rate is sufficiently low so that there will not be significant alteration of the material medium (transmutation) during any reasonable interaction time. In this case, there is said to be no target *depletion*, so that $N(\mathbf{r}, t) = N(\mathbf{r})$. In many instances of practical importance, the target material is of uniform composition as well, so that N can be treated as a constant.

3.7 One Speed Neutrons in a Slab

Let's now explore a particularly straightforward but somewhat restrictive case that exercises the flux and reaction rate concepts, as well as introducing some new probability ideas. Consider a one speed (monoenergetic) and monodirectional beam of neutrons incident on a macroscopic (thick) slab of material as illustrated in Fig. 3.7. Since we have not developed a theory for describing how neutrons travel in general in such a medium, we are restricted to this simplified case. For the same reason, we are also restricted to describing only how beam neutrons behave before they interact with atoms of target material, the so called "uncollided flux." Once beam particles interact in with target atoms, they are removed from the uncollided beam and are no longer guaranteed to be monoenergetic and monodirectional. Indeed, as long as neutrons remain monoenergetic and monodirectional, they are by definition without interaction. There is a one-to-one correspondence between interactions of any type and particles removed from the uncollided beam.

Figure 3.7 illustrates this physical situation. Beam neutrons are incident on a thick slab of target material with thickness L, where L is much greater than the thin target dimension Δx. As neutrons enter the medium,

Example 15: Interaction Rates and Target Depletion
A very large experimental nuclear reactor has a nearly uniform and steady neutron flux of $10^{12}\frac{\text{n}}{cm^2\ s}$ at the center of its nuclear core. Calculate the interaction rate and total number of interactions for a small vanadium target placed in this flux for 1 min.

Data:
Vanadium target: $V = 1\ \text{cm}^3$, $\rho = 6.1\ \text{g/cm}^3$, $M \sim 51$ amu, $\sigma = 10$ b

Solution:
Assuming no target depletion, the reaction rate expression (Eq.(3.23)) with uniform, steady flux reduces to $R(t) = \varphi\Sigma V = \varphi\sigma NV$. Given the physical density above, the atom density of vanadium in the target is $N = \rho N_A/M = 7.2 \times 10^{22}\frac{\text{atoms}}{cm^3}$. Then,

$$\begin{aligned} R(t) &= 10^{12}\frac{\text{n}}{cm^2\ s}(10^{-23}\ \text{cm}^2)7.2 \times 10^{22}\frac{\text{atoms}}{cm^3}(1\ \text{cm}^3) \\ &= 7.2 \times 10^{11}\frac{\text{interactions}}{s} \\ \mathbb{R} &= R(t)\Delta t = 4.32 \times 10^{13}\ \text{interactions} \end{aligned}$$

To check the reasonableness of our no depletion assumption, consider that the total number of depleted atoms is at most equal to the total number of interactions during the interaction time. (This number is less in reality since some of the interactions are scattering.) Then the depleted fraction is $\frac{4.32\times10^{13}\ \text{interactions}}{7.2\times10^{22}\ \text{atoms}} = 6 \times 10^{-10} \ll 1$. During the 60 s interaction interval, only a tiny fraction of the original vanadium atoms are consumed.

they begin to interact with atoms of target material in the slab and are subsequently removed from the uncollided flux. So long as there are no sources of neutrons in the slab itself, the uncollided flux will decrease monotonically with distance from the surface of incidence ($x = 0$). In a slab of uniform composition, the macroscopic cross section, and hence the interaction probability per unit path length, are constant. As we'll see shortly, this implies that the rate of uncollided flux decrease is proportional to the uncollided flux itself. In general, $\varphi(\mathbf{r})$ is a function of three-dimensional configuration space. The symmetry in this problem reduces the dependence to one dimension, *i.e.* our slab is of infinite extent in both directions perpendicular to $\hat{x}$.

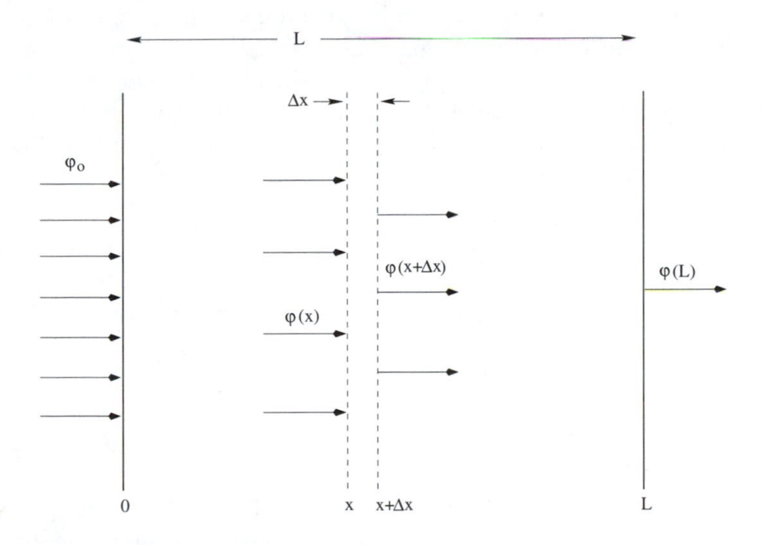

Figure 3.7: Uncollided neutron flux attenuation in a thick absorber of length L. Beam source neutrons are incident from the left at $x = 0$ with flux φ_o. The uncollided flux at any position within $[0, L]$ diminishes exponentially as $\exp(-\Sigma x)$.

We will track the uncollided flux through the thick slab since we can readily describe the behavior of uncollided neutrons and the rate of first interactions. The collided flux (flux that suffers at least one interaction in the target) cannot be so easily tracked. To be sure, the collided component of the neutron flux is significant, even dominant in most cases. Yet, again, this is well beyond the limitations of our model. For convenience, we'll also ignore target depletion, usually a very acceptable assumption, as we saw in example 15. The uncollided flux at arbitrary position x is then $\varphi(x)$, while that at position $x + \Delta x$, a tiny distance further, is $\varphi(x + \Delta x)$. Our objective will be to find $\varphi(x)$ as a function of the incident flux, φ_o and material properties. This will enable calculations of interaction rates and probabilities in the macroscopic medium.

To facilitate discussion, imagine focusing on a portion of the uncollided beam with vertical beam cross section of A (cm^2). Then, the number of uncollided neutrons crossing that area at position x per second is $\varphi(x)A$. Similarly, the number of uncollided neutrons passing through that area at position $x + \Delta x$ is $\varphi(x + \Delta x)A$. Then, the difference

$$[\varphi(x) - \varphi(x + \Delta x)]\, A$$

is the rate of neutron removal from the uncollided flux in Δx. From the definition of the uncollided flux, this must be the rate of first interactions in volume $\Delta x \times A$ of the slab. If we now shrink Δx small enough, our thin target approximation will hold so that the interaction probability is just $\Sigma\ \Delta x$. Then, we can also express the rate of first interactions as $\Sigma\ \Delta x\ \varphi(x)\ A$, and

$$\frac{\varphi(x) - \varphi(x + \Delta x)}{\Delta x} = \Sigma\ \varphi(x) \quad .$$

In the limit $\Delta x \to 0$, the left hand side is just $-\frac{d\varphi(x)}{dx}$. The solution of this differential equation, $\frac{d\varphi(x)}{dx} = -\Sigma\ \varphi(x)$ with the boundary condition $\varphi(x = 0) = \varphi_o$, is the familiar exponential attenuation law

$$\varphi(x) = \varphi_o \exp(-\Sigma x) \tag{3.25}$$

This expression provides the basis for some useful quantities. For example, one might ask at what position, $x_{1/2}$, is the uncollided flux exactly half its incident value. Expression (3.25) immediately yields, $x_{1/2} = \ln 2/\Sigma$. Similarly, we can find the e-folding distance, x_e, where $\varphi(x_e) = e^{-1}\varphi_o$, so that $x_e = \Sigma^{-1}$.

The above formalism has been developed by specifically describing neutron interactions. Other uncharged particles, like photons (of sufficiently high energy, *i.e.* γ particles), behave similarly, having constant interaction probability per unit path length. Reaction rate equations, fluxes, and cross sections can be prescribed in much the same way, although the types of reactions encountered are quite different. When treating γ interactions, the flux is sometimes replaced by the intensity, $I(\mathbf{r}, t) = E_\gamma \varphi_\gamma(\mathbf{r}, t)$, but has similar interpretation. As well, the macroscopic cross section is often replaced by the linear attenuation coefficient, μ, of identical dimension and interpretation (sec. 6.3). The reaction rate density expression for γs then becomes, $\mathcal{R}(\mathbf{r}, t) = \varphi_\gamma(\mathbf{r}, t)\mu(\mathbf{r}, t)$.

3.7.1 Probability Concepts

The description of uncollided flux attenuation lends naturally to some probability concepts that expand our understanding. To begin, consider the incident flux φ_o. Neutrons in the beam enter the slab at $x = 0$ with this flux. Everywhere in the slab, they have a constant probability of interaction per unit travel length of Σ. Since this represents only a probability of interaction, some neutrons will survive to any location in L without interacting. Some will escape all together without interacting for any L (so long as $L < \infty$).

The probability that a neutron will interact in any dx of the thick slab is then $\Sigma\ dx = -d\varphi(x)/dx$. Since $\varphi(x)$ represents the uncollided flux at any position, then $\varphi(x)/\varphi_o = \exp(-\Sigma x)$ is the probability of a neutron surviving to position x without having a single interaction. This probability is only zero at $L \to \infty$, so some neutrons will always traverse the entire slab without suffering a single interaction. The quantity $1 - \exp(-\Sigma x)$ is then the probability of having at least one interaction between 0 and x. We cannot say more. There is no information to be extracted from the above formalism that will tell us how many interactions neutrons will have once they have suffered at least one.

Combining the interaction and survival probabilities, we can identify the combination

$$\frac{\varphi(x)}{\varphi_o}\ \Sigma\ dx = \exp(-\Sigma x)\ \Sigma\ dx = p(x)\ dx$$

as the probability a neutron incident at $x = 0$ will have its first interaction in dx about x after having survived to x. The probability of having an interaction anywhere in the slab $(0 \leq x \leq L)$ then is

$$\int_o^L p(x)\ dx = \int_o^L \exp(-\Sigma x)\ \Sigma\ dx = 1 - \exp(-\Sigma x)$$

This is just $1 - \varphi(L)/\varphi_o$ as we saw above.

3.7.2 Mean Free Path and Average Flux

With the probability language developed above, we can now ask some meaningful questions about the average behavior of neutrons in the thick slab. For instance, it is important to quantify the average distance of travel for a neutron (incident at $x = 0$) in the slab before having its first interaction of any kind. This quantity is known as the "mean free path" and will allow us to give a quantitative description of exactly what we mean by a thick or thin target. Since $p(x)\ dx$ is the probability our neutron will first interact

Example 16: Two Slab Attenuation
Consider two slabs placed together with neutrons incident on the left side of slab 1, which has thickness L_1 and macroscopic cross section Σ_1. The second slab has thickness L_2 and macroscopic cross section Σ_2. Find:
a) the probability of a neutron having its first interaction in L_1,
b) the probability of a neutron having its first interaction in L_2,
c) the probability of a neutron escaping both slabs without interaction,
d) and the sum of all these probabilities.

Solution:
a) $p_a = 1 - \exp(-\Sigma_1 L_1)$
b) $p_b = \exp(-\Sigma_1 L_1)\,[1 - \exp(-\Sigma_2 L_2)]$
c) $p_c = \exp(-\Sigma_1 L_1)\exp(-\Sigma_2 L_2)$
d) sum $= p_a + p_b + p_c = 1$. Since these are the only possible eventualities, the sum must be 1.

in dx about x, the mean free path is then the average over all possible positions ($x = [0, L]$) weighted by $p(x)$

$$\langle x \rangle = \frac{\int_o^L x p(x)\, dx}{\int_o^L p(x)\, dx} = \frac{1/\Sigma - (L + 1/\Sigma)\exp(-\Sigma L)}{1 - \exp(-\Sigma L)}$$

In the limit $L \to \infty$, the mean free path for an infinite medium becomes $\langle x \rangle_\infty = 1/\Sigma$.

We can now quantify what we mean by a "thin" target. When introducing the concept of a thin target, we were interested in ensuring that neutrons suffered at best only one interaction in the target and that the incident flux is negligibly attenuated. These conditions are synonymous with the requirement that the interaction probability be very small. In this way, the probability of suffering a second interaction in the thin target is vanishingly small and all target atoms "see" the same projectile stream. The first interaction probability in any target of length L is $1 - \exp(-\Sigma L)$. This quantity is small ($\ll 1$) when $\Sigma L \ll 1$, or $L \ll 1/\Sigma$. In terms of mean free path, a target is thin when $L \ll \langle x \rangle_\infty$.

The volume averaged flux is also an important quantity. Often times, we are interested in only the average or integral behavior of a system rather than details regarding spatial dependence. The average flux over an interaction volume V is defined

$$\langle\varphi\rangle = \frac{\int_V \varphi(x)\, dV}{\int_V dV}$$

For our one-dimensional problem, $dV = A\, dx$ so

$$\langle\varphi\rangle = \frac{\int_0^L \varphi_o \exp(-\Sigma x)\, A\, dx}{AL} = \frac{\varphi_o}{L\Sigma}\left[1 - \exp(-\Sigma L)\right] \tag{3.26}$$

In the limit $\Sigma L \to 0$, then $\langle\varphi\rangle = \varphi_o$. This makes good physical sense. When $L \to 0$, the slab is thin, and the neutron flux is not appreciably altered from its incident value. When $\Sigma \to 0$, there are no interactions whatsoever in a slab of any thickness. The flux retains its incident value everywhere.

Now, we can write an expression for reaction rates in our macroscopic slab with $\varphi(x)$. With constant composition and steady flux, $\mathcal{R}_j(x,t) = \varphi(x)\Sigma_j$ is now the reaction rate density for interactions of type j. Then, the total rate of interaction is

$$\begin{aligned} R &= \int_V \mathcal{R}(x,t)\, dV = \Sigma \int_0^L \varphi(x) A\, dx \\ &= \varphi_o A\left[1 - \exp(-\Sigma L)\right] = \langle\varphi\rangle \Sigma V \end{aligned}$$

When $L \to \infty$ (a very large target), then $R(t) \sim \varphi_o A$, *i.e.* all neutrons interact at least once if the target is thick enough.

If we instead consider only reactions of type j (*e.g.* (n,γ) for example), then

$$R_\gamma = \varphi_o A \frac{\Sigma_\gamma}{\Sigma}\left[1 - \exp(-\Sigma L)\right]$$

is the rate of (n,γ) interactions in the entire slab of thickness L. This agrees with our intuition, *i.e.* $R_\gamma = \frac{\Sigma_\gamma}{\Sigma} R$, and reaffirms our interpretation of cross section as a relative probability for interaction. We interpret the quantity Σ_γ/Σ as the relative probability of (n,γ) interactions among all interactions. Again, when $L \to \infty$, then $R_\gamma \sim \varphi_o A \Sigma_\gamma/\Sigma$.

Our results thus far lead to some very natural questions about stopping particles of radiation with material objects like slabs, and the generation of heat in such media as these particles are stopped. The answers to these questions comprise the subject of radiation shielding. An introduction to some of the subjects important to radiation shielding is presented in chapter 6 under the more fundamental subject of radiation interaction with matter. The reader is referred to [Shultis96] for a more advanced treatment. Indeed, what we have done in this chapter is the first step in describing

Example 17: Thick Target Interaction Rates and Probabilities
A thick ^{235}U slab target is irradiated by a uniform monoenergetic neutron beam of $\varphi = 10^{12} \frac{\text{n}}{\text{cm}^2\ \text{s}}$. The total interaction cross section is $\sigma = 689.7$ b, while that for fission alone is $\sigma_f = 582.2$ b. The mass density of uranium is 19.1 g/cm^3. Determine the interaction rate and average reaction rate density for fission and total interactions if the target is 1 cm thick and the beam illuminated area on target is 1 cm^2.

Solution:
The target interaction center density is

$$N = \frac{\rho N_A}{M_{\text{Al}}} = \frac{19.1\ \text{g/cm}^3 (6.022 \times 10^{23}\ \text{at/mole})}{235.0439\ \text{g/mole}} \sim 4.9 \times 10^{22}\ \text{at/cm}^3$$

so that $\Sigma = \sigma N = 33.8\ \text{cm}^{-1}$ and $\Sigma_f = \sigma_f N = 28.5\ \text{cm}^{-1}$. Then, the total and fission interaction rates in the slab are

$$\begin{aligned} R &= \varphi_o A\left[1 - \exp(-\Sigma L)\right] = 10^{12}\ \text{interactions/s} \\ R_f &= \frac{\Sigma_f}{\Sigma} R(t) = 8.43 \times 10^{11}\ \text{interactions/s} \end{aligned}$$

By the definition of average flux, the average interaction rate density is $\langle \mathcal{R}_j(x,t) \rangle = \langle \varphi \rangle \Sigma_j = R_j / V$.

Discussion:
The probability for a neutron in this beam to reach L without an interaction $= \exp(-\Sigma L) \sim 2.1 \times 10^{-15}$. There is high probability for neutrons to interact somewhere in this target. Even though L is only 1 cm, this is a very thick target since Σ is large.

neutron and γ particle interaction with matter. The scattering of neutrons (or γ rays) by slab molecules transfers projectile energy to target molecules which quickly distributes and becomes heat energy. As well, neutron absorption reactions may lead to the production of energetic charged particles through radioactive decay (chpt. 4). These charged particles stop suddenly and deposit their energy of motion as heat energy in the target. The latter is a localized heating mechanism and is, therefore, proportional to $\varphi(x)\Sigma_j$. Neutron-induced γ interactions may also deposit energy that heats the slab material. This is a much more difficult problem to quantify since energetic photons rarely deposit all of their energy locally. Then, in addition to tracking neutron transport to initiate such reactions, we are compelled to track photon transport to quantify heat deposition.

The slab attenuation problem reveals a natural mechanism to reduce radiation hazards. Providing flux attenuating material between a source of intense radiation and humans constitutes a physical barrier for hazard reduction. Increasing the thickness of this barrier provides exponential benefit from the reduction of uncollided flux. The impact of collided flux is almost always dominant, however, and cannot be ignored. Yet, the overall benefit is almost always enhanced by increasing barrier thickness.

3.8 Reaction Cross Sections for Homogeneous Mixtures

Our preceding discussions and calculations have all involved so called *pure* media, those that consist of only one chemically pure substance or element, or only one nuclide. We have implicitly insisted on this condition so that the neutron interaction properties of the system could be described by a single value of σ_j or Σ_j determined from some experiment like that described in sec. 3.3. Real life situations often present more complicated combinations of substances that we'll call *mixtures* for which we must find a representative (or aggregate) cross section to correctly determine reaction rates.

Recall that in the definition of the macroscopic cross section, $\Sigma = \sigma N$, one is required to determine the atom number density of the target material. In all cases we've already examined, the target composition was the simplest imaginable, a single isotope like ^{27}Al. In this case, it is a trivial matter to determine the macroscopic cross section since N is found directly from the physical density, $N = \rho N_A/M$, where M is the gram molecular weight (g/mol). For a single isotope, the gram molecular weight is identical to the atomic mass in amu (sec. 1.3.5). For ^{27}Al, $M = 26.981538$ amu/molecule $= 26.981538$ g/mol. (Of course, this is just the definition of N_A.) Returning to our Al thin target example (sec. 3.3), recall that we found $\sigma \sim 1.7$ b.

With $\rho = 2.7$ g/cm^3 and M from above, this yields $N \sim 6 \times 10^{22}$ at/cm^3, so $\Sigma = 0.1$ cm^{-1}, in good agreement with $p/\Delta x$, as it should be.

To examine more complicated systems, let's return to the reaction rate density expression (Eq.(3.22)) for reactions of type j

$$\mathcal{R}_j(\mathbf{r},t) = \varphi(\mathbf{r},t)\Sigma_j$$

where we consider a spatially and temporally constant cross section, *i.e.* a local region where material properties do not vary and the *no depletion* assumption is valid. The reaction rate density for neutrons interacting with atoms of a particular isotope in a mixture of isotopes can be found by extension of the above

$$\mathcal{R}_j^i(\mathbf{r},t) = \varphi(\mathbf{r},t)\Sigma_j^i \tag{3.27}$$

where the superscript refers to isotope i. We should interpret this expression as determining the reaction rate density for interactions of type j with isotopes of type i in our mixture. The reaction rate density for the entire mixture can be recovered by summation over isotopes, $\mathcal{R}_j(\mathbf{r},t) = \sum_i \mathcal{R}_j^i(\mathbf{r},t)$. This extended description introduces the new quantity

$$\Sigma_j^i = N^i\sigma_j^i = f_n^i N\sigma_j^i$$

where N^i is the atom density for isotope i only, and f_n^i is the atom fraction or fractional abundance, familiar from chapter 1.

Example 18: Mixture Cross Section: 1

What is the macroscopic absorption cross section for ^{235}U, Σ_a^{235}, in natural uranium?

Solution:

The fractional abundance for ^{235}U in natural U is $f_n^{235} = 0.0072$, the mass density of U is $\rho_{\mathrm{U}} = 19.1$ g/cm^3, and the microscopic absorption cross section for ^{235}U is $\sigma_a^{235} = 680.8$ b. Then $N^{\mathrm{U}} = \rho_{\mathrm{U}} N_A/M = 19.1(0.6022 \times 10^{24})/238.0289 = 0.0483 \times 10^{24}$ at/cm^3. So

$$\Sigma_a^{235} = 0.0072(0.0483)680.8 = 0.237 \text{ cm}^{-1}$$

Our mixture in the above example consists of the isotopes of natural uranium, 99.9945% of which is ^{235}U and ^{238}U. To determine the absorption reaction rate density in natural uranium (example 19), we need to calculate

$$\mathcal{R}_a^{\mathrm{U}}(\mathbf{r},t) = \varphi(\mathbf{r},t)\Sigma_a^{\mathrm{U}} \quad = \quad \varphi(\mathbf{r},t)\left[N^{235}\sigma_a^{235} + N^{238}\sigma_a^{238}\right]$$

$$= \varphi(\mathbf{r},t)N^{\mathrm{U}}\left[f_n^{235}\sigma_a^{235} + f_n^{238}\sigma_a^{238}\right]$$

For the general problem of this type, we find

$$\Sigma_j^m = \sum_i N^i \sigma_j^i = N^m \sum_i f_n^i \sigma_j^i \qquad (3.28)$$

where the superscript m refers to the "mixture." When the individual f_n^i's and N^i's are prescribed, the solution is quite straightforward (example 19). This procedure extends naturally to chemical compounds as well

---

Example 19: Mixture Cross Section: 2

What is the macroscopic absorption cross section for natural uranium?

Solution:

The fractional abundance for ^{235}U in natural U is $f_n^{235} = 0.0072$, that for ^{238}U is $f_n^{238} = 0.9928$, the mass density of U is $\rho_U = 19.1$ g/cm^3, and the microscopic absorption cross section for ^{235}U is $\sigma_a^{235} = 680.8$ b, while that for ^{238}U is $\sigma_a^{238} = 2.7$ b. Then,

$$\Sigma_a^{\mathrm{U}} = \frac{\rho_{\mathrm{U}} N_A}{M_{\mathrm{U}}}\left[0.0072(680.8) + 0.9928(2.7)\right] = 0.366 \text{ cm}^{-1}$$

a bit higher than that of ^{235}U in natural U.

Discussion:

Although ^{238}U has a much smaller cross section, its abundance is so much greater than ^{235}U, it has a significant contribution to Σ_a^{U} for natural U.

(example 20). For example, we may require the macroscopic cross section for substances like ordinary water. This mixture is ubiquitous in nuclear systems.

For many cases of interest, the atom number density or atom fraction may not be among the information available. Instead, we may be given the mixture density and mass (or weight) fraction, f_w^i, of particular isotopes. Our general expression for the mixture cross section (Eq.(3.28)) is still valid, although it will need to be expanded a bit differently. By definition $f_w^i = \rho_i/\rho$ is the contribution to mass density from isotope i relative to the mixture. Since $\rho \propto M$ and $\rho_i \propto f_n^i M_i$, then

$$f_w^i = \frac{M_i}{M} f_n^i$$

Example 20: Mixture Cross Section: 3

Find the macroscopic absorption cross section for ordinary water.

Solution:

Equation (3.28) requires us to write down explicitly all the contributions to $\Sigma_a^{H_2O}$ from all the stable isotopes of its constituents.

$$\begin{aligned}\Sigma_a^{H_2O} &= N^{^1H}\sigma_a^{^1H} + N^{^2H}\sigma_a^{^2H} + N^{^{16}O}\sigma_a^{^{16}O} + N^{^{17}O}\sigma_a^{^{17}O} + N^{^{18}O}\sigma_a^{^{18}O} \\ &= f_n^{^1H}N^H\sigma_a^{^1H} + f_n^{^2H}N^H\sigma_a^{^2H} + f_n^{^{16}O}N^O\sigma_a^{^{16}O} + f_n^{^{17}O}N^O\sigma_a^{^{17}O} \\ &\quad + f_n^{^{18}O}N^O\sigma_a^{^{18}O}\end{aligned}$$

but, $N^H = 2N^{H_2O}$, $N^O = N^{H_2O}$, and $M_{H_2O} = 18.0153$ g/mol, so

$$\begin{aligned}\Sigma_a^{H_2O} &= N^{H_2O}\left[2f_n^{^1H}\sigma_a^{^1H} + 2f_n^{^2H}\sigma_a^{^2H} + f_n^{^{16}O}\sigma_a^{^{16}O} + f_n^{^{17}O}\sigma_a^{^{17}O}\right. \\ &\quad \left. + f_n^{^{18}O}\sigma_a^{^{18}O}\right] \\ &= \frac{1.0(0.6022)}{18.0153}[2(0.99985)0.332 + 2(0.00015)0.00053 \\ &\quad + (0.99759)0.000178 + (0.00037)0.235 + (0.00204)0.00016] \\ &= 0.0222\ \text{cm}^{-1}\end{aligned}$$

Discussion:

Notice that the overwhelming majority of the contribution to the mixture cross section comes from a single isotope, ^{1}H, that has both large abundance and cross section.

and our mixture cross section equation can be rewritten

$$\Sigma_j^m = \rho N_A \sum_i f_w^i \frac{\sigma_j^i}{M_i} \tag{3.29}$$

Example 21: Mixture Cross Section: 4
Instead of finding natural uranium in our reactor, we are given an enriched mixture so that the isotopic abundance of ^{235}U is increased well in excess of its natural abundance. Since ^{235}U has a much larger cross section for absorption (and fission) than does ^{238}U, we expect such enrichment to yield a higher fission rate, all other factors being equal. Consider a uranium mixture enriched to 10 w/o in ^{235}U. What is the net macroscopic absorption cross section?

Solution:
Equation (3.29) now requires the weight percent and atomic mass of each isotope.

$$\begin{aligned}
\Sigma_a^{\mathrm{U}} &= N^{235}\sigma_a^{235} + N^{238}\sigma_a^{238} \\
&= \rho N_A \left[f_w^{235} \frac{\sigma_a^{235}}{M_{235}} + f_w^{238} \frac{\sigma_a^{238}}{M_{238}} \right] \\
&= 19.1(0.6022) \left[0.1 \frac{680.8}{235.0439} + 0.9 \frac{2.7}{238.0509} \right] \\
&= 3.45 \text{ cm}^{-1}
\end{aligned}$$

Discussion:
Notice that this is much greater (by about an order of magnitude) than that of natural uranium (cf: example 19). This is reasonable since ^{235}U is the majority of the influence on Σ^{U}, and we have increased the abundance of this isotope by about an order of magnitude over its natural abundance.

Finally, let's consider a more involved case that requires elements of the two different types of solutions outlined above, *i.e.* our mixture contains an assortment of constituents for which we know atom fractions for some and weight fractions for others. A perfect example of this more general case is real fission reactor fuel consisting of uranium dioxide (UO_2) ceramic fuel pellets. Let's imagine (although high for a real reactor) that the fuel is

Example 22: Mixture Cross Section: 5

Find the macroscopic absorption cross section, $\Sigma_{\mathrm{a}}^{304}$, of 304 stainless steel, which has a density of $\rho = 7.86$ g/cm^3 and has the composition given in the table below:

material	σ_a^i (b)	M_i	f_w^i
C	0.0035	12.011	0.0008
Cr	3.1	51.996	0.19
Ni	4.5	58.69	0.10
Fe	2.56	55.847	0.7092

Solution:

$$\begin{aligned}
\Sigma_a^{304} &= 7.86(0.6022)\left[0.0008\frac{0.0035}{12.011} + 0.19\frac{3.1}{51.996} + 0.1\frac{4.5}{58.69}\right. \\
&\quad \left. + 0.7092\frac{2.56}{55.847}\right] \\
&= 0.244 \text{ cm}^{-1}
\end{aligned}$$

enriched to 10 w/o in ^{235}U to compare to example 21. We wish to determine macroscopic cross sections for this mixture, *e.g.* $\Sigma_a^{UO_2}$. For simplicity, we'll consider O to be comprised entirely of ^{16}O. We've already seen that this makes up the overwhelming majority of the contribution. The mass density for this mixture is 10.5 g/cm^3, so

$$\begin{aligned}\Sigma_a^{UO_2} &= N^{235}\sigma_a^{235} + N^{238}\sigma_a^{238} + N^{16}\sigma_a^{16} \\ &= f_n^{235}N^{U}\sigma_a^{235} + f_n^{238}N^{U}\sigma_a^{238} + f_n^{16}N^{O}\sigma_a^{16}\end{aligned}$$

but

$$N^U = N^{UO_2} \;, \quad N^O = 2N^{UO_2} \;, \quad f_n^{235} = f_w^{235}\frac{M_U}{M_{235}}$$

This reduces to the final expression

$$\Sigma_a^{UO_2} = \frac{\rho N_A}{M}\left[f_w^{235}\frac{\sigma_a^{235}}{M_{235}}M_U + f_w^{238}\frac{\sigma_a^{238}}{M_{238}}M_U + 2\sigma_a^{16}\right]$$

However, we cannot solve this yet. We do not know the new mixture molecular weight since it is not naturally occurring. We must find it from the composition, *i.e.* $M = M_{O_2} + M_U$, where $M_{O_2} = 2M_O$. M_U is different from that naturally occurring since we have changed the composition. To find it, return to the definition of weight fraction, $f_w^i = f_n^i M_i/M$. Summing this over i gives an expression for M_U

$$\sum_i \frac{f_w^i}{M_i} = \sum_i \frac{f_n^i}{M_U} = \frac{1}{M_U}\sum_i f_n^i = \frac{1}{M_U}$$

Using this to determine M_U,

$$\frac{1}{M_U} = \frac{0.1}{235.0439} + \frac{0.9}{238.0509}$$

so $M_U = 237.7747$ g/mol. Employing this in the calculation of $\Sigma_a^{UO_2}$, we first find $M = M_{UO_2} = 269.765$ g/mol and finally, $\Sigma_a^{UO_2} = 1.66$ cm^{-1}. Note that this is substantially smaller (about a factor of two) than that calculated for uranium metal enriched to 10 w/o in ^{235}U (cf: example 21). This is to be expected. The addition of oxygen to the mixture dilutes the uranium concentration. Indeed, the UO_2 mixture density is about a factor of two below that of uranium metal.

3.9 Homogeneity

In the mixture cross section calculations we have been performing, it has been implicitly assumed that as far as our interacting neutron is concerned, the mixture of nuclides in the system is somehow evenly distributed. We have, further, treated the cross section problem as though the neutron were able to "see" all available isotopes (in their respective abundances) during its lifetime in the material media. In doing so, we have constructed a *homogeneous medium.*

To define this more specifically, consider a system including two different substances with concentrations N_1 and N_2, respectively. Upon mixing, we define the composition of the mixture as the ratio N_1/N_2. This mixture is uniform if N_1 and N_2 are independently $\neq f(\mathbf{r})$. We say the mixture is homogeneous if $N_1/N_2 \neq f(\mathbf{r})$, a somewhat less restrictive condition. A homogeneous mixture need not be uniform as long as the ratio N_1/N_2 does not vary. Yet, a uniform mixture is always homogeneous. If our system is comprised of a monatomic substance only (single atom or isotope), then it is guaranteed to be homogeneous, but need not be uniform.

On some scale, all physical systems are both inhomogeneous and non-uniform. If we magnify our perspective to the material grain or, even more, to the atomic scale, there are always large spatial gradients. Yet, what is important here is the degree of homogeneity and uniformity on the scale of the physical process under scrutiny. For neutron interactions, the relevant scale is the interaction mean free path, $\langle x \rangle_\infty$.

Consider the physical system depicted below (Fig. 3.8) comprised of a mixture of two substances. Particles of substance 1 are interspersed among a those of substance 2 in either a regular or random way. It matters little for our neutron interactions. A good example might be sugar or salt crystals dissolved in water. Perhaps a more appropriate example of a nuclear system might be UO_2 powder mixed in water or some other moderator. Let us then define

$\langle x \rangle_{\infty_1}$ = neutron mean free path in a sample of pure substance 1

$\langle x \rangle_{\infty_2}$ = neutron mean free path in a sample of pure substance 2

$\langle r \rangle$ = average distance between substance 1 particles in the mixture

d = diameter of particles in substance 1

This mixture is homogeneous if both

$$\langle x \rangle_{\infty_2} \gg \langle r \rangle$$
$$\langle x \rangle_{\infty_1} \gg d$$

i.e. the neutron must sample many interaction sites of all substances in the mixture during its characteristic transport scale, $\langle x \rangle_\infty$.

Since $\langle x \rangle_\infty$ can be relatively small (a few cm or less), many nuclear systems, including nuclear power reactors, are not homogeneous mixtures

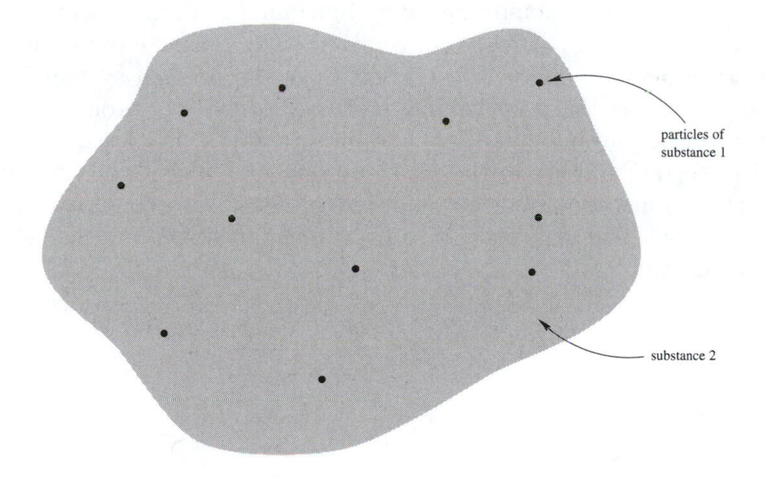

Figure 3.8: Mixing of particles of substance 1 in a reservoir of substance 2. The new substance forms a homogeneous mixture for neutron interactions when the mean free path for neutrons in substance 1 is much greater than the average diameter of substance 1 particles and when the mean free path for neutrons in substance 2 is much greater than the average distance between substance 1 particles. Neutrons then "select" atoms to interact with randomly and in proportion to the mixing ratio.

of substances according to the definition above. They instead consist of long metallic fuel rods that contain the UO_2 fuel, which are immersed in a water moderator. While the separation between these rods may be less than the neutron mean free path in water, the dimensions of the fuel rods themselves ($d \sim 0.5$ in) are not smaller than the neutron mean free path in the fuel. Neutron flux and reaction rate calculations for such systems become quite difficult and are usually performed numerically on high speed computers, even then only after considerable approximation.

Under certain limited conditions, it may be acceptable to homogenize even such a system, *i.e.* construct an analytic model that replaces the real system with one that is homogeneous by imagining the composition to be decomposed into small particles and remixed. The remixed system maintains average neutron interaction properties of the original system if we insist on conserving the total volume and total number of atoms of each substance. In performing a homogenization this way, some of the interesting and important physics in the problem may be removed, and we may now be able to answer only a very limited subset of questions, like those regarding average flux and reaction rates, rather than exploring the details of local variations.

References

[Dickerson79] Dickerson, R. E., Gray, H. B., and Haight, G. P., *Chemical Principles*, 3^{ed}, Benjamin Cummings, Pub. Co. Inc, Menlo Park, CA, 1979.

[Duderstadt76] Duderstadt, J. J., and Hamilton, L. J., *Nuclear Reactor Analysis*, John Wiley & Sons, NY, 1976.

[GE96] General Electric Co., *Nuclides and Isotopes*, 15^{ed}, 1996.

[Krane88] Krane, K. S., *Introductory Nuclear Physics*, John Wiley & Sons, NY, 1988.

[Lamarsh66] Lamarsh, J. R., *Introduction to Nuclear Reactor Theory*, Addison-Wesley, Reading, MA, 1966.

[Lamarsh83] Lamarsh, J. R., *Introduction to Nuclear Engineering*, 2^{ed}, Addison-Wesley Publishing Co. Inc., Reading, MA, 1983.

[Shultis96] Shultis, J. K., and Faw, R. E., *Radiation Shielding*, Prentice-Hall, Inc., Upper Saddle River, NJ, 1996.

Problems and Questions

1. Find the number of ^{235}U nuclei in 5 g of natural uranium.

2. COMPUTER EXERCISE: Construct a simple FORTRAN source code to evaluate the Lab System neutron scattered energies and angles as per Eqs. (3.4) and (3.6) of the text.
   Your code should:

   (a) compile and run on a workstation with f77,

   (b) have at least one subprogram and two nested loops, and

   (c) output clearly identifiable ASCII data for your results to incorporate into your favorite plotting package.

   You should present your results graphically in the form E_y/E_x vs. the lab scattering angle, ϕ, as a family of curves for $A = 1, 2, 7, 10, 12, 16, 27, 56, 90, 208$, and 235.

3. Consider a system of two free, non-relativistic charged particles undergoing an elastic collision

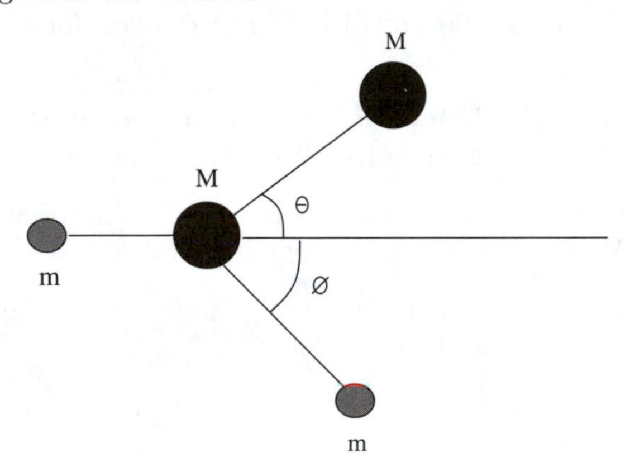

Figure P.(3): Two body elastic collision.

   Figure P.(3) describes the interaction in which m is an α particle with an initial velocity v_x and M is a stationary ^{7}Li atom. In addition, consider the case where both particles have identical speeds after the collision.

   (a) Determine the ratio of the initial velocity of the α particle to its velocity after the collision.

   (b) Determine the angle ϕ with which the α particle leaves the interaction.

(c) Determine the angle θ which the ^{7}Li atom leaves the interaction.

(Assume that there is no Coulomb interaction between the two particles before and after the collision when they are very distant. The interaction physics is then identical to elastic neutron scattering.)

4. A 10 MeV neutron elastically scatters with an ^{16}O nucleus, which is stationary in the lab system at a center of mass angle of 45°. What are the energies of the scattered neutron and the recoil nucleus in the lab and center of mass systems? What are the center of mass kinetic energy and center of mass velocity?

5. Consider the fusion reaction

$$\mathrm{d} + \mathrm{D} \rightarrow \mathrm{Y} + \mathrm{n}$$

Determine the product Y identity and product energies assuming the reactants are initially at rest. The energetic neutron leaving this interaction encounters a stationary ^{56}Fe nucleus and undergoes an elastic scattering interaction with the neutron emerging at a center of mass scattering angle of $\phi_c = 43.2^\circ$. Determine the product energies and lab scattering angle.

6. Consider a head on ($\phi_c = \pi$) elastic scattering collision between a neutron and a proton, stationary in the lab system. Describe the collision process by evaluating all of the following parameters: E_x^c, E_X, E_X^c, KE_c, E_Y, E_Y^c, E_y, and E_y^c as functions of E_x and masses.

7. Neutrons are produced for scattering experiments by bombarding a thin ^{9}Be target with 10 MeV α particles from an accelerator. Energetic neutrons are extracted along the α direction ($\phi = 0$). On a particular experiment, neutrons are detected at $\phi = \pi/6$ from the incident beam after striking a thin lithium target and scattering elastically with one ^{7}Li atom. What is the resultant neutron energy after the elastic scattering event?

8. For a neutron elastic backscatter event ($\phi_c = \pi$) with a stationary ($\vec{v}_2 = 0$) ^{56}Fe nucleus, determine the C and L system energies, E_x^c, E_X^c, E_Y^c, E_y^c, KE_c, E_Y, and E_y, as functions of E_x.

9. A 4 MeV neutron elastically scatters with a graphite nucleus at a lab angle of 30°. What are the energies of the scattered neutron and the recoil graphite nucleus in the lab and center of mass system? What is the velocity of the center of mass system? What is the center of mass kinetic energy?

10. A 1 MeV neutron elastically scatters through a lab angle of 45° in a collision with a ^{2}H nucleus. What is the energy of the scattered neutron and the recoil deuteron in both the C and L systems? What is the center of mass kinetic energy, KE_c?

11. A 2 MeV neutron traveling in water has a head-on elastic collision with a ^{16}O nucleus. What are the energies of the neutron and the nucleus after the collision (in the lab system)? Would you expect the water molecule involved in the interaction to remain intact after the collision (and why or why not)?

12. Plot the average fractional energy loss in neutron elastic scattering as a function of mass number of the target nucleus.

13. For a head on collision, determine the final neutron kinetic energy when the target is a proton and when it is a very heavy atom from the general expression, Eq.(3.4). What are the target energies (E_Y) after the collision in the two cases above.

14. Consider a 0.5 MeV fission neutron incident on a very large pool of heavy water (D_2O). Assuming all interactions to be elastic scattering on deuterons, determine:

    (a) the average number of collisions required to thermalize this fast neutron,

    (b) the mean distance traveled before the first interaction of this neutron ($\Sigma = 0.00985$ cm^{-1}), and

    (c) the probability of traveling the mean distance without having an interaction.

15. Using binary, two-product reaction kinematics in the center of mass, derive Eqs.(3.8), (3.9), and (3.10) for the more general inelastic scattering reaction. Show that these reduce to the specific case of elastic scattering when $E^* \to 0$.

16. Show the steps involved in arriving at the final form of expression (3.13).

17. Determine the average number of elastic collisions required to thermalize (0.025 eV) a fission neutron (2 MeV) in potassium ($A = 39$).

18. How many elastic collisions are required on average to slow down a 4 MeV fission neutron to 0.0253 eV if the moderator is: (i) hydrogen, (ii) deuterium, (iii) graphite, or (iv) sodium?

19. How many elastic collisions are required (on the average) to slow a 2 MeV fission neutron to 1 eV if the moderator is hydrogen? Graphite?

20. A beam of 1 eV neutrons with beam density 10^3 cm^{-3} strikes a 1 micron thick pure gold target. If the beam cross section projection on the target is 1 mm^2 and the total interaction cross section is 99 b, what is the interaction rate in the target? What are the probabilities of interaction and non-interaction? If the beam density were increased ten fold, what are the new reaction rate and probabilities?

21. An experimenter designs a neutron beam experiment to provide five independent beams converging on a single target at the center of a circle. Four of the beams produce neutrons at the constant rate of 10^4 n/s, and the fifth at 10^5 n/s. If all neutrons are monoenergetic at 100 eV, what is the "one speed" flux of neutrons at the convergence if all beams focus to 0.1 mm^2 at the target? If the fifth beam can be independently tuned to 200 eV while the others remain at 100 eV, what are the "one speed" fluxes on target?

22. What is the probability per unit length of a neutron interacting in a pure He gas at STP (0.05 b) and in a pure ^{238}U target (2.7 b)?

23. Consider two infinite slabs L and R in contact with each other, with thicknesses a and b and arranged with L to the left of R.

    (a) What is the probability that a neutron entering the slab from the left will have an interaction in L?

    (b) What is the probability that a neutron entering from the left will have its first interaction in R?

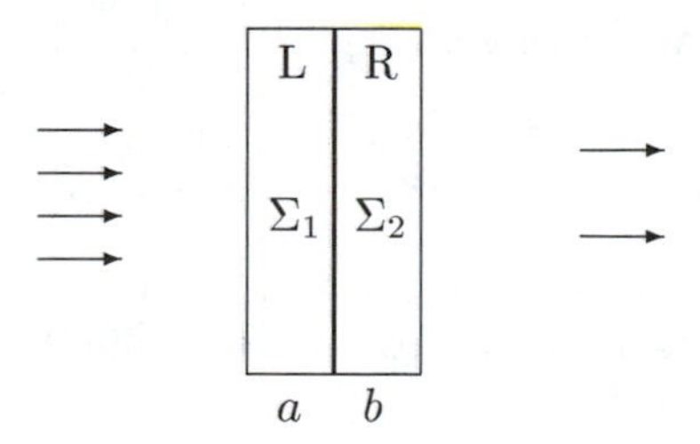

24. Consider an extension of question 23 in which we now have a mono-energetic neutron beam incident on a target comprised of three separate layers of different materials with macroscopic cross sections, Σ_1, Σ_2 and Σ_3, in layers of length L_1, L_2, and L_3, respectively. Determine:

(a) the probability of a neutron having its first interaction in L_1,

(b) the probability of a neutron having its first interaction in L_2,

(c) the probability of a neutron having its first interaction in L_3, and

(d) the probability of a neutron escaping the entire target without an interaction.

Verify that the sum of the probabilities in parts (a)–(d) is in fact = 1.

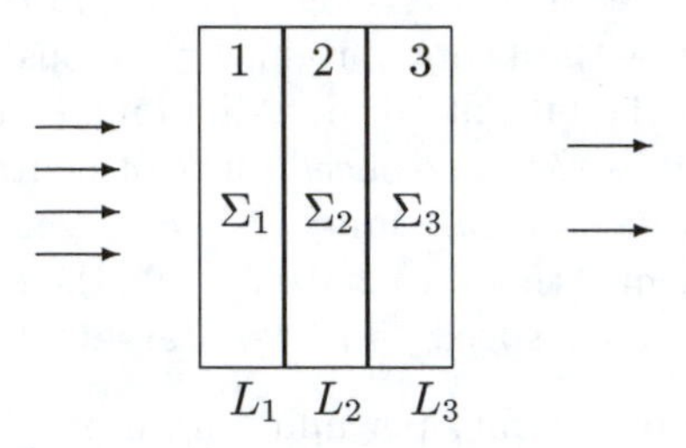

25. Consider neutrons incident normal to an infinite slab of thickness $3L$ consisting of two regions. The first region of thickness $2L$ is filled with ordinary water ($\rho = 1$ g/cm^3, $\sigma^{^1\text{H}} = 38$ b, $\sigma^{^2\text{H}} = 6.8$ b, $\sigma^{^{16}\text{O}} =$ 0.178 mb, $f_n^{^2\text{H}} = 0.015\%$, $f_n^{^{16}\text{O}} = 99.76\%$, all other cross sections can be assumed to be zero.). The second region is filled with uranium metal enriched to 95 w/o in ^{235}U ($\rho_U = 19.1$ g/cm^3, $\sigma^{235} = 680.8$ b, $\sigma^{238} = 2.7$ b). (a) Determine the macroscopic cross section in each region. (b) Find the uncollided flux as a fraction of the incident flux, φ_o, at $0, L, 2L$, and $3L$, for $L = 0.2$ cm. (c) Determine the probability of first interaction in each region.

26. Show that the infinite moderator mean free path, $< x >_\infty$, is equal to the inverse of the macroscopic total cross section, Σ.

27. We determined the mean free path for a neutron in an infinite medium by integrating $xp(x)\,dx$ from zero to infinity. The general definition for $< x >$ is as follows:

$$< x > \equiv \frac{\int_o^\infty x\,p(x)\,dx}{\int_0^\infty p(x)\,dx}$$

where the integral in the denominator goes to 1 for the infinite medium case. Now, find the mean free path for a finite medium of thickness L. Evaluate this expression in the limit $L \to \infty$ to check your solution.

28. Determine the macroscopic absorption cross section for a uranium fuel rod which has been enriched to 15 w/o ^{235}U and the remainder is ^{238}U. The density, ρ, of this fuel rod is equal to 19.1 g/cm^3 and the microscopic absorption cross sections are $\sigma_{235} = 680.8$ b and $\sigma_{238} = 2.7$ b.

29. Concrete has a density of 2.35 g/cm^3 and is composed of ten different elements. Determine the macroscopic absorption and scattering cross sections for concrete using the weight fractions f_w^i and the microscopic cross sections listed below.

material	M_i	f_w^i	σ_a^i (b)	σ_s^i (b)
O	15.9994	0.4983	0.00028	4.2
Si	28.0855	0.3158	0.168	1.7
Ca	40.078	0.0826	0.43	3.0
Al	26.981538	0.0456	0.23	1.4
K	39.0983	0.0192	2.1	1.5
Na	22.98977	0.0171	0.53	4.0
Fe	55.845	0.0122	2.56	11.0
H	1.00794	0.0056	0.333	38.0
Mg	24.305	0.0024	0.066	3.6
S	32.066	0.0012	0.52	1.1

30. Zircaloy is a corrosion resistant material used in nuclear reactors as fuel cladding. A particular sample of this material is comprised of 1.5 w/o in Sn, 0.15 w/o in Fe, 0.1 w/o in Cr, and the remainder is Zr. Using the data below, determine the macroscopic absorption cross section for this material if the physical density is 6.5 g/cm^3.

material	σ_a^i (b)	M_i (g/mole)
Sn	0.61	118.71
Fe	2.56	55.845
Cr	3.1	51.9961
Zr	0.184	91.224

31. Consider “one speed” neutrons incident perpendicular to a three region slab. The neutron beam is incident on region 1 which is comprised of an ordinary water layer 0.3 cm thick, followed by a 1.5 cm thick iron layer, and finally by a 0.5 cm thick 5 w/o solution of boric acid (H_3BO_3). Using the data below, calculate:

    (a) the uncollided flux at the exit of region 3,

    (b) the first interaction probability in each region,

(c) the probability of escaping each region without and interaction, and

(d) the sum of first interaction probabilities and the probability of escape from region 3.

material	σ_a^i(b)	σ_s^i(b)
H	0.333	38.0
O	0.00028	4.2
Fe	2.56	11.0
B	760.0	4.0

(ρ(water and boric acid solution) $\simeq$ 1 g/cm^3, ρ_{Fe} = 7.87 g/cm^3, $\varphi_o = 10^{10}$ n/(cm^2 s)).

32. Using the data and results in problem 31 above:

    (a) Plot the uncollided flux in each region on the same graph.

    (b) Find the average flux in each region.

    (c) Find the rate (reactions/s) of absorption reactions in each region.

33. A certain nuclear power reactor burns uranium dioxide (UO_2) pellets in which the uranium is enriched to 3.2 w/o in ^{235}U. The fuel mass density is 10.5 g/cm^3, the fission cross sections are $\sigma_f^{235} = 585$ b and $\sigma_f^{238} = 5 \times 10^{-6}$ b, and the steady uniform flux is $\varphi_o = 7.8 \times 10^{12} \frac{\text{n}}{\text{cm}^2\ \text{s}}$. What is the fission rate ($\frac{\text{fiss}}{\text{cm}^3\ \text{s}}$) for this reactor? Assuming a cylindrical reactor of 10 foot diameter and 10 foot height, what is the thermal fission power for the above conditions?

Chapter 4

Decay of Radionuclides

The overwhelming majority of the more than 2930 known isotopes are unstable and spontaneously emit energy and sub-atomic particle(s) from their atomic nuclei. In fact, most elements contain only a few stable nuclides and some only one. All isotopes of the elements with $Z > 83$ (bismuth) are unstable. Nuclides exhibiting this emission behavior we call "radionuclides" (chpt. 1), and the particles emitted from the nucleus we call "nuclear radiation" or simply "radiation." To be sure, we must take care to distinguish these nuclear radiations from other phenomena often described as radiation (*i.e.* microwaves, visible light, X-rays, ...) which have other than nuclear origin. These latter radiations originate in atomic, molecular, or electronic de-excitation.

The study of radioactive materials may be motivated by any of several intentions. The structure of the atomic nucleus is revealed in radionuclide decay. As well, the great wealth of nuclear applications described in chapter 1 are made possible only by the existence of radioisotopes and nuclear decay. In addition, large concentrations of radionuclides may pose a potential hazard to man.

Of the almost 2700 known radioisotopes, only 65 are naturally occurring. All others are man-made in nuclear reactors, accelerators, and nuclear weapons detonation. Naturally occurring radioisotopes are present in nature either by virtue of primordial existence and painfully slow nuclear decay, or by continuous production in the atmosphere by cosmic radiation. This latter group contains the nuclides ^{3}H, ^{7}Be, and ^{14}C, while the former includes familiar radionuclides like ^{40}K, and ^{238}U and its decay products.

Man-made or "artificial" radioactivity was discovered in 1934 by Pierre Joliot and Irene Curie who shared the 1935 Nobel prize in chemistry for their pioneering work. In their experiment, they bombarded boron atoms with α particles in the reaction

$$^4_2\text{He} + \,^{10}_5\text{B} \longrightarrow \,^{13}_7\text{N} + \text{n}$$

thereby producing the radioactive nuclide ^{13}N. Following this great legacy, physicists all over the world have created hundreds of artificially radioactive isotopes in nuclear reactors and particle accelerators.

In what follows, we'll discuss details of the principle nuclear decay modes important in nuclear power and radioisotope applications. To provide a quantitative approach, we'll employ the now familiar conservation laws of mass/energy and linear momentum as our primary analysis tools. The remainder of the chapter will be devoted to a detailed description of the kinetics of radionuclide decay chains.

4.1 Decay Modes

Radioactive decay is said to occur when energy and particle(s) are spontaneously emitted from the atomic nucleus. Symbolically, we can describe this process much as we did for nuclear reactions

$$\text{X} \longrightarrow \text{Y} + y_1 + y_2 + \cdots$$

where the light products, $y_1, y_2, \ldots$, represent emitted particles of radiation. For our purposes, we may consider this as a particular type of nuclear reaction, one in which there is only one reactant, and make use of much the same analysis formalism as we had developed in chapter 2 for the more general case. The lone reactant for radionuclide decay reactions is called the "parent" nucleus, while the heavy product, Y, is often referred to as the "daughter" nucleus. The daughter nucleus may itself be a radionuclide as may its daughter products, giving rise to a decay chain. Many cases of interest involve only two products (Y and y_1). When this is the case, the simplified form of the simultaneous momentum and mass/energy conservation expression may be used (Eq.(2.8)), resulting in monoenergetic products. Several important cases, however ($\beta^{\pm}$ and fission, for example), involve three or more decay products. For these important cases, our two-product analysis of chapter 2 is not valid. We will find no unique solution for product energies. Products then take on a distribution of energies.

Regardless of the number of products, spontaneity is always a property of nuclear decay. Since there exists only one reactant, allowed solutions to mass/energy conservation include only $Q > 0$. This is sufficient to imply spontaneity as we saw in chapter 2.

Examining the chart of the nuclides reveals the following decay modes (table 4.1) and the approximate number of radionuclides exhibiting each.

The decay modes internal conversion (IC) and electron capture (EC) are not shown in this list. The important decay mode spontaneous fission (SF) is also not included in this list, nor is it considered in our discussion of decay modes below. We will, instead, devote the majority of chapter 5 to the discussion of fission, both spontaneous and induced.

Table 4.1: Decay Modes

decay modes	approximate number of examples
α particle ($^4_2\mathrm{He}^{++}$ nucleus)	450
β^- particle (electron)*	1100
β^+ particle (positron)*	250
p (proton)	~ 40
n (neutron)	~ 80
γ ray	2000

*neutrinos also are emitted

Many radionuclides decay by more than one mode but never simultaneously. Multiple modes are either competing parallel or sequential decay modes. The most common example of a sequential decay mode is β^- followed by γ decay. Examples of competing parallel decay modes include γ and IC, and β^- and β^+. Shortly, we'll discuss the energetics of each of the above.

4.2 Conservation Laws for Radionuclide Decay

As has been done throughout, we employ the conservations laws to provide order and a quantitative description.

4.2.1 Conservation of Nucleons

For an arbitrary two-product decay reaction,

$$^{A}_{Z}\mathrm{X}^{N} \longrightarrow \, ^{A'}_{Z'}\mathrm{Y}^{N'} + \, ^{A''}_{Z''}y^{N''}$$

total nucleon conservation always holds, *i.e.* $A = A' + A''$. However, proton and neutron numbers are not always independently conserved. For most decay modes

$$Z = Z' + Z'' \qquad \text{and} \qquad N = N' + N''$$

The exceptions are the EC and $\beta^{\pm}$ decay modes. These do not independently conserve nucleons. Instead, the EC and $\beta^{\pm}$ modes involve the transformation of one type of nucleon to the other.

4.2.2 Conservation of Mass-Energy and Momentum

Mass/energy and linear momentum, however, must always be conserved. Let's examine each decay mode in turn.

4.2.2.1 Gamma Decay

The simplest to consider is γ decay. In this decay mode, the radionuclide, ${}^{A}_{Z}\mathrm{X}^{N}$, exists initially in some nuclear excited state denoted by the superscript $*$. Symbolically, this is written ${}^{A}_{Z}\mathrm{X}^{*N}$. We do not consider, at this point, how this nuclide came to be in this state. Perhaps this was its fate as the product of a nuclear reaction, a consequence of nucleosynthesis, or more likely the product of another decay event.

Regardless of how it got to be in this excited state, the nuclear excitation may decay in the following way

$${}^{A}_{Z}\mathrm{X}^{*N} \longrightarrow {}^{A}_{Z}\mathrm{X}^{N} + \gamma$$

This decay process is analogous to the emission of atomic radiation in atomic de-excitation events. The nuclear excited state will likewise de-excite by photon emission to a lower excited state or to the ground state. The emitted nuclear photon (γ ray) is monoenergetic with energy, E_γ, equal to that of the energy level difference, less a small fraction acquired by the recoiling nucleus. Mass/energy and linear momentum balance will allow us to quantify. Applying mass/energy before and after the interaction yields

$$m_{\mathrm{X}}c^2 + E^* = m_{\mathrm{X}}c^2 + \mathrm{KE}_{\mathrm{X}} + E_\gamma$$

where $E_\gamma = h\nu$ and E^* represents the nuclear excitation energy level. Since there is no mass change in the interaction, $E^* = \mathrm{KE}_{\mathrm{X}} + E_\gamma$. The excitation energy is shared in proportion determined by momentum conservation. Here, it is also clear that E^* represents the liberated energy, so we may identify it with the Q-value,

$$Q_\gamma = E^* = \mathrm{KE}_{\mathrm{X}} + E_\gamma \tag{4.1}$$

As both KE_{X} and E_γ are always positive definite quantities, $Q_\gamma > 0$ always.

With a single reactant (radionuclide) at rest before the interaction, linear momentum balance yields the co-linear result, $p_X = p_\gamma (= E_\gamma/c)$ from which we may find by squaring

$$\mathrm{KE_X} = \frac{E_\gamma^2}{2m_X c^2} \tag{4.2}$$

Combining this with our expression for Q_γ reveals the quadratic expression for E_γ

$$E_\gamma \left(1 + \frac{E_\gamma}{2m_X c^2}\right) = Q_\gamma \tag{4.3}$$

where the entirety of the difference, $Q_\gamma - E_\gamma$, is due to the small recoil contribution (Eq.(4.2)). Since the recoil contribution is always small (*i.e.* $E_\gamma \sim$ few MeV and $2m_X c^2 > 4000$ MeV), then $E_\gamma \sim Q_\gamma = E^*$.

Nuclear excited states that decay by photon emission generally do so promptly after formation, usually $\lesssim 10^{-9}$ s. Occasionally, however, we find excited states that have much longer lifetimes, perhaps seconds or even days. Such states are called "metastable states" or "isomeric states." The latter term is borrowed from chemistry, where it is used to distinguish among chemical compounds of the same makeup but different chemical properties. The nuclear decay transition that marks the return of the excited, metastable state to the ground state is called an "isomeric transition," and the radionuclides that participate in such decay are called "isomers" or metastable atoms. Metastable states are indicated by the superscript m, *i.e.* ^{137m}Ba is used to represent the metastable state of barium-137 that decays to the ground state, ^{137}Ba, by emission of a 0.662 MeV γ-ray.

4.2.2.2 Alpha Decay

Alpha particle (α) decay is also straightforward to describe since, again, there are only two reaction products,

$${}^A_Z\mathrm{X}^N \longrightarrow {}^{A-4}_{Z-2}\mathrm{Y}^{N-2} + {}^4_2\,\mathrm{He}$$

The radionuclide ${}^A_Z\mathrm{X}^N$ exists before the interaction as an unstable configuration of nucleons, which attains a more stable configuration upon decay by release of an α (^{4}He nucleus) from the nucleus. The parent nucleus, ^{A}X is considered unstable, but not an excited state, as the latter terminology is reserved for nuclear level transitions that are only allowed by photon emission. The daughter product, $^{A-4}$Y, may be unstable and be part of a decay chain. It may also be produced in an excited state, $^{A-4}\mathrm{Y}^*$, and decay by γ emission before continuing in a decay chain or remaining as a stable product.

Mass/energy balance reveals

$$m_X c^2 = m_Y c^2 + \mathrm{KE}_Y + m_{^4\mathrm{He}} c^2 + \mathrm{KE}_{^4\mathrm{He}}$$

which indicates a liberated energy of

$$\begin{aligned} Q_\alpha &= \mathrm{KE}_Y + \mathrm{KE}_{^4\mathrm{He}} \\ &= c^2[m_X - m_Y - m_{^4\mathrm{He}}] \end{aligned} \tag{4.4}$$

The first line in this expression shows the spontaneity condition, $Q_\alpha > 0$. In the nuclide chart, we find $4 \leq Q_\alpha \leq 10$ MeV. Equation (4.4) may also be used to write a mass condition for spontaneous alpha decay

$$m_X > m_Y + m_{^4\mathrm{He}} \tag{4.5}$$

So long as this condition is satisfied, spontaneous decay by α emission is energetically possible. This is a necessary, but not sufficient condition. In the nuclide chart, this condition roughly corresponds to that region above the $Z^*(N)$ curve (sec. 1.6.6.1) for $A \gtrsim 47$. Most heavy nuclides are α emitters, especially those with a high Z/N ratio.

In the above analysis leading to the determination of Q_α, it has been implicitly assumed that all reaction participants are always in the ground, neutral state. This is not the case for intermediate steps in the decay process. The α particle is emitted in the decay process as doubly ionized helium, *i.e.* ${}^4_2\mathrm{He}^{++}$, while the daughter product will immediately possess an excess of two electrons. The excess electrons are quickly shed. As well, the α will eventually neutralize by acquiring two atomic electrons from the material media in which it interacts. The net result is that all sub-atomic particles, including electrons, are conserved, and that the same Q_α expression results to within the small difference in electron binding energy difference between product and parent. The electron binding energy difference, however, is an extremely small energy difference on our nuclear scale.

4.2.2.3 Beta Minus Decay

Many unstable nuclei decay by the emission of energetic negative electrons from the nucleus in β^- decay. In fact, β^- decay is very common in the nuclide chart. The terminologies, negative beta decay, beta minus decay, negatron decay, negative electron decay, and simply beta decay, all refer to this decay mode. Though the result of nuclear β^- decay is an energetic electron from the nucleus, electrons do not exist in the nucleus under normal circumstances. We know this because of several observations. Several radionuclides can decay by both positron (β^+) and negatron emission. Were

Example 23: Alpha Decay Energy Release
What is the Q_α distribution among reaction products released from alpha decay in the following reaction?

$$^{238}\text{U} \longrightarrow {}^{234}\text{Th} + {}^{4}\text{He}$$

Solution:
From our Q_α expression, Eq.(4.4), we immediately find $Q_\alpha \sim 4.2$ MeV. Since there are only two products in α decay and there is no net momentum, the products are monoenergetic and co-linear. We may then use the simplified form of the kinetic energy sharing solution, Eq.(2.8).

$$\text{KE}_\alpha = Q_\alpha \frac{m_{\text{Th}}}{m_{\text{Th}} + m_{^4\text{He}}} = \frac{234}{238} Q = 4.13\,\text{MeV}$$

and

$$\text{KE}_{\text{Th}} = Q_\alpha \frac{m_{^4\text{He}}}{m_{\text{Th}} + m_{^4\text{He}}} = \frac{4}{238} Q = 0.07\,\text{MeV}$$

As with general nuclear reactions, the lighter product carries most of the decay energy.

both particles to exist in the nucleus, they would be expected to annihilate (sec. 1.4). As well, electrons with β^- decay energies (few MeV) have de Broglie wavelengths much larger than nuclear dimensions. Several quantum mechanical arguments[Evans55] also preclude the existence of nuclear electrons.

The decay process instead involves nucleon transformation. Symbolically, the β^- decay process can be written

$$ {}^{A}_{Z}\mathrm{X}^{N} \longrightarrow \left[{}^{\ A}_{Z+1}\mathrm{Y}^{\mathrm{N}-1}\right]^{+1} + \beta^- + \overline{\nu} $$

where β^- is written to indicate the negative electron from the nucleus and distinguish it from electrons of other origin e^-, $\overline{\nu}$ is an anti-neutrino that always accompanies β^- emission, and the $[\,]^{+1}$ notation indicates that the daughter atom is immediately left in an ionized state by the nuclear transformation process. An example is the common laboratory ^{137}Cs source

$$ {}^{137}_{55}\mathrm{Cs} \longrightarrow \left[{}^{137}_{56}\mathrm{Ba}\right]^{+1} + \beta^- + \overline{\nu} $$

The nuclear transformation in β^- decay converts nucleon form while preserving the total nucleon number, A. Since the transformation reduces neutron number while increasing that of nuclear protons, the exchange can be visualized

$$ \mathrm{n} \longrightarrow \mathrm{p} + \beta^- + \overline{\nu} $$

Indeed, a free neutron (not bound to the nucleus) would so decay with an average life time of about 15 min.

The decay process is completed with atomic recombination of the daughter atom leaving all products in their ground, neutral state

$$ \left[{}^{\ A}_{Z+1}\mathrm{Y}^{N-1}\right]^{+1} + e^- \longrightarrow {}^{\ A}_{Z+1}\mathrm{Y}^{N-1} + \gamma_x $$

where γ_x represents the emission of characteristic atomic photon(s) in the recombination process, ranging from X-rays to visible light. The overall β^- decay process is then summarized

$$ {}^{A}_{Z}\mathrm{X}^{N} \longrightarrow {}^{\ A}_{Z+1}\mathrm{Y}^{N-1} + \beta^- + \overline{\nu} + \gamma_X - e^- \tag{4.6} $$

for which a mass/energy balance yields

$$ m_{\mathrm{X}}c^2 = m_{\mathrm{Y}}c^2 + \mathrm{KE}_{\mathrm{Y}} + \mathrm{KE}_{\beta^-} + E_{\overline{\nu}} + E_{\gamma_X} \tag{4.7} $$

since the β^- particle is identical in mass (and charge) to an atomic electron.

The liberated energy can then be quantified

$$ \begin{aligned} Q_{\beta^-} &= \mathrm{KE}_{\mathrm{Y}} + \mathrm{KE}_{\beta^-} + E_{\overline{\nu}} + E_{\gamma_X} \\ &= c^2\left[m_{\mathrm{X}} - m_{\mathrm{Y}}\right] \end{aligned} \tag{4.8} $$

from which we obtain the mass condition

$$\boxed{m_{\mathrm{X}} > m_{\mathrm{Y}}} \tag{4.9}$$

for spontaneous decay. A review of the semi-empirical binding energy equation (Eq.(1.23)) suggests that this condition is satisfied (*i.e.* β^- decay is possible) for all nuclei with $Z \lesssim Z^*$. In the nuclide chart, $Q_{\beta^-} \leq 3$ MeV. The daughter product recoil energy, $\mathrm{KE_Y}$, and the recombination photon energy, E_{γ_X}, (the latter being equal in magnitude to the ionization potential) are on the order of a few eV to perhaps keV, and always much smaller than the net nuclear energy released. The β^- particle and anti-neutrino carry almost all of the liberated energy

$$Q_{\beta^-} \simeq \mathrm{KE}_{\beta^-} + E_{\overline{\nu}} \tag{4.10}$$

Since more than two products result from β^- decay (Eq.(4.6)), we cannot use mass/energy and momentum conservation to identify unique values for product energies. The result instead is a statistical sharing of KE_{β^-} and $E_{\overline{\nu}}$ among decay events, *i.e.* although these receive a unique and measurable fraction of Q_{β^-} on a single decay event, we cannot predict this value. Only the relative probability that the β^- particle will attain a certain value can be predicted. Measuring product energies on many such decay events maps out the distribution of β^- particle energies. Such a compilation of data is called an energy spectrum. An example β^- spectrum from the decay of ^{210}Bi is shown in Fig. 4.1 as a normalized distribution function of β^- particles vs. β^- particle energy ($n(\mathrm{KE}_{\beta^-})$ vs. KE_{β^-}). It was data such as this, in fact, that lead to W. Pauli's postulate in 1931 on the existence of the then unknown anti-neutrino in β^- decay. He reasoned that there must be a third, undetected particle involved in the decay process that carries some energy and momentum, and produces the measured distribution. This unknown and very difficult to detect particle that we now call the anti-neutrino was discovered by Reines and Cowen in the 1950s.

From the β^- distribution, we can identify some important features. Firstly, since β^- and $\overline{\nu}$ share Q_{β^-}, the maximum β^- kinetic energy is

$$\mathrm{KE}_{\beta^-}|_{\max} \sim Q_{\beta^-}$$

For the example in Fig. 4.1, $\mathrm{KE}_{\beta^-}|_{\max} \sim 1.16$ MeV. These spectral data are usually normalized so that

$$n_{\beta^-} = \int_o^{Q_{\beta^-}} n(\mathrm{KE}_{\beta^-})\, d(\mathrm{KE}_{\beta^-}) = 1$$

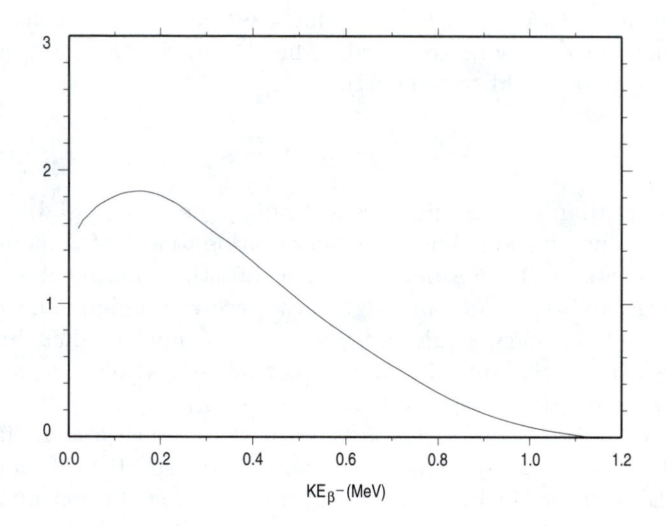

Figure 4.1: β^- particle energy spectrum from ^{210}Bi. Normalized β^- distribution function vs. β^- particle kinetic energy. Solid line smoothed through the data taken from ***Quantum Physics of Atoms, Molecules, Solids, Nuclei, and Particles*** by Robert Eisberg and Robert Resnick, Copyright©1985 by John Wiley & Sons, Inc. Reprinted by permission of John Wiley & Sons, Inc.

The distribution average yields the average β^- kinetic energy

$$\begin{aligned}\langle \mathrm{KE}_{\beta^-} \rangle &= \frac{\int_o^{Q_{\beta^-}} \mathrm{KE}_{\beta^-}\, n(\mathrm{KE}_{\beta^-})\, d(\mathrm{KE}_{\beta^-})}{\int_o^{Q_{\beta^-}} n(\mathrm{KE}_{\beta^-})\, d(\mathrm{KE}_{\beta^-})} \\ &= \int_o^{Q_{\beta^-}} \mathrm{KE}_{\beta^-}\, n(\mathrm{KE}_{\beta^-})\, d(\mathrm{KE}_{\beta^-})\end{aligned}$$

and is typically in the range, $\frac{1}{4}Q_{\beta^-} \lesssim \langle \mathrm{KE}_{\beta^-} \rangle \lesssim \frac{1}{3}Q_{\beta^-}$. For the example in Fig. 4.1, $\langle \mathrm{KE}_{\beta^-} \rangle \sim 0.34$ MeV, a bit less than $\frac{1}{3}Q_{\beta^-}$. The most probable value of β^- energy is found where $n(\mathrm{KE}_{\beta^-})$ is maximized. Since the distribution is always skewed in favor of low energy, the most probable value is always less than $\langle \mathrm{KE}_{\beta^-} \rangle$. The Fig. 4.1 example of ^{210}Bi has a most probable β^- kinetic energy of ~ 0.155 MeV, about half $\langle \mathrm{KE}_{\beta^-} \rangle$.

4.2.2.4 Positron Decay

Energetic positrons can be emitted in the decay of unstable nuclei. When such particles originate from the atomic nucleus, they are given the symbol β^+. For the same reasons as discussed in the context of β^- decay, positrons do not exist in the nucleus. They are the result of nuclear transformation. The positron decay process for an arbitrary radionuclide can be expressed

$${}^A_Z\mathrm{X}^N \longrightarrow \left[{}_{Z-1}^{\;\;A}\mathrm{Y}^{N+1}\right]^{-1} + \beta^+ + \nu$$

Like β^- decay, a third particle always accompanies β^+ decay. In this case, it is the neutrino, ν. The daughter product from decay is, again, left in an ionized state immediately following nuclear transformation. This time, it is a negative ion which quickly neutralizes by electron emission

$$\left[{}_{Z-1}^{\;\;A}\mathrm{Y}^{N+1}\right]^{-1} \longrightarrow {}_{Z-1}^{\;\;A}\mathrm{Y}^{N+1} + e^-$$

so that the net decay reaction is

$${}^A_Z\mathrm{X}^N \longrightarrow {}_{Z-1}^{\;\;A}\mathrm{Y}^{N+1} + \beta^+ + \nu + e^- \tag{4.11}$$

A common example is the β^+ decay of ^{22}Na

$${}^{22}_{11}\mathrm{Na} \longrightarrow {}^{22}_{10}\mathrm{Ne} + \beta^+ + \nu + e^-$$

Again the total nucleon number, A, is preserved while individual nucleon numbers are not. For β^+, we can visualized the net exchange

$$\mathrm{p} \longrightarrow \mathrm{n} + \beta^+ + \nu$$

Mass/energy is always conserved, however. This will again allow the quantification of liberated energy. For the net β^+ decay reaction of Eq.(4.11), mass/energy conservation yields

$$m_X c^2 = m_Y c^2 + \mathrm{KE_Y} + 2m_e c^2 + \mathrm{KE}_{\beta+} + E_\nu + \mathrm{BE}_e$$

where BE_e represents the binding energy of the last electron in Y^{-1}. Then the liberated energy becomes

$$\begin{aligned} Q_{\beta+} &= \mathrm{KE_Y} + \mathrm{KE}_{\beta+} + E_\nu + \mathrm{BE}_e \\ &= c^2 \left[m_X - m_Y - 2m_e\right] \end{aligned} \tag{4.12}$$

In the nuclide chart, we find $Q_{\beta+} \lesssim 3$ MeV. Since $Q_{\beta+}$ must always be greater than zero for spontaneous decay, we can identify

$$\boxed{m_X > m_Y + 2m_e} \tag{4.13}$$

as the necessary (but not sufficient) mass condition for β^+ decay. This condition is more demanding than that for β^- decay. In β^+ decay, two electron masses are ejected from the atom. In β^- decay, no net electron masses are expelled.

The recoil, $\mathrm{KE_Y}$, and electron binding, BE_e, contributions to $Q_{\beta+}$ are small in comparison to the sum $\mathrm{KE}_{\beta+} + E_\nu$ so that, without much error, we can write

$$Q_{\beta+} \sim \mathrm{KE}_{\beta+} + E_\nu$$

As with β^- decay, there is a statistical sharing of $Q_{\beta+}$ among β^+ and ν resulting in a continuous energy spectrum for $\mathrm{KE}_{\beta+}$. Regardless of its birth energy, however, the fate of β^+ is always the same. It will eventually lose energy in collisional interactions with electrons and atoms in the surrounding medium. When β^+ slows sufficiently, it will combine with an ordinary electron. The two will temporarily exist as the pseudo-atom positronium before annihilating with each other. The annihilation event usually produces two energetic photons of $m_e c^2 \sim 0.511$ MeV each (chpt. 6).

4.2.2.5 Electron Capture

The electron capture (EC) decay mode for unstable nuclei is marked by the capture of an atomic electron by the unstable nucleus. Symbolically, we write for EC,

$${}^{A}_{Z}\mathrm{X}^{N} \longrightarrow \left[{}_{Z-1}^{\;\;A}\mathrm{Y}^{N+1}\right]^* + \nu$$

revealing a transformation which looks very much like β^+ decay without positron or atomic electron emission. In fact, EC competes with β^+ decay for certain nuclei. The nuclear transformation can be visualized as

$$\mathrm{p} + e^- \longrightarrow \mathrm{n} + \nu$$

Since it is most often the lowest lying (K-shell) atomic electron that is assimilated into the nucleus, this decay mode is often called K-capture. In the nuclide chart, EC is given the symbol ε. As with $\beta^\pm$ decay, EC produces daughter nuclei that are isobaric with the parent. An example of EC decay is

$${}^{54}_{25}\mathrm{Mn} \longrightarrow {}^{54}_{24}\mathrm{Cr} + \nu$$

The [ ]* notation above indicates atomic excitation. Following electron capture, the daughter atom is left with electron shells that are not in their stable configuration. Emission of a characteristic photon cascade (usually X-rays) quickly follows

$$\left[{}_{Z-1}^{\;\;A}\mathrm{Y}^{N+1}\right]^* \longrightarrow {}_{Z-1}^{\;\;A}\mathrm{Y}^{N+1} + \sum_j \gamma_{x_j}$$

where $\sum_j \gamma_{x_j}$ represents the characteristic photon cascade as the daughter atom readjusts to the ground state.

The energy liberated in this process is found in the usual way

$$\begin{aligned} Q_{\mathrm{EC}} &= \mathrm{KE}_{\mathrm{Y}} + E_\nu + \sum_j E_{x_j} \\ &= c^2\left[m_{\mathrm{X}} - m_{\mathrm{Y}}\right] \end{aligned} \tag{4.14}$$

In the nuclide chart, $Q_{\mathrm{EC}} \leq 3$ MeV. From Eq.(4.14), we directly find the necessary mass condition for spontaneous EC decay

$$\boxed{m_{\mathrm{X}} > m_{\mathrm{Y}}} \tag{4.15}$$

Reviewing the mass conditions for the three isobaric decay modes ($\beta^\pm$, EC) suggests an interesting consequence. Adjacent isobars cannot both be stable. If the $Z+1$ isobar is more massive than its nearest isobaric neighbor (Z), then β^+ or EC are possible. If, instead, the Z isobar is more massive, β^- decay is possible. This conclusion is well supported by the nuclide chart. There are no known examples in nature of adjacent isobars that are both stable, and a scant few examples of more than one stable isobar.

4.2.2.6 Proton Decay

Radioactive decay by proton emission leaves the daughter atom with an extra atomic electron (negative ion) which is subsequently emitted from the atomic electron cloud. The net proton decay reaction becomes

$$ {}^{A}_{Z}\mathrm{X}^{N} \longrightarrow {}^{A-1}_{Z-1}\mathrm{Y}^{N} + \mathrm{p} + e^{-} $$

so the daughter is an isotone of the unstable parent nuclide, ${}^{A}_{Z}\mathrm{X}^{N}$. An example of a proton decay reaction is

$$ {}^{73}_{35}\mathrm{Br} \longrightarrow {}^{72}_{34}\mathrm{Se} + \mathrm{H} $$

Mass/energy conservation for the net reaction in this case looks like

$$ m_{\mathrm{X}}c^2 = m_{\mathrm{Y}}c^2 + \mathrm{KE}_{\mathrm{Y}} + m_{\mathrm{H}}c^2 + \mathrm{KE}_{\mathrm{H}} $$

where we have ignored the tiny difference between the emitted electron binding energy in Y^{-1} and that in H. The energy release in proton decay is then

$$ \begin{aligned} Q_{\mathrm{p}} &= \mathrm{KE}_{\mathrm{Y}} + \mathrm{KE}_{\mathrm{H}} \\ &= c^2\left[m_{\mathrm{X}} - m_{\mathrm{Y}} - m_{\mathrm{H}}\right] \end{aligned} \tag{4.16} $$

so that for spontaneous emission ($Q_{\mathrm{p}} > 0$), this implies

$$ \boxed{m_{\mathrm{X}} > m_{\mathrm{Y}} + m_{\mathrm{H}}} \tag{4.17} $$

In the nuclide chart, $Q_{\mathrm{p}} \lesssim 10$ MeV.

4.2.2.7 Neutron Decay

Decay by neutron emission from the atomic nucleus is an important decay mode. Symbolically, it can be described

$$ {}^{A}_{Z}\mathrm{X}^{N} \longrightarrow {}^{A-1}_{\;\;\;Z}\mathrm{X}^{N-1} + \mathrm{n} $$

The product nuclide and reactant are isotopes in this case. Since charged particle emission is not involved and Z is constant, there need not be any perturbation of the electron shell structure.

Mass/energy conservation for this case yields

$$ m_{\mathrm{X^N}}c^2 = m_{\mathrm{X^{N-1}}}c^2 + \mathrm{KE}_{\mathrm{X^{N-1}}} + m_{\mathrm{n}}c^2 + \mathrm{KE}_{\mathrm{n}} $$

so that the reaction Q-value becomes

$$ \begin{aligned} Q_{\mathrm{n}} &= \mathrm{KE}_{\mathrm{X^{N-1}}} + \mathrm{KE}_{\mathrm{n}} \\ &= c^2\left[m_{\mathrm{X^N}} - m_{\mathrm{X^{N-1}}} - m_{\mathrm{n}}\right] \end{aligned} \tag{4.18} $$

and the mass condition for spontaneous decay is

$$\boxed{m_{\mathrm{X}^{\mathrm{N}}} > m_{\mathrm{X}^{\mathrm{N}-1}} + m_{\mathrm{n}}} \tag{4.19}$$

An example of a neutron decay reaction is

$$^{138}_{54}\mathrm{Xe} \longrightarrow {}^{137}_{54}\mathrm{Xe} + \mathrm{n}$$

Although rather rare, neutron decay reactions are extremely important. The control of nuclear reactors relies on the maintenance of very sensitive neutron balance. Some of the products of fission reactions are neutron-emitting radionuclides that contribute to this balance. The slow release of neutrons from the decay of such nuclides allows control of the neutron inventory and, hence, the reactor.

4.2.2.8 Internal Conversion

Internal conversion (IC) concludes our list of radioactive decay modes. In this mode, nuclear excitation energy is released by ejection of an atomic electron (usually K-shell) as a competing mode to γ decay. Symbolically we write for IC

$$^{A}_{Z}\mathrm{X}^{*N} \longrightarrow \left[{}^{A}_{Z}\mathrm{X}^{N}\right]^{+1} + e^{-}$$

where the ejected electron, e^{-}, is fast and monoenergetic. The latter property of IC electrons is in stark contrast to β^{-} released electrons. Just like γ decay, there is no rest mass change in IC. The symbol e^{-} is used for internal conversion instead of "IC" in the nuclide chart.

The daughter atom is left in an ionized state that will subsequently neutralize with an ambient electron and emit recombination photons.

$$\left[{}^{A}_{Z}\mathrm{X}^{N}\right]^{+1} + e^{-} \longrightarrow {}^{A}_{Z}\mathrm{X}^{N} + \sum_{j} \gamma_{\mathrm{x}_j}$$

When recombination occurs directly to the vacancy left by the IC electron, there is only one emitted characteristic photon of $E_{\gamma\mathrm{X}} = \mathrm{BE}_e$, the atomic binding energy of the IC electron. In this situation, mass/energy balance yields

$$Q_{\mathrm{IC}} = E^{*} = \mathrm{KE}_{\mathrm{X}} + \mathrm{KE}_{e} + \mathrm{BE}_{e} \tag{4.20}$$

The atom recoil energy, KE_{X}, is usually very small so that

$$\mathrm{KE}_{e}^{\mathrm{K}} \sim Q_{\mathrm{IC}} - \mathrm{BE}_{e}^{\mathrm{K}} \tag{4.21}$$

where the superscript K refers to a K-shell electron. L, M, and higher shell electrons can also be seen with monotonically decreasing likelihood.

Figure 4.2 summarizes the Z, N transformations induced by nuclear decay along with a few others that may be produced in nuclear reactions. Blocks marked "out" surrounding the "original nucleus" indicate the relative position of the decay product nucleus with respect to that of the parent. Not shown are the modes γ and IC, since they do not alter the Z, N identity of the original nucleus. Nuclear decay reactions, being spontaneous

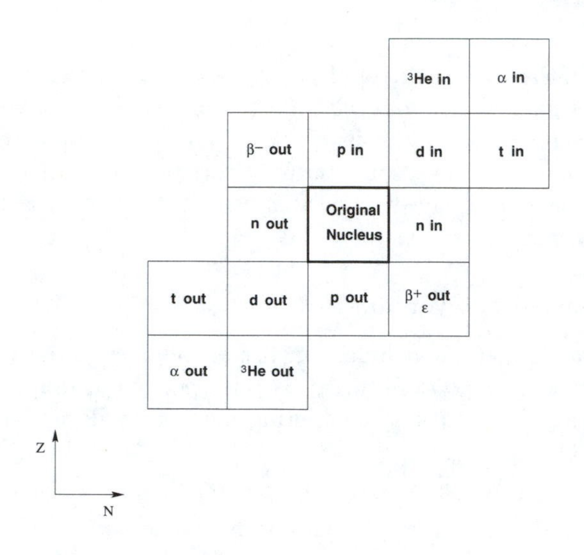

Figure 4.2: Relative position of daughter nuclei from nuclear decay in the Nuclide Chart. (From *Nuclides and Isotopes*, Copyright©1996 by the General Electric Co. Reprinted by permission of the General Electric Co.)

($Q > 0$), all produce products of lesser mass. Indeed, a three-dimensional plot of nuclide masses vs. Z and N reveals that stable nuclei all reside at the trough of this curve in the "valley of stability." The two-dimensional representation of these data was displayed in Fig. 1.3.

4.2.3 Energy Level Diagrams

A convenient tool for organizing nuclear decay data and nuclear energy level structure is the energy level diagram. In its most common form, the energy level diagram is, in essence, a plot of nuclear energy level vs. atomic number, Z, showing nuclear transitions (decay modes) among energy levels. Besides identifying decay modes and energies, many other useful data are also included, like *branching ratio* (relative probability of decay by more than one route), nuclear isomerism, and nuclear lifetime in units of half-life (sec. 4.3.3). The nuclide chart and *Table of Isotopes*[Firestone96] provide the data for these diagrams.

Several examples will illustrate the utility of such diagrams. Let's first consider the simplest case of a radionuclide that decays by one mode directly to the ground state of its daughter nuclide. The β^- decay of ^{32}P to the ground state of ^{32}S is such an example (Fig. 4.3). The energy scale indicates that the β^- emitter, ^{32}P, lies 1.71 MeV above the ground state (0 MeV) of ^{32}S, so that $Q_{\beta^-} = 1.71$ MeV. Arrows point to the right in this energy level diagram since β^- decay increases Z.

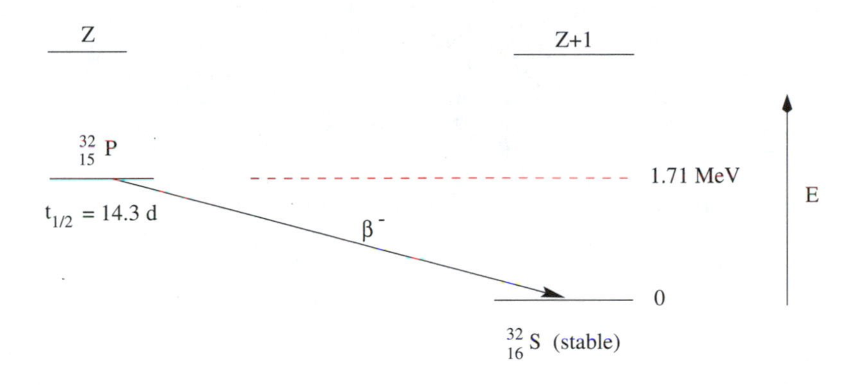

Figure 4.3: Energy level diagram for ^{32}P β^- decay. An energy level diagram shows nuclear transitions between energy levels with energy scale referenced to the ground state of the daughter nucleus.

More common in β^- decay is the production of the daughter nucleus in an excited state. Figure 4.4 illustrates this situation for ^{60}Co, which β^- decays (with $Q_{\beta^-} = 0.31$ MeV) more than 99% of the time to the excited state 2.5 MeV above the ground state of ^{60}Ni. This excited state immediately decays by the emission of two monoenergetic photons of 1.17 and 1.33 MeV, respectively. A tiny fraction of decay events ($\sim$ 0.1%) result in β^- decay directly to the state 1.33 MeV above ground, which subsequently decays by single photon emission.

Figure 4.5 illustrates an example that includes nuclear isomerism and internal conversion (IC). In 8% of the ^{137}Cs decays, β^- decay is directly to the ground state of ^{137}Ba. The remaining 92% of ^{137}Cs decays are by β^- to the long-lived isomeric state ^{137m}Ba, 0.662 MeV above the ground state. This metastable state decays predominantly by 0.662 MeV γ emission. However, about 10% of the time, ^{137m}Ba decays by K-shell IC. The frequency is less for L-shell ($\sim$ 2%) and much less for M-shell ($\ll$ 1%).

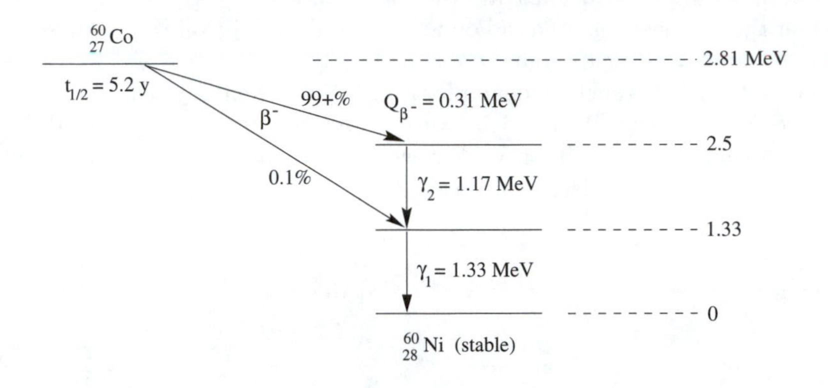

Figure 4.4: Energy level diagram for ^{60}Co β^- decay.

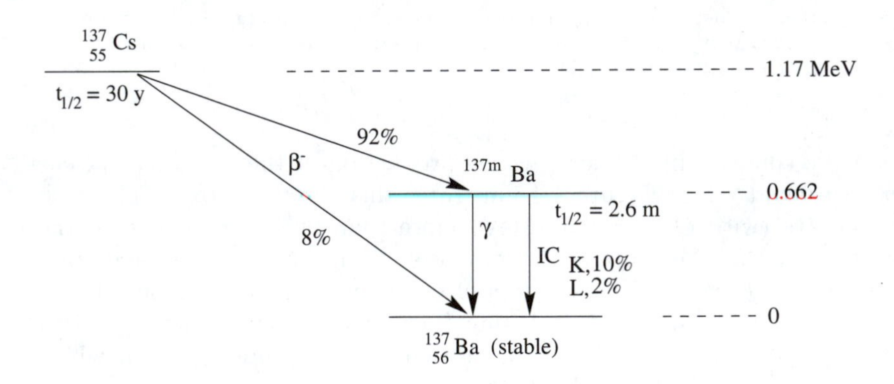

Figure 4.5: Energy level diagram for ^{137}Cs β^- decay showing γ and IC transitions from metastable ^{137m}Ba. For the K and L internal conversion transitions, $\mathrm{BE}_e^{\mathrm{K}} \sim 38$ keV and $\mathrm{BE}_e^{\mathrm{L}} \sim 6$ keV, respectively.

An α decay example is shown in Fig. 4.6. Here, the arrows point to the left since Z is reduced in α decay. This example shows that 94.5% of the ^{226}Ra α decays are immediately to the ground state of ^{222}Rn. The remainder are to the excited state, 0.186 MeV above the ^{222}Rn ground state. As with excited states following β^- decay, this one too will decay by γ emission. Note that although ^{222}Rn is unstable and itself decays by α emission, there is no difficulty in identifying it as a ground (but unstable) state and referencing the nuclear states of ^{226}Ra to this state.

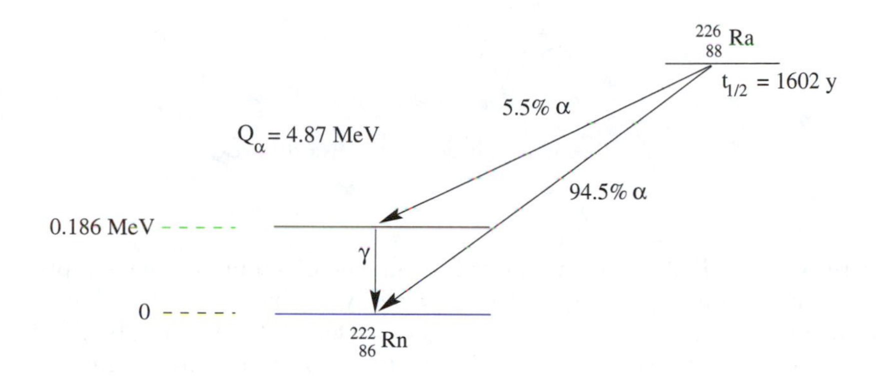

Figure 4.6: Energy level diagram for ^{226}Ra α decay.

Our final example illustrates β^+ and EC decay. Figure 4.7 depicts the decay of ^{22}Na of which 90% are β^+ decays ($Q_{\beta+} = 0.55$ MeV) to the 1.27 MeV level above ^{22}Ne. Competing with β^+ about 10% of the time is EC decay. Again, single γ emission leaves ^{22}Ne in its ground state. A tiny fraction, $\sim$ 0.05%, of ^{22}Na decays are β^+ directly to the ^{22}Ne ground state.

4.3 Radioactive Decay Law

Our objective is to determine the temporal behavior or kinetics of system(s) of radionuclide(s). Consider first the simplest case, a pure sample of radioactive material so that this sample consists only of a single radionuclide. Upon decay, daughter product atoms are removed so that at any time there are only radioisotope atoms present. For a sample of mass $m(t)$, the number of radioactive atoms in the sample at time t is

$$\mathcal{N}(t) = m(t)\frac{N_A}{M} \tag{4.22}$$

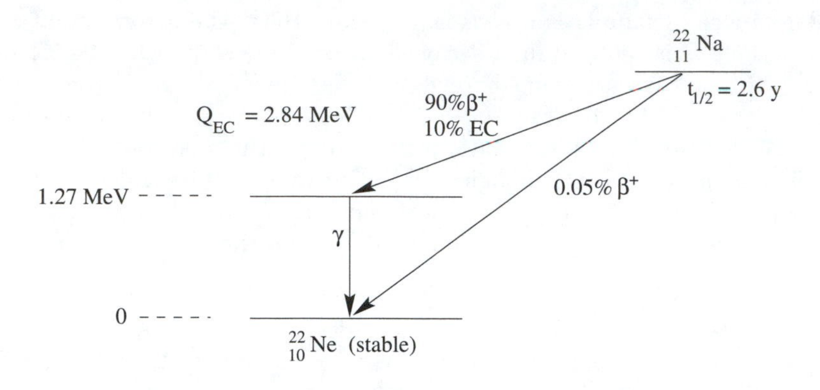

Figure 4.7: Energy level diagram for ^{22}Na β^+ decay.

where $\mathcal{N}(t) = \int_V N(t)\ dV$ is the total number of atoms in the sample. Atoms in this sample may spontaneously decay by one or more of the decay modes described in the previous section (sec. 4.2). Obviously, $\mathcal{N}(t)$ is a monotonically decreasing function of time. Just how it decreases is our concern here. The answer resides in the observation that radioactive decay is a statistical process. While it is impossible to predict when an individual decay event will occur, it is possible to predict the average decay behavior for a statistically large sample ($\mathcal{N} \ggg 1$). This situation is very much like that for neutron interactions discussed in chapter 3. While we cannot determine *a priori* where an individual neutron will interact, we can predict the average behavior of a large sample of neutrons. For neutron interactions, the average behavior was quantified by the constant probability per unit length of interaction, Σ. In radioactive decay, we find that the probability per unit time for decay is a constant. Since this probability is constant, the average decay rate in our sample is proportional to the number of radioactive atoms in the sample

$$A(t) \propto \mathcal{N}(t)$$

where $A(t)$ is the nuclear activity, or just the "activity" of the sample. This is the decay analog of the neutron reaction rate being proportional to the neutron flux. The activity, $A(t)$, represents the instantaneous number of nuclear decays or disintegrations per second in the sample. It also indicates the instantaneous number of radionuclide atoms disappearing per second, and hence the number of daughter atoms appearing per second. Though not as popular as dis/s (disintegrations per second), the accepted SI unit for decay is Bq (becquerel) = 1 dis/s in honor of the discoverer of nuclear radiation, Henri Becquerel, who shared the 1903 Nobel prize in physics.

The proportionality constant relating the activity to the radioactive inventory is the "decay constant," given the symbol λ so that

$$\boxed{A(t) \equiv \lambda \mathcal{N}(t)} \tag{4.23}$$

This expression constitutes the most fundamental form of the "radioactive decay law." This form is always applicable regardless of the purity of the sample or the complexity of the decay problem. Since it can always be applied, it is an extremely important and useful expression, as we'll see shortly. The decay constant, λ, has units of inverse time and is a strict constant. It has a definite and immutable value, different for each radioisotope. The β^- emitter ^{3}H has a decay constant of $\lambda = 1.79 \times 10^{-9}$ s^{-1} so that in a sample of 10^{20} tritium atoms, we would expect an instantaneous activity of 1.79×10^{11} dis/s.

Activity has such large dimension, as seen in the previous example, that a readjustment in scale is once again in order. To honor the radiochemists Pierre and Marie Curie, who did much of the early experimental work describing natural radioactivity and shared in the 1903 Nobel prize with Becquerel, activity is often quoted in units of the curie (Ci). One curie is defined as that activity equal to the activity of 1 g of pure ^{226}Ra. Since ^{226}Ra has a decay constant of $\lambda_{226} \sim 1.37 \times 10^{-11}$ s^{-1} and $\mathcal{N}_{226} \sim 2.7 \times 10^{21}$ atoms, then $A_{226}(1g) \sim 3.7 \times 10^{10}$ dis/s and

$$1\text{Ci} \equiv 3.7 \times 10^{10} \text{ dis/s}$$

Familiar multiples include pCi, nCi, and μCi, most often applicable to small laboratory radiation sources, to kCi and MCi, which are representative of nuclear reactor fission product inventories.

With a very large sample ($\mathcal{N} \ggg 1$), there is little difficulty in using the continuum mathematics to describe the discrete decay process. Then, we can write the activity

$$A(t) = \lim_{\Delta t \to 0} \left[-\frac{\Delta \mathcal{N}(t)}{\Delta t} \right] = -\frac{d\mathcal{N}(t)}{dt}$$

since it represents the instantaneous time rate of decrease in number of radioactive atoms in our pure sample. Combining this with our primitive form of the radioactive decay law (Eq.(4.23)), we arrive at

$$\frac{d\mathcal{N}(t)}{dt} = -\lambda \mathcal{N}(t)$$

the differential form of the radioactive decay law. This form only describes the decay of a pure sample or the first component of a closed decay chain (sec. 4.4.1) since we allow no other means of introduction or removal of atoms of this type.

Furnishing a boundary condition (or more appropriately, an initial condition since this is a time dependent problem) completes the physical description of the system. The problem then becomes a mathematical exercise to arrive at an explicit, time dependent solution. The most usual boundary condition is to specify the radioisotope inventory at $t = 0$, *i.e.* $\mathcal{N}(t = 0) = \mathcal{N}_o$. Then, the complete solution is

$$\mathcal{N}(t) = \mathcal{N}_o e^{-\lambda t} \tag{4.24}$$

the exponential form of the radioactive decay law, which also only applies to a pure sample or the first component of a closed decay chain. Since the activity is proportional to $\mathcal{N}(t)$, we may write

$$A(t) = A_o e^{-\lambda t}$$

where $A_o = \lambda \mathcal{N}_o$ is the initial activity.

The exponential form of the radioactive decay law allows us to predict the future state of our large sample. The decay constant, λ, controls the rate of decay. For the exponential decay in our pure sample, $1/\lambda$ is the e-folding time, *i.e.* at $t = 1/\lambda, \mathcal{N}(t = 1/\lambda) = e^{-1}\mathcal{N}_o = 0.368\mathcal{N}_o$ just as $1/\Sigma$ is the e-folding distance for uncollided neutron flux.

The interesting, yet entirely academic situation arises when our initially large sample decays through many, many e-folding times to a small and statistically insignificant sample where $\mathcal{N} \sim 1$. Our continuum description is no longer valid here, and decay may only be described by a statistical model. Such situations are of little practical importance, however, since the activity is extremely low.

As a practical application, since the decay of each individual radioactive atom releases energy Q, the rate of energy release or decay power in a large sample is

$$P(t) = QA(t) = QA_o e^{-\lambda t} \tag{4.25}$$

The recoverable portion of this decay power can be put to practical use for applications like space satellite power sources or remote lighting, etc., as discussed in chapter 1. Decay power also represents a significant fraction ($\sim 10\%$) of nuclear reactor power and all of the decay heating power in fuel removed from the reactor.

Finally, before moving to more complex system of radionuclides, let's use the radioactive decay law to describe detailed balance of nuclei in our simple system. Assuming the immediate daughter product, Y, to be stable in the decay reaction X $\longrightarrow$ Y + y + ..., then all atoms in our sample are either of type X or Y at all times. Our exponential form of the radioactive decay law should confirm this. The total number of atoms of type Y produced to time t in our sample is exactly equal to the total number of nuclear transformations to time t

$$\begin{aligned} \mathcal{N}_{\mathrm{Y}}(t) &\equiv \int_o^t A(t')\,dt' \\ &= \mathcal{N}_o\left(1 - e^{-\lambda t}\right) \\ &= \mathcal{N}_o - \mathcal{N}(t) \end{aligned}$$

so that $\mathcal{N}(t) + \mathcal{N}_{\mathrm{Y}}(t) = \mathcal{N}_o$. All atoms at any time are either of type X or Y, and the sum is always the initial inventory, $\mathcal{N}_o$. The total number of atoms is indeed conserved. When $t \to \infty$, we find $\mathcal{N}_{\mathrm{Y}}(\infty) = \mathcal{N}_o$ and $\mathcal{N}(\infty) = 0$, *i.e.* all atoms of type X eventually decay and then exist as the stable daughter, Y.

4.3.1 Branching Decay

Often, we find particular radionuclides that may decay by more than one decay mode, especially in heavy elements. One such example is ^{238}U, which may decay by α emission or spontaneous fission. The transuranic ^{241}Pu may decay by β^- or α emission. There are many such examples. Though each decay event chooses one route or the other, the probability per unit time of any radioactive atom in the sample decaying by either route is constant. Consider the following general case where radionuclide X can decay by either of two routes to stable daughter nuclei

$$\mathrm{X} \begin{array}{l} \overset{\lambda_1}{\nearrow} \quad \mathrm{Y}_1\ (\text{stable}) + y_1 + \ldots \\ \\ \underset{}{\overset{\lambda_2}{\searrow}} \quad \mathrm{Y}_2\ (\text{stable}) + y_2 + \ldots \end{array}$$

Here, X can be considered an unstable compound nucleus with more than one exit channel (sec. 2.3). Each exit channel has a distinct decay probability and, hence, decay constant λ_j.

Since the decay constant is a nuclear property of the individual radioisotope and independent of the number of radioactive atoms, presence of other radioactive atoms, or the presence of other decay channels, the decay probabilities for each channel are independent and additive. To see this, write

$$A(t) = \sum_j A_j(t) = A_1(t) + A_2(t)$$

as the total or composite activity for this radioisotope. By the radioactive decay law, $A(t) = \lambda_1 \mathcal{N}(t) + \lambda_2 \mathcal{N}(t)$. We can then immediately identify $\lambda = \lambda_1 + \lambda_2$ as the composite decay constant.

The explicit product inventories then become

$$\begin{aligned} \mathcal{N}_1(t) &= \int_o^t A_1(t')\,dt' = \frac{\lambda_1}{\lambda}\mathcal{N}_o\left(1 - e^{-\lambda t}\right) \\ \mathcal{N}_2(t) &= \int_o^t A_2(t')\,dt' = \frac{\lambda_2}{\lambda}\mathcal{N}_o\left(1 - e^{-\lambda t}\right) \end{aligned}$$

where it is clear that the ratio λ_j/λ is the relative probability for decay by channel j. Since these probabilities are inalterable, the daughter concentration ratio, $\mathcal{N}_1(t)/\mathcal{N}_2(t) = \lambda_1/\lambda_2$, is also unchanging.

4.3.2 Specific Activity

At times, it may be more appropriate to consider activity concentrations rather than absolute activities. The concentration of nuclear activity may be expressed in units of Ci/g or Ci/l (curie per liter). The latter is often used when trace quantities of radionuclides are diluted in carrier fluids like air or water, and is often used to set regulatory limits. The total activity of the radionuclide inventory is considered to be homogenized over the entire fluid volume. Hence, the concentration (Ci/l) is very sensitive to changes in physical density of the carrier fluid and further dilution.

The specific activity, expressed in Ci/g, is most often used to characterize the nuclear hazard of solid compounds containing radioisotopes. If the sample is pure so that it contains only atoms of the radionuclide in question, then it is said to be "carrier free." In this case, the specific activity

$$A_s(t) = \frac{A(t)}{m(t)} = \frac{\lambda \mathcal{N}(t)}{\mathcal{N}(t) M / N_A} \neq f(t)$$

is a constant. This, however, requires that all daughter product atoms be immediately removed from the sample, a condition that is difficult to guarantee, except perhaps for some small sources with only gaseous decay products.

For the more general case where carriers cannot be removed,

$$A_s(t) = \frac{\lambda \mathcal{N}(t)}{\mathcal{N}_o M / N_A}$$

is time dependent. This expression ignores the small decay mass loss, Q/c^2.

The β^- emitter ^{14}C is not carrier free in the following samples:

1. $^{14}CO_2$
2. $^{14}CH_4$
3. ^{14}CO and ^{12}CO mixture
4. ^{14}C in elemental carbon (^{12}C, ^{13}C)

4.3.3 Half-Life

The familiar terminology when reporting nuclear lifetimes of unstable nuclides is the half-life, the time required for the activity of a pure sample to decay to one half its original activity, *i.e.* $A(t = t_{1/2}) = \frac{1}{2}A_o$. From the exponential form of the radioactive decay law (Eq.(4.24)), the half-life is related to the decay constant by

$$t_{1/2} = \frac{\ln 2}{\lambda}$$

This introduces a convenient way to rewrite the exponential form of the radioactive decay law as

$$\mathcal{N}(t) = \mathcal{N}_o \left(\frac{1}{2}\right)^{t/t_{1/2}}$$

which naturally reveals the halving of the pure sample radioactive inventory or activity on each successive $t_{1/2}$.

4.3.4 Decay Probability Revisited

The decay language developed thus far allows us to express some additional probability concepts much as we did for neutron interactions and the uncollided flux (sec. 3.7.1). We have already identified λ as the probability per unit time of decay, so the quantity $\lambda\ dt$ is the probability of decay in dt, just as $\Sigma\ dx$ is the probability of neutron interaction in dx. Since $\mathcal{N}(t)/\mathcal{N}_o = e^{-\lambda t}$ is the fraction of radionuclides that survive to time t, then $e^{-\lambda t}$ is the survival probability to t. Combining these, we find

$$e^{-\lambda t} \times \lambda\ dt = p(t)\ dt$$

is the probability of decay in dt about t. The fraction of nuclei transformed in time $[0, t]$ is then

$$\mathcal{F}(t) = \int_o^t p(t')\, dt' = 1 - e^{-\lambda t}$$

which is just $1 - \mathcal{N}(t)/\mathcal{N}_o$ as we expect. Letting $t \to \infty$, we find $\mathcal{F}(\infty) = 1$.

The probability function, $p(t)\, dt$, allows us to find the mean lifetime τ of a radionuclide, the nuclear lifetime, as

$$\tau = \lim_{t\to\infty} \int_o^t t' p(t')\, dt' = \frac{1}{\lambda} \tag{4.26}$$

identical to the e-folding time.

4.3.5 Measuring Half-Life

There is no lack of applications that validate accurate measure of nuclear lifetimes. The study of nuclear structure requires precise estimates. Decay constants, known to reasonable accuracy, are required to estimate the absolute activity of diagnostic radiotracers.

In the laboratory, radionuclide lifetimes can be measured with an experimental system like that shown schematically in Fig. 4.8. Particles of radiation from a radioactive source are incident on a "radiation detector" and induce some physical change within the detector, usually the liberation of electric charge (chpt. 6), which is manifest in an electrical signal pulse. Electrical pulses can be easily detected and recorded with a pulse counter circuit. Operating such a system for time interval, Δt, provides a total number of pulse counts, C, which is proportional to the total number of radioactive decays, $\Delta\mathcal{N}$, from the source during that time interval. If the source $t_{1/2}$ is long with respect to Δt, then $C/\Delta t$ is proportional to the instantaneous source activity. Such a system may be used then to measure λ, and hence $t_{1/2}$, yet the measurement and data analysis procedure may differ depending on the absolute magnitude of $t_{1/2}$. Two such methods are discussed below.

The most obvious analysis suggests observing the exponential decay of the radiation source, the "decay source method." A plot of $\Delta\mathcal{N}/\Delta t$ vs. t on a semi-log scale should appear as a straight line with slope $-\lambda$, as illustrated by the sample data shown in Fig. 4.9. This technique is not without limitations, however. To convince ourselves that we are indeed observing an exponential decay, we require source reduction through at least one $t_{1/2}$, and preferably up to $4t_{1/2}$. This practical requirement limits the set of available radionuclides for this method to those with $t_{1/2}$ less than several hours to, perhaps, several days.

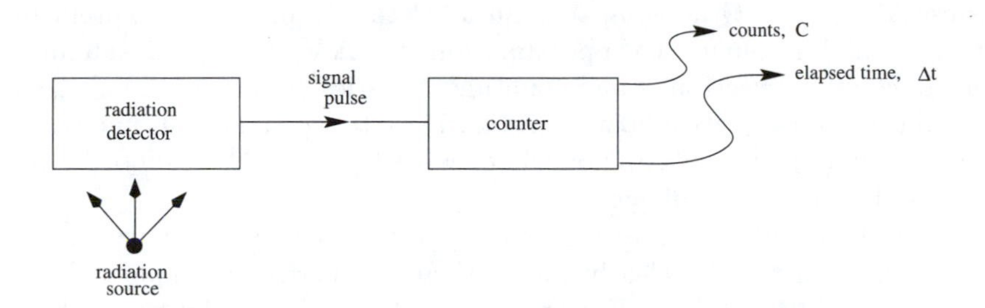

Figure 4.8: Schematic of a simple radiation counting system. A radiation detector generates electronic pulses when particles of radiation from a radioisotope source interact with atoms in the detector. These pulses are counted in counting time interval Δt to provide an estimate of the source activity.

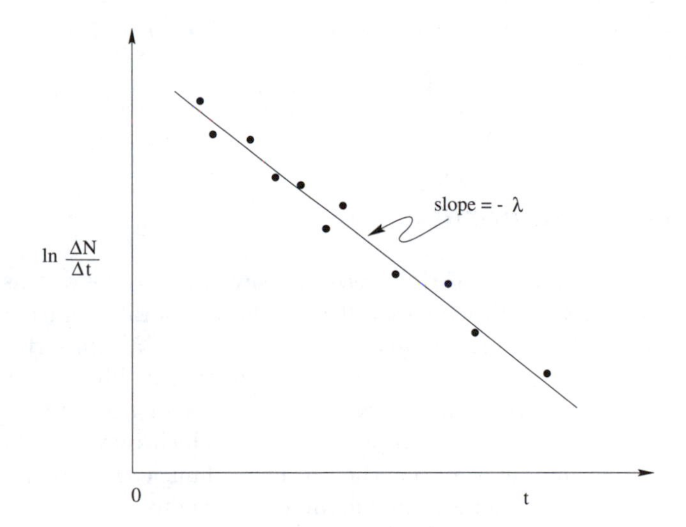

Figure 4.9: Sample data from a decay source method experiment for determining half-life.

The decay source method is also bounded on the low end by $t_{1/2}$. In order for $\Delta\mathcal{N}/\Delta t$ to be representative of the instantaneous activity, we must ensure $\Delta t \ll t_{1/2}$. Otherwise, substantial decay during the measurement interval would invalidate the approximation $A \sim \Delta\mathcal{N}/\Delta t$. Practical limitations in counting electronics and counting statistics (*i.e.* we need $\Delta\mathcal{N} \gg 1$ for good statistics) place a bound on counting interval and dwell time (time between counts) such that radionuclides with $t_{1/2} \lesssim 10^{-3}$ s cannot be investigated with this technique.

For radionuclides with very long $t_{1/2}$, years to millions of years, the decay source method is clearly impractical. A perfect example is ^{235}U with $t_{1/2} = 7.04 \times 10^8$ y. For any reasonable counting duration, there is no noticeable source activity reduction. That is not to imply there is no radioactive decay. Indeed, a 1 g source of ^{235}U emits about 80,000 α particles per second. However, a day's or even a year's worth of decay would not show a noticeable activity reduction.

Instead, an "absolute" method may be used to estimate the decay constant of such nuclides. With the same experimental setup, we can record the total number of counts from a ^{235}U sample decay in an interval of say $\Delta t = 1$ h. Careful measure of the pure sample mass determines $\mathcal{N}$ before counting. Since we can safely assure $\Delta t \ll t_{1/2}$, the source inventory after counting is also $\mathcal{N}$ and, to excellent approximation, $\Delta\mathcal{N}/\Delta t$ is the source activity. We can then use the radioactive decay law in its primitive form (Eq.(4.23)) to estimate

$$\lambda = \frac{A}{\mathcal{N}} \sim \frac{\Delta\mathcal{N}/\Delta t}{\mathcal{N}}$$

4.4 Decay Kinetics

Through the developments of the preceding several sections of this chapter, the tools have been provided to discuss the temporal evolution of radionuclide decay chains. The study of decay chains is an important one. Some naturally occurring, long-lived radioisotopes give birth to product nuclei that are themselves radioactive. They may, in turn, produce radionuclide products, giving rise to decay chains, which may be quite long in many cases. As another example, the products of nuclear fission reactions (chpt. 5) can give rise to long radionuclide decay chains.

Decay chain problems are classified as belonging to one of two types of systems, open and closed systems. Closed decay systems are characterized by a constant total atom inventory in the system. There is no introduction or removal of atoms, into or out of the system after $t = 0$ for closed systems. Mathematically, such cases are initial value problems. Physically, we imagine being provided an initial inventory of atoms in some distribution among

Example 24: Absolute Method of Half-Life Measurement
1 g of pure ^{235}U contains $\mathcal{N} \sim 2.56 \times 10^{21}$ atoms. In a 1 h count, we infer from the number of counts a total number of disintegrations, $\Delta\mathcal{N} \sim 2.88 \times 10^{8}$ decays. What is the estimated value of $t_{1/2}$ from these data?

Solution:
From the fundamental form of the radioactive decay law, Eq.(4.23), we can estimate

$$\lambda \sim \frac{\Delta\mathcal{N}/\Delta t}{\mathcal{N}} = \frac{2.88 \times 10^{8}/3600}{2.56 \times 10^{21}} \sim 3.125 \times 10^{-17} \text{ s}^{-1}$$

This implies a $t_{1/2}$ of 2.2×10^{16} s or 7.03×10^{8} y, very close to the accepted value.

decay chain components and observing the sequential decay. Nuclides in this system change identity though decay, yet the total nuclide inventory remains constant for all time. Enclosed samples of naturally occurring radioisotopes, or fission products that remain contained after removal from a reactor, are excellent examples of closed systems.

By contrast, an open decay system features radionuclide production and removal mechanisms other than decay. Fission production of radioisotopes in an operating reactor, radionuclide production by neutron irradiation of a material sample in a reactor, ground water transport of contaminated soils, isotope separation, radiotracers carried by blood flow, and bioelimination of ingested radionuclides are perfect examples of open systems. Such systems are rich in diversity of solution and application. The price to pay for this diversity, however, is often the difficulty of solution. Time dependent production functions must be obtained independent of the nuclear decay physics, often involving chemical and biological mechanisms. Nonlinear differential equations are often introduced in modeling open systems, which may have no analytic solution. In what follows, we'll develop some general solutions and some examples for both categories of problems that are chosen to be relevant and interesting, yet analytically tractable.

4.4.1 Closed Systems

A closed system decay chain can be described symbolically by the following:

$$\mathrm{X}_1 \xrightarrow{\lambda_1} \mathrm{X}_2 \xrightarrow{\lambda_2} \mathrm{X}_3 \xrightarrow{\lambda_3} \cdots \xrightarrow{\lambda_{j-1}} \mathrm{X}_j \xrightarrow{\lambda_j} \cdots \xrightarrow{\lambda_{n-1}} \mathrm{X}_n \text{ (stable)}$$

where X_j is a daughter product with associated decay constant λ_j at some arbitrary position in the decay chain. Every decay chain terminates with a stable product, X_n. Here, we focus on daughter product atom concentrations and ignore the fate of the radiation particle. Our interest is only in the temporal behavior of the radionuclide chain. We'll shift attention back to the radiation particles later (chpt. 6) where we discuss their fate in the context of radiation interactions with matter.

An example of a long, naturally-occurring decay chain is the actinium series

$$\begin{array}{l} {}^{235}\mathrm{U} \xrightarrow{\alpha} {}^{231}\mathrm{Th} \xrightarrow{\beta^-} {}^{231}\mathrm{Pa} \xrightarrow{\alpha} {}^{227}\mathrm{Ac} \xrightarrow{\beta^-} {}^{227}\mathrm{Th} \xrightarrow{\alpha} \\ {}^{223}\mathrm{Ra} \xrightarrow{\alpha} {}^{219}\mathrm{Rn} \xrightarrow{\alpha} {}^{215}\mathrm{Po} \xrightarrow{\alpha} {}^{211}\mathrm{Pb} \xrightarrow{\beta^-} {}^{211}\mathrm{Bi} \xrightarrow{\alpha} \\ {}^{207}\mathrm{Tl} \xrightarrow{\beta^-} {}^{207}\mathrm{Pb} \text{ (stable)} \end{array}$$

This is a closed system when there are no sources or sinks for atoms in the chain other than decay. The total nuclide inventory is then conserved. The two other naturally-occurring decay chains include the thorium series beginning with $^{232}\mathrm{Th}$ and terminating in stable $^{208}\mathrm{Pb}$, and the uranium series beginning with $^{238}\mathrm{U}$ and terminating in stable $^{206}\mathrm{Pb}$ (cf: [GE96]).

Of course, the simplest example of such a problem is the simple decay system ($n = 2$) where the immediate daughter product of X_1 is the stable X_2. We have already seen the differential equation describing this system, $d\mathcal{N}_1(t)/dt = -\lambda_1\mathcal{N}_1(t)$. With the initial condition $\mathcal{N}_1(t = 0) = \mathcal{N}_{1_o}$, the solution is the familiar exponential form of the radioactive decay law (Eq.(4.24)), $\mathcal{N}_1(t) = \mathcal{N}_{1_o}e^{-\lambda_1 t}$. Next in line of complexity is the binary decay chain ($n = 3$) for which there are two consecutive radionuclide decays to stable X_3,

$$\mathrm{X}_1 \xrightarrow{\lambda_1} \mathrm{X}_2 \xrightarrow{\lambda_2} \mathrm{X}_3 \text{ (stable)}$$

The set of differential decay equations for this chain are

$$\begin{aligned} \frac{d\mathcal{N}_1(t)}{dt} &= -\lambda_1\mathcal{N}_1(t) \\ \frac{d\mathcal{N}_2(t)}{dt} &= \lambda_1\mathcal{N}_1(t) - \lambda_2\mathcal{N}_2(t) \\ \frac{d\mathcal{N}_3(t)}{dt} &= \lambda_2\mathcal{N}_2(t) \end{aligned} \tag{4.27}$$

The first in this set is the familiar differential form of the radioactive decay law for X_1. The remaining two equations are not forms of the radioactive decay law, thereby validating our earlier assertion that the differential and exponential forms of the radioactive decay law apply only to a pure sample or to the first component of a decay chain, *i.e.* $A_j(t) = \lambda_j \mathcal{N}_j(t)$ for all j, $A_j(t) = -d\mathcal{N}_j(t)/dt$ only for $j = 1$.

Equations (4.27) represent a set of three coupled, differential equations which must be solved sequentially beginning with $\mathcal{N}_1(t)$, and they require three initial conditions to obtain an explicit time dependent solution. We may choose to specify the inventories $\mathcal{N}_1(t_1), \mathcal{N}_2(t_2), \mathcal{N}_3(t_3)$ arbitrarily and at arbitrary times. Any non-negative inventories are physically acceptable. The choice of making all initial inventories zero, however, leads only to the trivial solution. The most usual specification, and easiest to solve, is

$$\begin{aligned} \mathcal{N}_1(t=0) &= \mathcal{N}_{1_o} \\ \mathcal{N}_2(t=0) &= 0 \\ \mathcal{N}_3(t=0) &= 0 \end{aligned} \tag{4.28}$$

i.e. we begin with an initially pure finite sample of radionuclide X_1 and observe its decay through X_2 to the stable X_3.

Providing the set of differential equations and boundary conditions completes the physical description of the system, the physical model. Obtaining solutions for these is a mathematical exercise. As the coupling in this set of equations suggests, sequential solution is required. The explicit time dependent function $\mathcal{N}_1(t)$ is a source term for $\mathcal{N}_2(t)$, and $\mathcal{N}_2(t)$ is a source term for $\mathcal{N}_3(t)$, ... The problem setup dictates that the first component solution take the exponential form of the radioactive decay law. Then, the differential equation for $\mathcal{N}_2(t)$ can be explicitly written

$$\frac{d\mathcal{N}_2(t)}{dt} = \lambda_1 \mathcal{N}_{1_o} e^{-\lambda_1 t} - \lambda_2 \mathcal{N}_2(t)$$

which may be solved by introducing the integrating factor $e^{\lambda_2 t}$ so that

$$\frac{d}{dt}\left[\mathcal{N}_2(t) e^{\lambda_2 t}\right] = \lambda_1 \mathcal{N}_{1_o} e^{(\lambda_2 - \lambda_1)t}$$

and the left side of the equation is a complete differential. We insist, for the time being, that $\lambda_1 \neq \lambda_2$. We'll examine $\lambda_1 = \lambda_2$ as a special case in sec. 4.4.2. With $\mathcal{N}_2(t=0) = 0$, this integrates to

$$\boxed{\mathcal{N}_2(t) = \frac{\lambda_1 \mathcal{N}_{1_o}}{\lambda_2 - \lambda_1}\left[e^{-\lambda_1 t} - e^{-\lambda_2 t}\right]} \tag{4.29}$$

which can now be used as the explicit source for $\mathcal{N}_3(t)$. The $\mathcal{N}_3(t)$ equation can be directly integrated with $\mathcal{N}_3(t=0)=0$ to find

$$\begin{aligned}\mathcal{N}_3(t) &= \mathcal{N}_{1_o}\left[1-e^{-\lambda_1 t}-\frac{\lambda_1}{\lambda_2-\lambda_1}\left(e^{-\lambda_1 t}-e^{-\lambda_2 t}\right)\right] \\ &= \mathcal{N}_{1_o}\left(1-e^{-\lambda_1 t}\right)-\mathcal{N}_2(t)\end{aligned} \tag{4.30}$$

Using the radioactive decay law (Eq.(4.23)), we find the activity of the two radionuclides, X_1 and X_2, in the chain

$$\begin{aligned}A_1(t) &= A_{1_o}e^{-\lambda_1 t} \\ A_2(t) &= A_{1_o}\frac{\lambda_2}{\lambda_2-\lambda_1}\left[e^{-\lambda_1 t}-e^{-\lambda_2 t}\right]\end{aligned} \tag{4.31}$$

4.4.2 Properties of Binary Decay

The following properties of the binary decay solutions (Eqs.(4.24), (4.29), (4.30), (4.31)) provide consistency check, examine special cases, and describe interesting features. Firstly, since the binary system is closed, we should check nuclide conservation in the final solutions. An examination of expressions (4.29) and (4.30) verifies

$$\mathcal{N}_1(t)+\mathcal{N}_2(t)+\mathcal{N}_3(t)=\mathcal{N}_{1_o}$$

for all t. For a closed system, nuclide conservation dictates that all radionuclides in the systems eventually decay

$$\int_o^\infty \lambda_1\mathcal{N}_1(t)\,dt=\int_o^\infty \lambda_2\mathcal{N}_2(t)\,dt=\mathcal{N}_{1_o}$$

and reside at $t=\infty$ as the stable end product X_3, $\mathcal{N}_3(t=\infty)=\mathcal{N}_{1_o}$, *i.e.* all nuclei in the system eventually pass through all states, X_1, X_2, X_3.

There are several special times of interest. We identify the time $t_\mathcal{N}$, as that at which the X_1 and X_2 inventories are identical. By equating $\mathcal{N}_1(t_\mathcal{N})=\mathcal{N}_2(t_\mathcal{N})$, we find

$$t_\mathcal{N}=\frac{1}{\lambda_1-\lambda_2}\ln\left[\frac{2\lambda_1-\lambda_2}{\lambda_1}\right]$$

having a physically realizable solution (finite, positive) only when $2\lambda_1>\lambda_2$. Otherwise, the two radionuclide inventory histories never cross paths. Since the inventory of X_2 is initially zero, this implies that when $2\lambda_1\leq\lambda_2$, the daughter inventory $\mathcal{N}_2(t)<\mathcal{N}_1(t)$ for all t.

The activities are equal at t_a. By equating $A_1(t_a)=A_2(t_a)$, we find

$$t_a=\frac{1}{\lambda_2-\lambda_1}\ln\frac{\lambda_2}{\lambda_1}$$

which always has a physical solution, *i.e.* the radionuclide activities always cross. Since both $\mathcal{N}_2(t)$ and $A_2(t)$ are zero at $t = 0$ and $t = \infty$, they must pass through a maximum in the interval $0 \leq t_{\max} \leq \infty$. By differentiating either expression (4.29) or (4.31), we find that both $\mathcal{N}_2(t)$ and $A_2(t)$ pass through a maximum at $t_{\max} = t_a$. So, the activities $A_1(t)$ and $A_2(t)$ cross at $A_2\,|_{\max} = A_2(t_a)$.

The special case $\lambda_1 = \lambda_2$ is an interesting limit, though not very practical. There are no known cases in nature where the parent nuclide has exactly the same decay constant as its immediate product. Nonetheless, our solution formalism provides an answer, should such a case ever be found. The $\mathcal{N}_2(t)$ expression above (Eq.(4.29)) is undefined in the limit $\lambda_1 \to \lambda_2$. By use of L'Hospital's rule or by reverting to the original differential equation for $\mathcal{N}_2(t)$, one readily finds

$$\lim_{\lambda_1 \to \lambda_2} \mathcal{N}_2(t) = \mathcal{N}_{1_\circ}\,(\lambda t)\,e^{-\lambda t}$$

where $\lambda = \lambda_1 = \lambda_2$. Should such a solution be desired, the recalculation of $\mathcal{N}_3(t)$ by integration is compelled.

Returning to the more general case $\lambda_1 \neq \lambda_2$, we investigate special cases for decay constants varying in ratio. First, consider $\lambda_2 > \lambda_1$. This obviously implies $t_{1/2}^{(1)} > t_{1/2}^{(2)}$; the parent nuclide is longer lived than the daughter. The general $A_2(t)$ solution (Eq.(4.31)) yields an interesting result at long times, $t > 3/(\lambda_2 - \lambda_1)$,

$$A_2(t) \simeq \frac{\lambda_2}{\lambda_2 - \lambda_1} A_1(t)$$

Since this implies that the ratio $A_2(t)/A_1(t)$ is a constant under these conditions, this special case is referred to as "transient equilibrium." Although this terminology is an oxymoron, it does reveal the essence of the decay scenario. While both $A_1(t)$ and $A_2(t)$ are decaying and, hence, transient, they are decaying together (in equilibrium with each other) with the decay constant of X_1. The activity ratio is always greater than one in transient equilibrium. The inventory ratio, $\mathcal{N}_2(t)/\mathcal{N}_1(t) \sim \lambda_1/(\lambda_2 - \lambda_1)$, however, only exceeds unity when $2\lambda_1 > \lambda_2$, in agreement with our inventory crossing condition.

The decay chain

$$^{58}\text{Cr} \xrightarrow{\beta^-} {}^{58m}\text{Mn} \xrightarrow{\beta^-} {}^{58}\text{Fe (stable)}$$

is an example of transient equilibrium. One possible decay path of ^{58}Cr is by β^- emission (with $t_{1/2} \sim 7$ s, $\lambda_1 \sim 0.099$ s^{-1}) to the isomeric state ^{58m}Mn which can β^- decay (with $t_{1/2} \sim 3$ s, $\lambda_2 \sim 0.231$ s^{-1}) directly to stable ^{58}Fe. A plot of the two radionuclide activities normalized to A_{1_o} and with $\mathcal{N}_2(t=0)=0$ is shown in Fig. 4.10. The activity crossing time, $t_a \sim 6.42$ s, is clearly present at the peak daughter activity. This chain is in transient equilibrium at $t \sim 3/(\lambda_2 - \lambda_1) \sim 22.7$ s, after which time the ^{58m}Mn decay parallels that of ^{58}Cr.

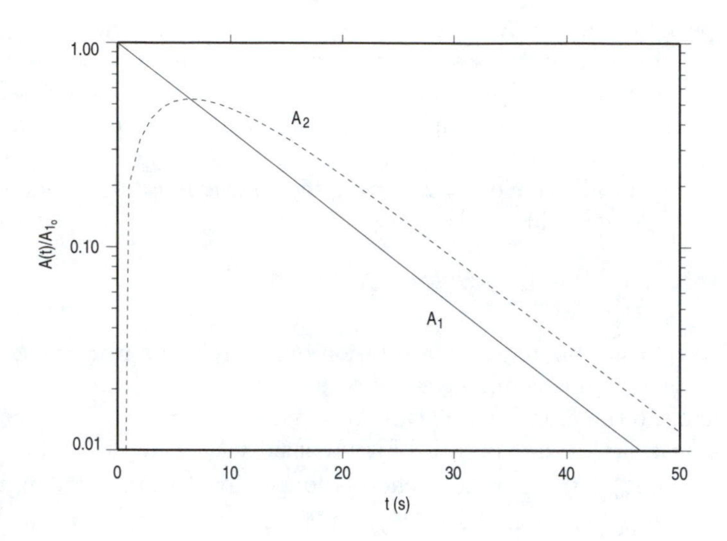

Figure 4.10: Transient equilibrium for the binary decay chain $^{58}\text{Cr} \xrightarrow{\beta^-} {}^{58m}\text{Mn} \xrightarrow{\beta^-} {}^{58}\text{Fe}$ (stable). The solid line indicates the normalized activity of ^{58}Cr. The dashed line indicates the normalized activity of ^{58m}Mn.

The more extreme case, $\lambda_2 \gg \lambda_1$, called "secular equilibrium," is a special case of transient equilibrium. In this limit, the inventory and activity ratios become $\mathcal{N}_2(t)/\mathcal{N}_1(t) \sim \lambda_1/\lambda_2$ and $A_2(t)/A_1(t) \sim 1$ for $t \gtrsim 3/\lambda_2$. Not only is the activity ratio a constant for this case, but it is unity in equilibrium. The daughter and parent decay together with the same decay rate and activity so that after $t \gtrsim 3/\lambda_2$, the total activity is well represented by $2A_1(t)$. The chain

$$^{247}\text{Pu} \xrightarrow{\beta^-} {}^{247}\text{Am} \xrightarrow{\beta^-} {}^{247}\text{Cm} \cdots$$

is a good example of secular equilibrium since the half-lives for ^{247}Pu and ^{247}Am are 2.3 d ($\lambda_1 \sim 0.0126\,\text{h}^{-1}$) and 23 min ($\lambda_2 \sim 1.81\,\text{h}^{-1}$), respectively, with an activity crossing time of $t_a \sim 2.74$ h. The activities of the first two components of this chain are shown from $0 \le t \le 20$ h in Fig. 4.11. The terminology "secular equilibrium" is derived from decay chains with even more disparate decay rates that remain in equilibrium for ages. Consider the chain

$$^{238}\text{U} \xrightarrow{\alpha} {}^{234}\text{Th} \xrightarrow{\beta^-} {}^{234}\text{Pa} \cdots$$

which exhibits the decay constants $\lambda_1 \sim 1.43 \times 10^{-10}$ y^{-1} and $\lambda_2 \sim 0.0288$ d^{-1}, respectively. This system is in equilibrium from $t_a \sim 866$ d. To within about 5%, both ^{238}U and ^{234}Th retain the initial activity of the parent, A_{1_o}, until $\frac{1}{20}\frac{1}{\lambda_1} \sim 3.5 \times 10^8$ y.

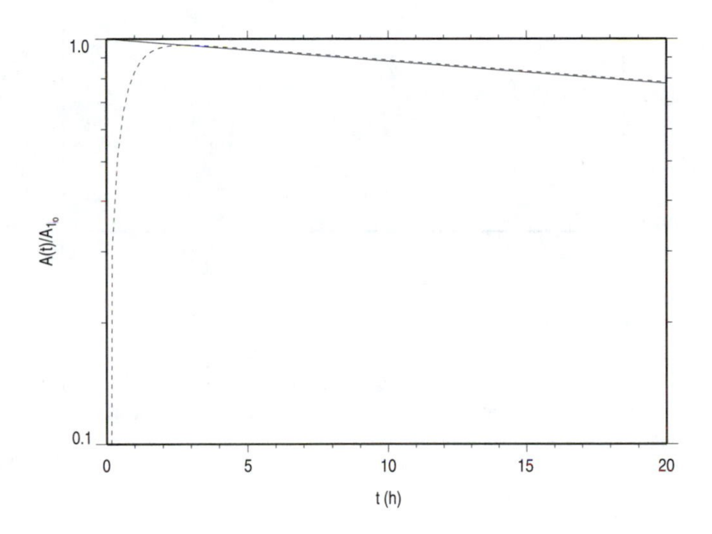

Figure 4.11: Secular equilibrium for the binary decay chain $^{247}\text{Pu} \xrightarrow{\beta^-} {}^{247}\text{Am} \xrightarrow{\beta^-} {}^{247}\text{Cm} \cdots$. The solid line indicates the normalized activity of ^{247}Pu. The dashed line indicates the normalized activity of ^{247}Am.

To complete the discussion, consider the case $\lambda_1 > \lambda_2$. Again, at long times ($t \gtrsim 3/(\lambda_1 - \lambda_2)$), simplification of the binary decay expression (Eq.(4.31)) can be made

$$A_2(t) \sim \frac{\lambda_2}{\lambda_1 - \lambda_2} A_{1_o} e^{-\lambda_2 t}$$

The daughter product decays with its own decay constant. The activity ratio, $A_2(t)/A_1(t)$, is time dependent for this case. Among the many examples of this case is the decay chain

$$^{233}\text{Pu} \xrightarrow{\varepsilon} {}^{233}\text{Np} \xrightarrow{\varepsilon} {}^{233}\text{U} \cdots$$

The activity of the first two components, ^{233}Pu ($\lambda_1 \sim 0.033$ min^{-1}) and ^{233}Np ($\lambda_2 \sim 0.0191$ min^{-1}), is show in Fig. 4.12 for $0 \leq t \leq 200$ min.

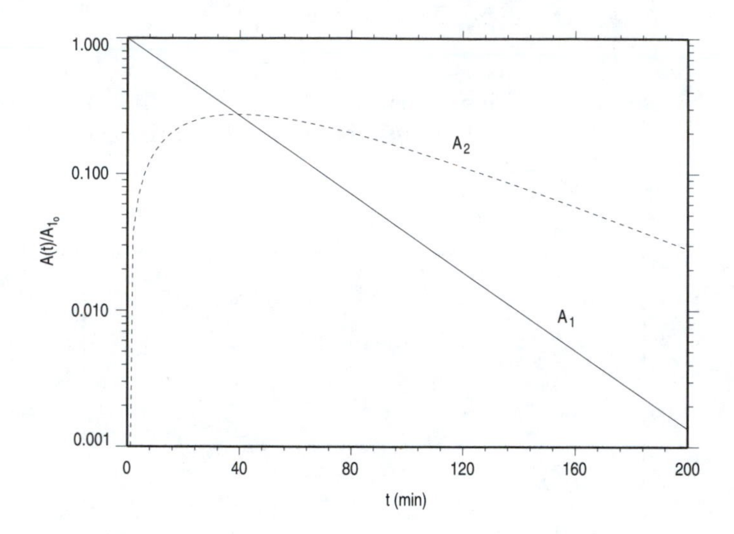

Figure 4.12: Normalized activities for the binary decay chain $^{233}\text{Pu} \xrightarrow{\varepsilon} {}^{233}\text{Np} \xrightarrow{\varepsilon} {}^{233}\text{U} \cdots$. The solid line indicates the normalized activity of ^{233}Pu. The dashed line indicates the normalized activity of ^{233}Np.

4.4.3 General Decay Chain Solutions

The general solution to an arbitrarily long decay chain

$$\text{X}_1 \xrightarrow{\lambda_1} \text{X}_2 \xrightarrow{\lambda_2} \text{X}_3 \xrightarrow{\lambda_3} \cdots \xrightarrow{\lambda_{j-1}} \text{X}_j \xrightarrow{\lambda_j} \cdots \xrightarrow{\lambda_{n-1}} \text{X}_n \text{ (stable)}$$

with the boundary conditions

$$\mathcal{N}_j(t=0) = \begin{cases} \mathcal{N}_{1_o} & , j = 1 \\ 0 & , 2 \leq j \leq n \end{cases}$$

is given by the Bateman equations. The activity of the jth member of the chain is

$$\begin{aligned} A_j(t) &= \mathcal{N}_{1_o} \sum_{i=1}^{j} C_i e^{-\lambda_i t} \qquad (4.32) \\ &= \mathcal{N}_{1_o} \left(C_1 e^{-\lambda_1 t} + C_2 e^{-\lambda_2 t} + \cdots + C_j e^{-\lambda_j t} \right) \end{aligned}$$

The coefficients, C_m, are found from the following product

$$C_m = \frac{\prod_{i=1}^{j} \lambda_i}{\prod_{i=1}^{j} (\lambda_i - \lambda_m)} = \frac{\lambda_1 \lambda_2 \lambda_3 \ldots \lambda_j}{(\lambda_1 - \lambda_m)(\lambda_2 - \lambda_m)\ldots(\lambda_j - \lambda_m)} \tag{4.33}$$

where $i = m$ is excluded from the product in the denominator only.

As an example, let's Use Eqs.(4.32) and (4.33) to find the activity of the first and second members in a chain. For $j = 1$,

$$A_1(t) = \mathcal{N}_{1_o} C_1 e^{-\lambda_1 t}$$

the now familiar exponential form of the radioactive decay law. The only coefficient to find here is C_1, which from Eq.(4.33) is $C_1 = \lambda_1$. For the second component, $j = 2$,

$$A_2(t) = \mathcal{N}_{1_o} C_1 e^{-\lambda_1 t} + \mathcal{N}_{1_o} C_2 e^{-\lambda_2 t}$$

The coefficients are again found from expression (4.33) with $m = 1$ and $m = 2$

$$C_1 = \frac{\lambda_1 \lambda_2}{\lambda_2 - \lambda_1}$$
$$C_2 = \frac{\lambda_1 \lambda_2}{\lambda_1 - \lambda_2}$$

Then, the activity reduces to

$$A_2(t) = A_{1_o} \frac{\lambda_2}{\lambda_2 - \lambda_1} \left[e^{-\lambda_1 t} - e^{-\lambda_2 t}\right]$$

in agreement with our earlier solution, Eq.(4.31).

4.4.4 Open Systems

In contrast to closed systems, open systems allow introduction and removal of nuclei in our decay chain by means other than decay. We can write a general open system decay chain symbolically as

$$
\begin{array}{ccccccccccc}
\mathcal{P}_1(t) & & \mathcal{P}_2(t) & & \mathcal{P}_3(t) & & & & \mathcal{P}_j(t) & & \\
\downarrow & & \downarrow & & \downarrow & & & & \downarrow & & \\
\mathrm{X}_1 & \xrightarrow{\lambda_1} & \mathrm{X}_2 & \xrightarrow{\lambda_2} & \mathrm{X}_3 & \xrightarrow{\lambda_3} & \cdots & \xrightarrow{\lambda_{j-1}} & \mathrm{X}_j & & \xrightarrow{\lambda_j} \\
\downarrow & & \downarrow & & \downarrow & & & & \downarrow & & \\
\mathcal{L}_1(t) & & \mathcal{L}_2(t) & & \mathcal{L}_3(t) & & & & \mathcal{L}_j(t) & &
\end{array}
$$

$$
\begin{array}{ccccc}
 & & \mathcal{P}_n(t) & & \\
 & & \downarrow & & \\
\cdots & \xrightarrow{\lambda_{n-1}} & \mathrm{X}_n & & \text{(stable)} \\
 & & \downarrow & & \\
 & & \mathcal{L}_n(t) & &
\end{array}
$$

where $\mathcal{P}_j(t)$ is the time dependent production function (atoms/s) for atoms of type j introduced into the system, and $\mathcal{L}_j(t)$ is the time dependent removal or loss function (atoms/s) for atoms of type j from the system. In general, production and loss functions are allowed for all components of the chain including the stable end product, X_n. The first order equation for arbitrary component j in this chain becomes

$$\frac{d\mathcal{N}_j(t)}{dt} = \mathcal{P}_j(t) + \lambda_{j-1}\mathcal{N}_{j-1}(t) - \lambda_j\mathcal{N}_j(t) - \mathcal{L}_j(t) \tag{4.34}$$

The border components, X_1 and X_n, have differential equations of slightly different form since $\lambda_{j-1} = 0$ when $j = 1$, and $\lambda_n = 0$. Mathematically, this system again represents a set of n coupled first order equations, which must be solved sequentially from $j = 1$ to n since the jth equation requires an explicit form of the immediately preceding inventory. To solve, we'll need n boundary conditions, the n functions $\mathcal{P}_j(t)$, and the n functions $\mathcal{L}_j(t)$. Even given these functions, the set may be analytically intractable since we are not guaranteed to obtain integrable forms with arbitrary, time dependent functions. In such a situation, one could always turn to a numerical solution, however.

The general solution to the jth component inventory is facilitated by employing the integrating factor $e^{\lambda_j t}$ in Eq.(4.34), just as we did in the closed system problem, to arrive at the new first order equation exhibiting a complete differential

$$\frac{d}{dt}\left[\mathcal{N}_j(t)e^{\lambda_j t}\right] = \left[\mathcal{P}_j(t) + \lambda_{j-1}\mathcal{N}_{j-1}(t) - \lambda_j\mathcal{N}_j(t) - \mathcal{L}_j(t)\right]e^{\lambda_j t}$$

Without explicit production and loss functions, this expression can be reduced only to the quadrature form

$$\mathcal{N}_j(t) = \mathcal{N}_{j_o} e^{-\lambda_j t} + e^{-\lambda_j t} \int_o^t \left[\mathcal{P}_j(t') + \lambda_{j-1}\mathcal{N}_{j-1}(t') - \mathcal{L}_j(t')\right] e^{\lambda_j t} \quad (4.35)$$

where $\mathcal{N}_{j_o} = \mathcal{N}_j(t=0)$. This may be only be integrable when the production and loss functions take on convenient forms. When all functions $\mathcal{P}_j(t)$ and $\mathcal{L}_j(t)$ are identically zero, the open system reverts to a closed system problem. In this way, we should consider a closed system as just a special case of the more general open system.

As an example, consider the three-isotope decay chain with a single production route

$$\begin{array}{lllll} \mathcal{P}_1(t) & & & & \\ \downarrow & & & & \\ X_1 & \xrightarrow{\lambda_1} & X_2 & \xrightarrow{\lambda_2} & X_3 \quad \text{(stable)} \end{array}$$

This is a production and decay problem, a more specialized case of which will be discussed in sec. 4.5. From the general solution (Eq.(4.35)), we have

$$\begin{aligned} \mathcal{N}_1(t) &= \mathcal{N}_{1_o} e^{-\lambda_1 t} + e^{-\lambda_1 t} \int_o^t dt'\, \mathcal{P}_1(t') e^{\lambda_1 t} \qquad (4.36) \\ \mathcal{N}_2(t) &= \mathcal{N}_{2_o} e^{-\lambda_2 t} + e^{-\lambda_2 t} \int_o^t dt'\, \lambda_1 \mathcal{N}_1(t') e^{\lambda_2 t} \\ \mathcal{N}_3(t) &= \mathcal{N}_{3_o} + \int_o^t dt'\, \lambda_2 \mathcal{N}_2(t') \end{aligned}$$

for which we need only $\mathcal{P}_1(t)$ and the boundary conditions, $\mathcal{N}_{1_o}, \mathcal{N}_{2_o}, \mathcal{N}_{3_o}$.

Some examples of integrable (and practically achievable) production functions are shown in Fig. 4.13. To provide a few explicit solutions, let's consider the step, ramp, exponential, and sinusoidal sources.

1. Step function at $t_o = 0$:

$$\mathcal{P}_1(t) = \begin{cases} 0 & , t < 0 \\ \mathcal{P}_o & , t \geq 0 \end{cases}$$

Here, we may choose $\mathcal{N}_{1_o} = \mathcal{N}_{2_o} = \mathcal{N}_{3_o} = 0$, no initial inventory of any member of the chain. This does not lead to the trivial solution, as it does for the closed system, as long as $\mathcal{P}_1(t) \neq 0$. Then, the inventory and activity for X_1 are

$$\mathcal{N}_1(t) = \frac{\mathcal{P}_o}{\lambda_1}\left(1 - e^{-\lambda_1 t}\right)$$

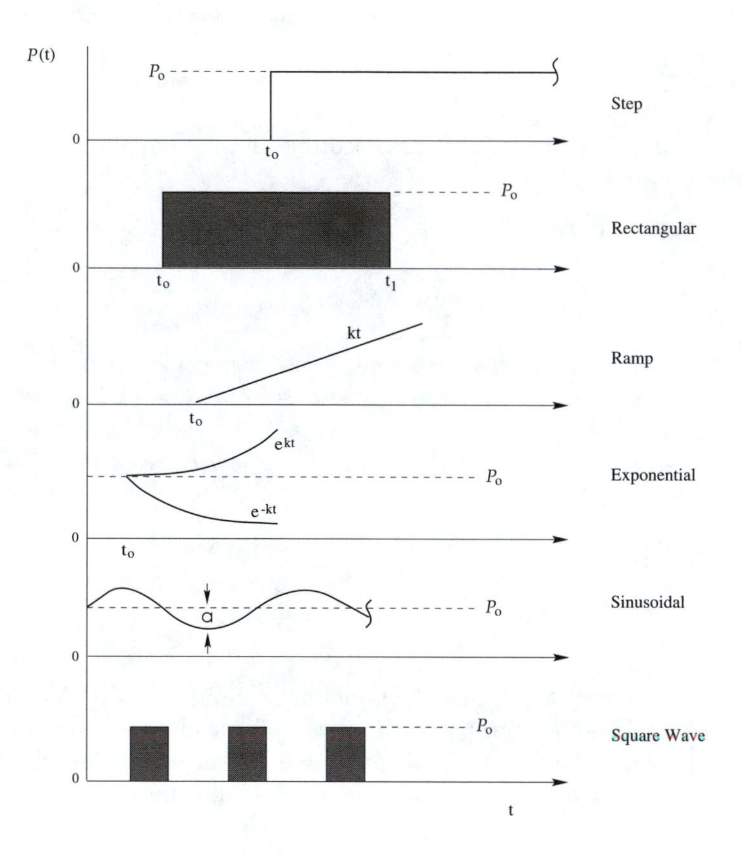

Figure 4.13: Some examples of potential production functions: $\mathcal{P}_j(t)$.

and

$$A_1(t) = \mathcal{P}_o\left(1 - e^{-\lambda_1 t}\right)$$

so that the asymptotic activity, or "saturation activity," is identical with the constant production rate. We'll revisit this important problem in sec. 4.5.

2. Ramp function at $t_o = 0$:

$$\mathcal{P}_1(t) = \begin{cases} 0 & , t < 0 \\ kt & , t \geq 0 \end{cases}$$

Again, we can allow $\mathcal{N}_{1_o} = 0$ to find

$$\mathcal{N}_1(t) = \frac{kt}{\lambda_1} - \frac{k}{\lambda_1^2}\left(1 - e^{-\lambda_1 t}\right)$$

This solution is obviously physical only for $t < \infty$. The source must be bounded. There is no physical means to provide a diverging source, linear or otherwise.

3. Exponential function at $t_o = 0$:

$$\mathcal{P}_1(t) = \begin{cases} 0 & , t < 0 \\ \mathcal{P}_o e^{-kt} & , t \geq 0 \end{cases}$$

The solution to this case is familiar

$$\mathcal{N}_1(t) = \frac{\mathcal{P}_o}{\lambda_1 - k}\left[e^{-kt} - e^{-\lambda_1 t}\right]$$

This is just the form of the solution for the second component of a closed binary system (Eq.(4.29)) with $k = \lambda_2$ and $\mathcal{P}_o = A_{1_o}$.

4. Sinusoidal function:

$$\mathcal{P}_1(t) = \mathcal{P}_o + a\sin(kt) \qquad , \text{all } t$$

With the initial condition $\mathcal{N}_1(t = 0) = 0$, the integral equation for $\mathcal{N}_1(t)$ then becomes

$$\mathcal{N}_1(t) = e^{-\lambda_1 t}\int_o^t dt'\ \left[\mathcal{P}_o + a\sin(kt')\right] e^{\lambda_1 t'}$$

Integrating by parts twice yields the explicit form

$$\mathcal{N}_1(t) = \frac{\mathcal{P}_o}{\lambda_1}\left(1 - e^{-\lambda_1 t}\right) + \frac{a}{\lambda_1^2 + k^2}\left[\lambda_1\sin(kt) - k\left(\cos(kt) - e^{-\lambda_1 t}\right)\right]$$

comprised of the constant production solution and an oscillatory component. At times long compared to both $1/\lambda_1$ and $1/k$, the solution reduces to

$$\mathcal{N}_{1_\infty} \simeq \frac{\mathcal{P}_o}{\lambda_1} + \frac{a}{\lambda_1^2 + k^2}\left[\lambda_1 \sin(kt) - k\cos(kt)\right]$$

and when $\lambda_1 \gg k$

$$\mathcal{N}_{1_\infty} \sim \frac{1}{\lambda_1}\left[\mathcal{P}_o + a\sin(kt)\right] \sim \mathcal{P}_1(t)/\lambda_1$$

Since the decay constant is so large, $\mathcal{N}_1$ reaches equilibrium quickly on each cycle of $\mathcal{P}_1(t)$ and, hence, follows the production. In the opposite extreme, at high frequency $k \gg \lambda_1$,

$$\mathcal{N}_{1_\infty} \sim \frac{\mathcal{P}_o}{\lambda_1} - \frac{a}{k}\cos(kt)$$

so the response is a small oscillation about a constant inventory.

4.5 Production of Radioactive Atoms with Neutrons

An important application of the general open system formalism involves the production of radioactive materials with neutrons in a process referred to as "neutron activation." Many stable isotopes readily absorb neutrons in (n,γ) reactions and are converted to radionuclei. Nuclear reactors are an ample source of free neutrons for such purposes. Radionuclei "activated" in this way can be employed in a vast number of important applications from the identification of trace elements in forensic pathology to the production of radiotracers for medical diagnosis.

Consider exposing a fixed sample of stable X_1 atoms with initial inventory $\mathcal{N}_{1_o}$ to neutron flux $\varphi(t)$ in a reactor. The rate at which (n,γ) reactions convert these to X_2 atoms

$$\mathrm{n} + \mathrm{X}_1 \text{ (stable)} \longrightarrow \mathrm{X}_2 + \gamma$$

can be found from the formalism developed in chapter 3, *i.e.*

$$\mathcal{P}_2(t) = R_\gamma = \sigma_{\gamma_1}\varphi(t)\mathcal{N}_1(t)$$

where σ_{γ_1} is the (n,γ) cross section for type X_1 atoms and we have assumed uniform flux over the sample volume. Since there is a one-to-one correspondence between interactions and the number of X_1 atoms consumed, then

$$\mathcal{L}_1(t) = \mathcal{P}_2(t)$$

Example 25: Biological Open System

Radionuclides are sometimes ingested into the body with food, water, or through inhalation, and may pose a hazard. Suppose $\mathcal{N}_{1_o}$ atoms of radionuclide X_1 are ingested instantaneously. Removal from the body occurs at the rate $-\lambda_1\mathcal{N}_1(t)$ from radioactive decay and at the rate $-k_b\mathcal{N}_1(t)$ from biological elimination. The latter represents a loss term of the form $\mathcal{L}_1(t) = k_b\mathcal{N}_1(t)$ where k_b is the biological removal rate constant. Then, the balance equation describing the inventory of radionuclide is

$$\frac{d\mathcal{N}_1(t)}{dt} = -\lambda_1\mathcal{N}_1(t) - k_b\mathcal{N}_1(t)$$

with the solution

$$\mathcal{N}_1(t) = \mathcal{N}_{1_o}e^{-(\lambda_1+k_b)t}$$

The measure of damage to the body during some time interval t is quantified by the total number of decays in the body during that interval, $\mathcal{N}_b(t)$, as

$$\mathcal{N}_b(t) = \int_o^t \lambda_1\mathcal{N}_1(t')\,dt' = \frac{\lambda_1}{\lambda_1 + k_b}\mathcal{N}_{1_o}\left[1 - e^{-(\lambda_1+k_b)t}\right]$$

The potential hazard can be estimated for long times ($t \to \infty$), so that

$$\mathcal{N}_{b_\infty} \sim \frac{\lambda_1}{\lambda_1 + k_b}\mathcal{N}_{1_o}$$

represents the total number of possible decays within the body. The eventual number of decays outside the body is then, $1 - \mathcal{N}_{b_\infty} = \frac{k_b}{\lambda_1+k_b}\mathcal{N}_{1_o}$. The hazard thus diminishes (and may be negligible regardless of $\mathcal{N}_{1_o}$) when $k_b \gg \lambda_1$. For some important radionuclides, however, the opposite is the case, $\lambda_1 \gg k_b$, and essentially all radionuclides in the sample decay in the body.

Tritium is a good example of the former where bioelimination is rapid and the hazard is mitigated. For ^{3}H, the biological rate constant $k_b \sim 0.069\ \mathrm{d}^{-1}$. Since $t_{1/2} = 12.3$ y ($\lambda_1 \sim 1.54\times10^{-4}\ \mathrm{d}^{-1}$), then indeed $k_b \gg \lambda_1$ and

$$\frac{\mathcal{N}_{b_\infty}}{\mathcal{N}_{1_o}} \sim \frac{\lambda_1}{k_b} \sim 0.0022$$

so only a small fraction of all ^{3}H decays occur in the body.

Atoms of nuclide X_2 are removed through decay to the stable X_3

$$X_2 \xrightarrow{\lambda_2} X_3 \text{ (stable)} + y + \cdots$$

at a rate $-\lambda_2 \mathcal{N}_2(t)$, and through neutron-induced interactions at a rate $\mathcal{L}_2(t) = \sigma_{\gamma_2}\varphi(t)\mathcal{N}_2(t)$ since X_2 atoms are born in the same neutron flux in which they are produced. An example of such an activation problem is the production of the β^- emitter tritium by neutron capture in deuterium

$$\begin{aligned} \mathrm{n} + \mathrm{D} &\longrightarrow \mathrm{T} + \gamma \\ \mathrm{T} &\xrightarrow{\lambda_\mathrm{T}} {}^3\mathrm{He} + \beta^- + \overline{\nu} \end{aligned}$$

Our differential balance equations for this system then become

$$\begin{aligned} \frac{d\mathcal{N}_1(t)}{dt} &= -\mathcal{L}_1(t) = -\sigma_{\gamma_1}\varphi(t)\mathcal{N}_1(t) \\ \frac{d\mathcal{N}_2(t)}{dt} &= \mathcal{P}_2(t) - \lambda_2\mathcal{N}_2(t) - \mathcal{L}_2(t) \\ &= \sigma_{\gamma_1}\varphi(t)\mathcal{N}_1(t) - \lambda_2\mathcal{N}_2(t) - \sigma_{\gamma_2}\varphi(t)\mathcal{N}_2(t) \\ \frac{d\mathcal{N}_3(t)}{dt} &= \lambda_2\mathcal{N}_2(t) - \mathcal{L}_3(t) = \lambda_2\mathcal{N}_2(t) - \sigma_{\gamma_3}\varphi(t)\mathcal{N}_3(t) \end{aligned}$$

since $\mathcal{P}_1(t) = \mathcal{P}_3(t) = 0$. Solutions are obtained only after $\varphi(t)$ is specified, and then analytic solutions are obtained only when the resulting functions are integrable. The most common and practical situation is to perform activation under constant reactor conditions and, hence, constant flux so that $\varphi(t) = \varphi_o$. Then

$$\boxed{\mathcal{N}_1(t) = \mathcal{N}_{1_o} e^{-\sigma_{\gamma_1}\varphi_o t}} \tag{4.37}$$

represents target sample consumption. For times short compared to $1/\sigma_{\gamma_1}\varphi_o$, the quantity $\sigma_{\gamma_1}\varphi_o t$ well represents the target sample depletion fraction.

The activated nuclide inventory may be readily obtained by employing the integrating factor $e^{(\lambda_2+\sigma_{\gamma_2}\varphi_o)t}$ to yield

$$\boxed{\mathcal{N}_2(t) = \frac{\sigma_{\gamma_1}\varphi_o\mathcal{N}_{1_o}}{\lambda_2 + \varphi_o(\sigma_{\gamma_2} - \sigma_{\gamma_1})}\left[e^{-\sigma_{\gamma_1}\varphi_o t} - e^{-(\lambda_2+\sigma_{\gamma_2}\varphi_o)t}\right]} \tag{4.38}$$

with the boundary condition $\mathcal{N}_{2_o} = 0$. A final integration provides the X_3 inventory

$$\boxed{\mathcal{N}_3(t) = \frac{\lambda_2 \sigma_{\gamma_1} \varphi_o \mathcal{N}_{1_o}}{\lambda_2 + \varphi_o(\sigma_{\gamma_2} - \sigma_{\gamma_1})} \left[\frac{e^{-\sigma_{\gamma_3}\varphi_o t} - e^{-\sigma_{\gamma_1}\varphi_o t}}{(\sigma_{\gamma_1} - \sigma_{\gamma_3})\varphi_o} - \frac{e^{-\sigma_{\gamma_3}\varphi_o t} - e^{-(\lambda_2 + \sigma_{\gamma_2}\varphi_o)t}}{\lambda_2 + (\sigma_{\gamma_2} - \sigma_{\gamma_3})\varphi_o} \right]} \tag{4.39}$$

when $\mathcal{N}_{3_o} = 0$.

These are the complete equations for the activation problem. In the limit $\varphi_o \sigma_{\gamma_{2,3}}/\lambda_2 \ll 1$ (most often the case), these expressions have the following intuitive properties. As it was for closed systems, nuclide conservation

$$\mathcal{N}_1(t) + \mathcal{N}_2(t) + \mathcal{N}_3(t) = \mathcal{N}_{1_o}$$

applies to this special open system problem since the irradiated sample is finite and not augmented during the activation. The inventory of X_3 atoms is identical to the total number of X_2 decays

$$\int_o^t A_2(t')\, dt' = \mathcal{N}_3(t)$$

for all t. Similarly, the production of X_2 atoms represents depletion of X_1 atoms

$$\int_o^t \mathcal{P}_2(t')\, dt' = \mathcal{N}_{1_o} - \mathcal{N}_1(t)$$

for all t. As $t \to \infty$, all atoms in the sample pass from type X_1 atoms to X_2 to X_3, so that $\mathcal{N}_1(\infty) = \mathcal{N}_2(\infty) = 0$, and $\mathcal{N}_3(\infty) = \mathcal{N}_{1_o}$.

While the target inventory, $\mathcal{N}_1(t)$, is monotonically decreasing by depletion, and the stable product inventory, $\mathcal{N}_3(t)$, is monotonically increasing, the radionuclide inventory, $\mathcal{N}_2(t)$, first increases and peaks before eventually returning to zero. The time of peak $\mathcal{N}_2(t)$ (and hence $A_2(t)$) is found by identifying the time, $t_{\max}$, at which $d\mathcal{N}_2(t)/dt = 0$

$$t_{\max} = \frac{1}{\lambda_2 - \varphi_o(\sigma_{\gamma_1} - \sigma_{\gamma_2})} \ln \left[\frac{\lambda_2 + \sigma_{\gamma_2}\varphi_o}{\sigma_{\gamma_1}\varphi_o} \right]$$

For the tritium production problem introduced earlier ($t_{1/2} = 12.3$ y, $\sigma_{\gamma_1} = 0.52$ mb), and assuming $\varphi_o = 10^{12}\ \frac{\mathrm{n}}{\mathrm{cm^2\,s}}$, we find $t_{\max} \sim 267$ y, and inordinate amount of time. The problem is not quite that bad, however. The reduction of $\mathcal{N}_\mathrm{T}(t)$ at long times is due largely to slow D target depletion. The buildup, however, occurs on the much faster decay time $1/\lambda_2$. The result is that the tritium inventory increases rapidly before gradually reaching its peak. It reaches 90% of peak activity in only $t_{90} \sim \frac{1}{\lambda_2} \ln \frac{1}{1-0.9} \sim 41$ y, a time that is largely independent of φ_o.

Example 26: Neutron Production of ^{38}Cl
A more practical laboratory example is the production of the β^- emitter ^{38}Cl ($t_{1/2} = 37.2$ min) from the stable ^{37}Cl ($\sigma_{\gamma_1} = 0.43$ b). Find the time to reach maximum activity and 90% of maximum activity if $\varphi = 10^{12}\ \frac{\text{n}}{\text{cm}^2\ \text{s}}$.

Solution:
Since $\lambda_2(\sim 3.1 \times 10^{-4}\ \text{s}^{-1}) \gg \sigma_{\gamma_1}\varphi_o(\sim 4.3 \times 10^{-13}\ \text{s}^{-1})$, then

$$t_{\text{max}} \sim \frac{1}{\lambda_2} \ln \frac{\lambda_2}{\sigma_{\gamma_1}\varphi_o} \sim 6.6 \times 10^4\ \text{s} = 18.3\ \text{h}$$

The time to reach 90% of maximum activity can be estimated from Eq.(4.38). Using the approximation above and recognizing that the first exponential term in Eq.(4.38) is always close to unity for any reasonable time (*i.e.* no depletion is an excellent assumption for $t \ll (\varphi_o\sigma_{\gamma_1})^{-1} \sim 2.3 \times 10^{12}\ \text{s} \sim 7.4 \times 10^4\ \text{y}$!!!), then

$$\frac{\mathcal{N}_2(t)}{\mathcal{N}_2|_{\text{max}}} = 0.9 \sim 1 - e^{-\lambda_2 t_{90}}$$

So, $t_{90} \sim \frac{1}{\lambda_2} \ln 10 \sim 7.4 \times 10^3\ \text{s} \sim 2.06\ \text{h}$.

4.5.1 Activation Equations

The approximations made above and in example 26 are almost universally applicable. There are exceptions, but for target and product nuclei with moderate absorption cross section (on the order of kb or less), then

$$\lambda_2 \gg \sigma_{\gamma_{1,2,3}} \varphi_o$$

As well, for any reasonable time

$$\sigma_{\gamma_1} \varphi_o t \ll 1$$

With these simplifications, the three inventory expressions, Eqs.(4.37)–(4.39), can be reduced as follows. For the target species X_1

$$\mathcal{N}_1(t) \simeq \mathcal{N}_{1_o} (1 - \sigma_{\gamma_1} \varphi_o t) \sim \mathcal{N}_{1_o}$$

again revealing the quantity $\sigma_{\gamma_1} \varphi_o t$ as the depletion fraction for small t. Since this quantity is negligible for very short times, the approximation $\sigma_{\gamma_1} \varphi_o t \ll 1$ is called the "no depletion" approximation.

Employing both approximations, the radionuclide inventory can be written

$$\boxed{\mathcal{N}_2(t) \simeq \frac{\sigma_{\gamma_1} \varphi_o \mathcal{N}_{1_o}}{\lambda_2} \left[1 - e^{-\lambda_2 t}\right]} \tag{4.40}$$

Without depletion, $\mathcal{N}_2(t)$ is monotonically increasing to the asymptote $\sigma_{\gamma_1} \varphi_o \mathcal{N}_{1_o} / \lambda_2$. The corresponding activity is now

$$\boxed{A_2(t) \simeq \sigma_{\gamma_1} \varphi_o \mathcal{N}_{1_o} \left[1 - e^{-\lambda_2 t}\right]} \tag{4.41}$$

and it too monotonically increases to the asymptotic or saturation activity

$$\boxed{A_\infty = \sigma_{\gamma_1} \varphi_o \mathcal{N}_{1_o}} \tag{4.42}$$

equal to the production rate in the absence of depletion. Here, $d\mathcal{N}_2(t)/dt = 0$, and the decay rate at saturation exactly balances production. The final stable product inventory becomes

$$\boxed{\mathcal{N}_3(t) \simeq \sigma_{\gamma_1} \varphi_o \mathcal{N}_{1_o} \left[t - \frac{1}{\lambda_2}(1 - e^{-\lambda_2 t})\right]} \tag{4.43}$$

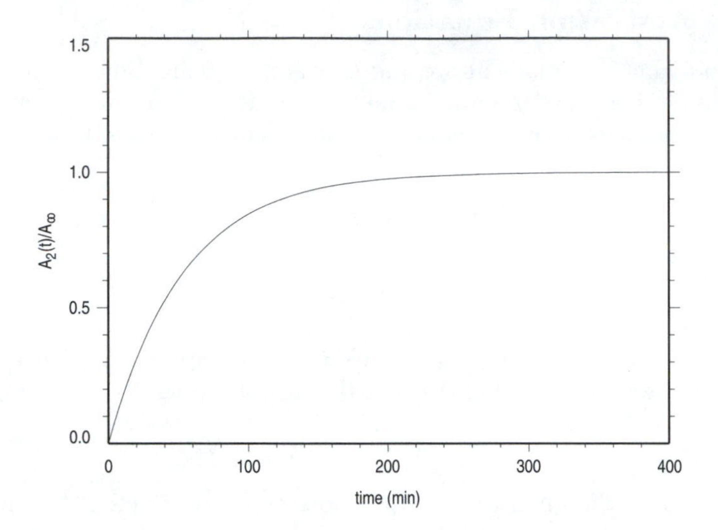

Figure 4.14: Approach to saturation for the β^- emitter ^{38}Cl as per example 26. Stable ^{37}Cl is irradiated in a research reactor to produce ^{38}Cl which decays with $t_{1/2} = 37.2$ min.

under these approximations. This set of equations is often called the "activation equations." A plot of $A_2(t)/A_\infty$ for the ^{38}Cl problem in example 26 is shown in Fig. 4.14 exploiting these approximations.

The practical question arises, "When do we terminate the irradiation?" Because of the exponential approach to saturation, there is obviously little merit in irradiating past a few half-lives of X_2. (Recall that even when allowing for depletion, $t_{90} \ll t_{\max}$.) The point of diminishing return is reached at about $t \sim 4t_{1/2}$ where

$$\frac{A_2}{A_\infty} = 1 - \left(\frac{1}{2}\right)^4 = \frac{15}{16} = 0.9375$$

There is little point in proceeding further.

4.5.2 Post-Irradiation Decay

To make use of radioactive materials produced in a reactor (or other neutron source), we must first remove the sample from the neutron flux. Once removed, activated sources will decay exponentially with characteristic time $1/\lambda_2$ since there is no longer a production or driving function for X_2. For an irradiation of time T, at time $(t-T) > 0$ after removal from the neutron source, the radionuclide activity will be

$$A_2(t) = A_2(T)e^{-\lambda_2(t-T)}$$

Combining this with our activation equation (Eq.(4.41)) for $A_2(T)$

$$A_2(t) = \sigma_{\gamma_1}\varphi_o\mathcal{N}_{1_o}\left[1 - e^{-\lambda_2 T}\right]e^{-\lambda_2(t-T)} \tag{4.44}$$

Figure 4.15 shows such a result for the ^{38}Cl example in Fig. 4.14. Here, we have removed the source from the reactor after $T = 3t_{1/2} \sim 111.6$ m. Making use of the source shortly after removal from the reactor is of obvious benefit.

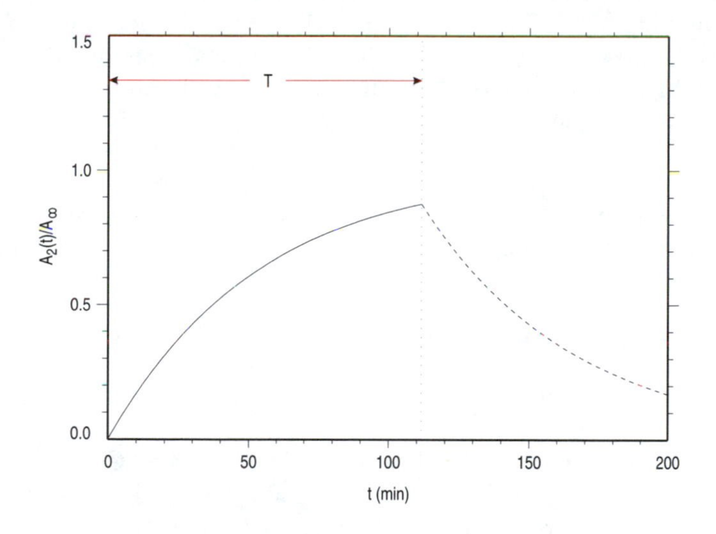

Figure 4.15: Post-irradiation decay after $3t_{1/2}$ for ^{38}Cl ($t_{1/2} = 37.2$ min).

4.6 Radiodating

Determining the age of relics and geological samples is yet another very important application of radioisotopes. The progression of a nuclear decay chain is straightforward to predict, as we have seen. Since the alteration of radionuclide inventories is governed by immutable rate constants, sample ages can be very accurately gauged.

Example 27: Neutron Production and Decay of ^{60m}Co

A researcher needs a ^{60m}Co (IT, IC) source for an X-ray scattering experiment. She has access to a research reactor with $\varphi_o = 10^{12}\ \frac{\mathrm{n}}{\mathrm{cm^2 s}}$. If the half-life of ^{60m}Co is 10.47 min and $\sigma_\gamma = 20.4$ b for stable ^{59}Co, what is the maximum specific activity for this photon source, the total activity for a 1 g sample of ^{59}Co after 30 min irradiation, and the activity of this sample 20 min after removal from the reactor?

Solution:

Since $\lambda_2 = 1.1 \times 10^{-3}\ \mathrm{s}^{-1}$ and $\sigma_\gamma \varphi_o = 2.04 \times 10^{-11}\ \mathrm{s}^{-1}$, we may use the simplified expressions. The maximum specific activity occurs at saturation

$$A_{s_\infty} = \sigma_\gamma \varphi_o \mathcal{N}_{1_o}/m = \sigma_\gamma \varphi_o N_A/M \sim 5.6\ \mathrm{Ci/g}$$

The saturation activity is $A_\infty \sim 5.6$ Ci for a 1 g sample. After 30 min

$$A_2(30\ \mathrm{min}) = A_{s_\infty}\left(1 - e^{-(30\ \mathrm{min})\lambda_2}\right) = 4.83\ \mathrm{Ci}$$

and at 20 min past removal

$$A_2(50\ \mathrm{min}) = A_2(30\ \mathrm{min})e^{-(20\ \mathrm{min})\lambda_2} = 1.285\ \mathrm{Ci}$$

Firstly, let's consider the simple closed system

$$X_1 \xrightarrow{\lambda_1} X_2 \text{ (stable)} + y + \cdots$$

With no initial inventory of daughter product X_2, nuclide conservation yields

$$\mathcal{N}_1(t) + \mathcal{N}_2(t) = \mathcal{N}_{1_o}$$

From this, it is a trivial matter to determine the sample age, t. Since $\mathcal{N}_{1_o} = \mathcal{N}_1(t)e^{\lambda_1 t}$ then

$$t = \frac{1}{\lambda_1} \ln \left[1 + \frac{\mathcal{N}_2(t)}{\mathcal{N}_1(t)} \right] \tag{4.45}$$

where the present concentration ratio, $\mathcal{N}_2(t)/\mathcal{N}_1(t)$ may be determined by mass spectroscopy or some chemical means. It is imperative that the system be closed so that only nuclear decay modifies the relative concentration.

In the more general case where $\mathcal{N}_{2_o} \neq 0$, nuclide conservation now reads

$$\mathcal{N}_1(t) + \mathcal{N}_2(t) = \mathcal{N}_{1_o} + \mathcal{N}_{2_o}$$

We now have too many unknowns in the problem to directly solve for the sample age. One unknown may be eliminated, however, if additional information on initial inventory is provided. Such information, in its most useful form, appears as a known concentration ratio between the immediate daughter product, X_2, and its stable isotope, X_2'. Such relative abundances are well known and extremely consistent. Any deviation then is a result of nuclear decay at a predictable rate. If X_2' is stable and not produced in the decay of some other radionuclide in the sample, then, for a closed system

$$\mathcal{N}_2'(t) = \mathcal{N}_{2_o}'$$

and we may write

$$\frac{\mathcal{N}_1(t) + \mathcal{N}_2(t)}{\mathcal{N}_2'(t)} = \frac{\mathcal{N}_{1_o} + \mathcal{N}_{2_o}}{\mathcal{N}_{2_o}'}$$

from which we can infer the sample age

$$t = \frac{1}{\lambda_1} \ln \left[1 + \frac{\mathcal{N}_2'(t)}{\mathcal{N}_1(t)} \left(\frac{\mathcal{N}_2(t)}{\mathcal{N}_2'(t)} - \frac{\mathcal{N}_{2_o}}{\mathcal{N}_{2_o}'} \right) \right] \tag{4.46}$$

The concentration ratios $\frac{\mathcal{N}_2'(t)}{\mathcal{N}_1(t)}$ and $\frac{\mathcal{N}_2(t)}{\mathcal{N}_2'(t)}$ are measured at the present time, while the primordial (or primitive) relative isotopic abundance $\frac{\mathcal{N}_{2_o}}{\mathcal{N}_{2_o}'}$ is assumed to be identical with that at present.

Example 28: Radiodating with ^{232}Th

Long-lived ^{232}Th ($t_{1/2} = 1.4 \times 10^{10}$ y) makes an excellent geological radiodating source. The thorium series (beginning with ^{232}Th) is a long decay chain to the stable ^{208}Pb. However, since all of the intermediate products are very short lived compared to ^{232}Th, secular equilibrium is quickly achieved between the parent and all radioactive daughter products. The final stable product can be treated as if it were the immediate daughter of ^{232}Th.

A geological sample (a rock) is found to contain 13.7 g of ^{232}Th and 3.1 g of ^{208}Pb. If we assume no initial lead inventory when the rock was formed, what is the age of the sample?

Solution:

We must further assume that the rock formation entraps all primordial components so that we are guaranteed a closed system, a good assumption. Since the sample contains at present 13.7 g of ^{232}Th or $\mathcal{N}_1(t) \sim 3.55 \times 10^{22}$ atoms and 3.1 g of ^{208}Pb or $\mathcal{N}_2(t) \sim 8.98 \times 10^{21}$ atoms, then we may use Eq.(4.45) with $\lambda_1 \sim 4.95 \times 10^{-11}$ y^{-1} to find

$$t \sim \frac{1}{4.95 \times 10^{-11}\ \mathrm{y}^{-1}} \ln\left[1 + \frac{8.98 \times 10^{21}}{3.55 \times 10^{22}}\right] \sim 4.56 \times 10^{9}\ \mathrm{y}^{-1}$$

This is very close to the accepted age of earth. This rock was formed with the earth.

Example 29: Radiodating with ^{14}C

^{14}C ($t_{1/2}$ = 5730 y) is an excellent radionuclide for dating the more recently deceased. It is continuously formed in the upper atmosphere by cosmic ray bombardment on atmospheric nitrogen. Organic matter takes up ^{14}C through injection of CO_2 and maintains an equilibrium concentration through biological elimination of about one ^{14}C atom for every 10^{12} atoms of ^{12}C. When the organism dies, the ^{14}C concentration diminishes through nuclear decay.

An ancient sample is found to have a present day ^{14}C specific activity of 1.6 pCi/g of carbon. If modern samples of similar origin have an equilibrium specific activity of 7.5 pCi/g, determine the sample age.

Solution:

If we assume that the ancient sample had the same equilibrium activity as do modern samples, then

$$A_s(t) = A_{s_o} e^{-\lambda t}$$

so that

$$t = \frac{1}{\lambda} \ln \frac{A_{s_o}}{A_s(t)} = \frac{5730}{\ln 2} \ln \frac{7.5}{1.6} \sim 12,774 \text{ y}$$

References

[Browne86] Browne, E., Firestone, R. B., and Shirley, V. S., *Table of Radioactive Isotopes*, John Wiley & Sons, Inc., NY, 1986.

[Eisberg85] Eisberg, R., and Resnick, R., *Quantum Physics of Atoms, Molecules, Solids, Nuclei, and Particles*, John Wiley & Sons, NY, 1985.

[Evans55] Evans, R. D., *The Atomic Nucleus*, McGraw Hill, Inc., NY, 1955.

[Firestone96] Firestone, R. B., Shirley, V. S., Baglin, C. M., Chu, S. Y. F., and Zipkin, J., *Table of Isotopes*, 8^{ed}, John Wiley & Sons, Inc., NY, 1996.

[GE96] General Electric Co., *Nuclides and Isotopes*, 15^{ed}, 1996.

[Knoll89] Knoll, G. F., *Radiation Detection and Measurement*, 2^{ed}, John Wiley & Sons, NY, 1989.

[Krane88] Krane, K. S., *Introductory Nuclear Physics*, John Wiley & Sons, NY, 1988.

Problems and Questions

1. For decay by proton emission:
   (a) Symbolically describe the decay process in an arbitrary radionuclide.
   (b) What is the mass condition for spontaneous decay (*i.e.* $Q_p > 0$)?
   (c) Determine the expressions for the reaction product kinetic energies.

2. Symbolically describe the positron decay process for an arbitrary positron emitting radionuclide. What is the mass condition for spontaneous decay (*i.e.* $Q_{\beta+} > 0$)? Repeat for electron capture and describe any similarities or differences in the mass conditions.

3. For electron capture (EC) decay:
   (a) Symbolically describe the decay process in an arbitrary radionuclide.
   (b) What is the mass condition for spontaneous decay in electron capture (*i.e.* $Q_{EC} > 0$)?

4. Consider the α decay of ${}^{238}_{92}\text{U}$:
   (a) Write an energy balance for the decay.
   (b) Determine the value of Q_α for this decay.

5. Determine the unknown in each of the following decay equations, and calculate the Q-value:
   (a) ${}^{36}_{17}\text{Cl} \longrightarrow \text{X} + \beta^- + \overline{\nu}$
   (b) ${}^{26}_{13}\text{Al} \longrightarrow \text{X} + \beta^+ + \nu$
   (c) ${}^{241}_{95}\text{Am} \longrightarrow {}^{237}_{93}\text{Np} + \text{y}$
   (d) ${}^{73}_{35}\text{Br} \longrightarrow {}^{72}_{34}\text{Se} + \text{y}$
   (e) ${}^{137}_{54}\text{Xe} \longrightarrow \text{X} + {}^{1}_{0}\text{n}$

6. Determine the Q_n-value and product kinetic energies for the following reaction
$$^{138}\text{Xe} \longrightarrow {}^{137}\text{Xe} + \text{n}$$

7. In the following α decay reaction
$$^{226}\text{Ra} \longrightarrow {}^{222}\text{Rn} + {}^{4}\text{He}$$
determine the Q_α value, the kinetic energies of the products, KE_{222} and KE_α, and the velocities of the products, v_{222} and v_α.

8. A plasma discharge device is invented to purify polluted air by ionizing and filtering it. This device requires an electrical input power of at least 40 kW continuously for its useful lifetime of 40 y. It is proposed to power this air cleaner with the disintegration energy from the following reaction

$$^{238}_{94}\text{Pu} \longrightarrow {}^{234}_{92}\text{U} + {}^{4}_{2}\text{He}$$

which has a half-life of 87.7 y. If this decay reaction is the only source of energy available to this device, determine the initial mass of $^{238}_{94}$Pu required so that it has sufficient power for its entire useful lifetime.

9. Consider the α decay of ^{238}Pu ($t_{1/2} = 87.7$ y). Determine the activity per gram of ^{238}Pu and if the Q-value is 5.6 MeV, determine the power output per gram of ^{238}Pu.

10. A space based nuclear power station employs the α decay of ^{244}Cm as its source of energy. Determine:

    (a) the α and product energies in each disintegration,

    (b) the original mass (in kg) and activity (in Ci) of the ^{244}Cm fuel required to provide continuous power at a level of at least 100 kW for 100 y, and

    (c) the mass (in kg) and activity (in Ci) of the ^{244}Cm daughter product after 100 y.

11. Your new space probe requires 10 kW of continuous electrical power to perform its exploration function. As the nuclear engineer on the project, you are asked to evaluate the suitability of employing a radioactive material as the power source. The β^- emitter tritium ($\text{T} \longrightarrow {}^{3}\text{He} + \beta^- + \overline{\nu}$) is a candidate since it can be produced in a nuclear reactor by irradiating deuterium with neutrons ($\text{D} + \text{n} \longrightarrow \text{T} + \gamma$). From the decay process, you may extract $\langle \text{KE}_{\beta^-} \rangle \sim \frac{1}{4} \text{KE}_{\beta^-_{\text{max}}}$. If the half life for tritium is 12.3 y and you have access to a research reactor that has agreed to irradiate your sample for 1 y continuously at constant conditions, how much pure deuterium mass (kg) must you irradiate so that your probe has enough continuous power for a 2 y mission? [Hint: you may employ the simplified form of the activation equation, and $\sigma = 5.2 \times 10^{-28}$ cm^2, $\varphi_o = 10^{15}$ cm^{-2} s^{-1}.]

12. Determine the original mass (in grams) and activity (in Ci) of a pure sample of ^{238}Pu ($t_{1/2} = 87.7$ y) required to power a 10 kW satellite for 100 y if all of the alpha particle disintegration energy can be converted to electrical power.

13. A 10 mg sample of radioactive material has an initial specific activity of 3 mCi/g. If its half life is 38 h:

    (a) How many radioactive nuclei are remaining after 3 d?

    (b) What is the total number of radioactive decays in this 3 d interval?

14. Calculate the activity of 1 mg of ^{226}Ra in mCi. Determine the mass needed to obtain the same amount of activity for the following nuclei: $^{241}_{95}$Am, $^{238}_{94}$Pu, $^{238}_{92}$U, $^{224}_{88}$Ra, and $^{3}_{1}$H.

15. Consider the following β^- decay

$$^{40}_{19}\mathrm{K} \overset{t_{1/2}=1.28\times 10^9\ \mathrm{y}}{\longrightarrow} {}^{40}_{20}\mathrm{Ca\ (stable)} + \beta^- + \overline{\nu}$$

    (a) Determine the initial activity (in curies) of $^{40}_{19}$K in 10 kg of natural potassium.

    (b) Determine the number of $^{40}_{19}$K and $^{40}_{20}$Ca nuclei present after $1.623t_{1/2}$. (Assume initially 10 kg of natural potassium and no $^{40}_{20}$Ca is initially present.)

16. For the following nuclei, find their half-lives and mean-lives: $^{131}_{54}$Xe, $^{173}_{72}$Hf, $^{181}_{74}$W, $^{210}_{84}$Po, and $^{237}_{92}$U.

17. Consider the following branching decay scheme

$$^{254}_{99}\mathrm{Es} \longrightarrow \left\{ \begin{array}{ll} \overset{t_{1/2}=1.64\ \mathrm{d}}{\longrightarrow} & ^{254}_{100}\mathrm{Fm} + \beta^- + \overline{\nu} \\ \overset{t_{1/2}=276\ \mathrm{d}}{\longrightarrow} & ^{250}_{97}\mathrm{Bk} + \alpha \end{array} \right.$$

    (a) Determine the composite decay constant.

    (b) Find $\mathcal{N}_{\mathrm{Fm}}(t)$ and $\mathcal{N}_{\mathrm{Bk}}(t)$ at $t = 3$ days.

    (c) Given an initial activity of 1.8×10^7 dis/s, plot the total activity and the activities due to $^{254}_{100}$Fm and $^{250}_{97}$Bk independently, as a function of time until the total activity is one tenth of its initial value.

18. Discuss conditions for the measurement (*i.e.* method, measurement times) of half-life for the following radioisotopes: (a) ^{27}Mg, (b) ^{3}H, and (c) ^{238}U.
    (Hint: examine the linear approximation to the exponential decay law for reasonable measurement times (minutes to hours).)
    EXTRA CREDIT: Discuss an experiment to measure the half-life of a very short-lived radioisotope like ^{8}Li.

19. Show that, for a binary decay problem, that the sum $A_1(t) + A_2(t)$ goes through a maximum, and calculate the time of the maximum activity.

20. Show that for the binary decay problem, $\mathcal{N}_1(t) + \mathcal{N}_2(t) + \mathcal{N}_3(t) = \mathcal{N}_1(0)$ for all t.

21. Explain what is meant by "secular equilibrium" and "transient equilibrium."

22. Consider the following decay scheme

$$^{233}_{94}\text{Pu} \overset{t_{1/2}=20.9 \text{ min}}{\longrightarrow} {}^{233}_{93}\text{Np} \overset{t_{1/2}=36.2 \text{ min}}{\longrightarrow} {}^{233}_{92}\text{U}$$

There are initially 10 g of $^{233}_{94}$Pu and no initial inventory of $^{233}_{93}$Np and $^{233}_{92}$U.

   (a) Find the time when the sum of the activities of $^{233}_{94}$Pu and $^{233}_{93}$Np is at a maximum.

   (b) How many grams of $^{233}_{92}$U will there be when the sum of the activities of $^{233}_{94}$Pu and $^{233}_{93}$Np reaches a maximum?

23. Consider the binary decay system

$$\text{X}_1 \xrightarrow{\lambda_1} \text{X}_2 \xrightarrow{\lambda_2} \text{X}_3 \text{ (stable)}$$

   (a) Write down the set of differential equations and boundary conditions required to solve this system.

   (b) Describe the behavior of these solutions in secular equilibrium. Namely, (i) what are the conditions on the decay constants and time interval such that secular equilibrium is achieved and (ii) what is the total activity in secular equilibrium?

24. For the binary decay chain

$$\text{X}_1 \xrightarrow{\lambda_1} \text{X}_2 \xrightarrow{\lambda_2} \text{X}_3 \text{ (stable)}$$

   (a) Find the maximum activity of X_2 in terms of the initial activity of X_1 ($A_{1_\circ}$) with the conditions $\mathcal{N}_1(t = 0) = \mathcal{N}_{1_\circ}$, $\mathcal{N}_{2_\circ} = 0$, and $\lambda_1 \neq \lambda_2$.

   (b) If $\lambda_2 = 3\lambda_1$ (transient equilibrium), what is the numerical value of the ratio of $A_2\,|_{\text{max}}\,/A_{1_\circ}$?

25. Consider the following open system where $\mathcal{P}_{1,2,3}(t)$ represent terms producing isotopes 1, 2, and 3 that are also related by decay.

$$\text{X}_1 \xrightarrow{\lambda_1} \text{X}_2 \xrightarrow{\lambda_2} \text{X}_3 \xrightarrow{\lambda_3} \text{X}_4 \text{ (stable)}$$

(a) Set up the differential decay equations for $\mathcal{N}_1(t)$, $\mathcal{N}_2(t)$, $\mathcal{N}_3(t)$ and $\mathcal{N}_4(t)$.

(b) If $\mathcal{N}_{1_o} = \mathcal{N}_{2_o} = \mathcal{N}_{3_o} = \mathcal{N}_{4_o} = 0$, $\mathcal{P}_1(t) = \mathcal{P}_{1_o}$ (a constant), $\mathcal{P}_2(t) = \mathcal{P}_{1_o} e^{-\lambda_o t}$ and $\mathcal{P}_3(t) = kt$ where k and λ_o are constants, find explicit expressions for $\mathcal{N}_1(t)$ and $\mathcal{N}_2(t)$.

26. Consider the following closed system

$$^{214}_{83}\mathrm{Bi} \overset{t_{1/2}=19.9\ \mathrm{min}}{\longrightarrow} {}^{210}_{81}\mathrm{Tl} \overset{t_{1/2}=1.3\ \mathrm{min}}{\longrightarrow} {}^{210}_{82}\mathrm{Pb}$$

with the following initial conditions: $\mathcal{N}_{\mathrm{Bi}}(t=0) = \mathcal{N}_{1_o}$, $\mathcal{N}_{\mathrm{Tl}}(t=0) = 0.8\mathcal{N}_{1_o}$, and $\mathcal{N}_{\mathrm{Pb}}(t=0) = 0$.

(a) Find expressions for $\mathcal{N}_{\mathrm{Bi}}(t)$, $\mathcal{N}_{\mathrm{Tl}}(t)$, and $\mathcal{N}_{\mathrm{Pb}}(t)$.

(b) Find the time that $\mathcal{N}_{\mathrm{Bi}}(t) = \mathcal{N}_{\mathrm{Pb}}(t)$. (Note: This will require root solving of a transcendental equation!)

27. Consider the following open system

$$\mathcal{P}_1(t) \longrightarrow \mathrm{X}_1 \overset{\lambda_1}{\longrightarrow} \mathrm{X}_2 \overset{\lambda_2}{\longrightarrow} \mathrm{X}_3 \text{ (stable)}$$

with the initial conditions: $\mathcal{N}_1(t=0) = \mathcal{N}_{1_o}$ and $\mathcal{N}_{2_o} = \mathcal{N}_{3_o} = 0$. Find expressions for $\mathcal{N}_1(t)$ and $\mathcal{N}_2(t)$ using the following production expressions:

(a) Step Function: $\mathcal{P}_1(t) = \mathcal{P}_o =$ constant.

(b) Ramp Function: $\mathcal{P}_1(t) = kt$, where k is a constant.

(c) Exponential Function: $\mathcal{P}_1(t) = \mathcal{P}_o e^{-kt}$, where $\mathcal{P}_o$ and k are constants and $k \neq \lambda_1$.

(d) Sinusoidal Input: $\mathcal{P}_1(t) = \mathcal{P}_o + a\sin(kt)$, where $\mathcal{P}_o$, a, and k are constants such that $\mathcal{P}_o \geq |a|$.

28. For the binary decay problem with $\mathcal{N}_{2_o} = \mathcal{N}_{3_o} = 0$, plot the following normalized quantities on a semi-log scale: $\frac{\mathcal{N}_1(t)}{\mathcal{N}_{1_o}}$, $\frac{\mathcal{N}_2(t)}{\mathcal{N}_{1_o}}$, $\frac{\mathcal{N}_3(t)}{\mathcal{N}_{1_o}}$, $\frac{A_1(t)}{A_{1_o}}$ and $\frac{A_2(t)}{A_{1_o}}$. These should be shown in a plot vs. $\lambda_2 t$ for the three cases below.

(a) $\lambda_1 = 2\lambda_2$

(b) $\lambda_2 = 2\lambda_1$ (transient equilibrium)

(c) $\lambda_2 = 100\lambda_1$ (secular equilibrium)

You can display all the activities on the same graph and all the inventories on the same graph (for a total of six plots). In the spirit of maintaining some sense of uniformity, please make $\lambda_1 = 1.0$ and run your plots from $0 < \lambda_2 t < 10$ for the first two cases and $0 < \lambda_2 t < 100$ for the third case. Also, calculate the times $t_a = t_{\max}$ and t_n, and identify these on your plots.

29. Consider the following open binary decay system

$$\mathcal{P}_1(t) \longrightarrow \mathrm{X}_1 \xrightarrow{\lambda_1} \mathrm{X}_2 \xrightarrow{\lambda_2} \mathrm{X}_3 \text{ (stable)}$$

    (a) Write down the set of differential equations and initial conditions required to solve this general system.

    (b) Find $\mathcal{N}_1(t)$ for $\mathcal{P}_1(t) = \mathcal{P}_o(kt)^2$ and an initial inventory $\mathcal{N}_{1_o}$.

30. Derive solutions for $\mathcal{N}_1(t)$, $\mathcal{N}_2(t)$, and $\mathcal{N}_3(t)$ for the binary decay problem when $\mathcal{N}_{1_o} = \frac{2}{3}\mathcal{N}_o, \mathcal{N}_{2_o} = \frac{1}{3}\mathcal{N}_o$, and $\mathcal{N}_{3_o} = 0$. Check nuclide conservation in your final result.

31. Using your nuclide chart, find one good example of a binary decay chain exhibiting: (a) transient equilibrium, (b) secular equilibrium. Plot $A_1(t)$ and $A_2(t)$ for each on a semi-log scale. For consistency, make $\mathcal{N}_{1_o} = 1.0 \times 10^{20}$ atoms and plot from $t = 0$ to $t = 6/\lambda_2$ s.

32. In a given neutron activation, we irradiate a sample for a time equal to four half-lives of the desired radioactive isotope. Following irradiation, the sample is removed from the reactor and is put to use only after another three half-lives have expired. How much activity remains in the sample as a fraction of the saturation activity? (Hint: you may use the simplified form of the activation equation.)

33. A particular neutron activation experiment involves irradiating a pure, thin ^{55}Mn target (1 mg) in the beam tube of a research reactor to produce the β^- emitter ^{56}Mn ($t_{1/2} = 2.6$ h). The thermal neutron flux at the thin target is $10^8 \frac{\mathrm{n}}{\mathrm{cm^2\,s}}$, and the (n,$\gamma$) cross section for ^{55}Mn is 13.3 b. (a) What is the saturation activity and inventory of ^{56}Mn, assuming no depletion and $\sigma_\gamma \varphi_o \ll \lambda_2$? (b) The experiment is run long enough so that the ^{56}Mn concentration is in saturation. Sometime later, the reactor flux suddenly jumps to five times its original value. Find the new ^{56}Mn inventory as a function of time after the jump. (Again assume no depletion and $\sigma_\gamma \varphi_o \ll \lambda_2$.) (c) What is the new saturation activity in dis/s? (d) Sketch a plot of the ^{56}Mn inventory as a function of time after the flux change on a linear scale showing the initial and final inventory, and time scale.

34. An initially pure 14 g sample of ^{134}Xe ($\sigma_\gamma = 0.2$ b) is irradiated in a constant neutron flux of $10^{13} \frac{\text{n}}{\text{cm}^2\,\text{s}}$ to produce the β^- emitter ^{135}Xe ($t_{1/2} = 9.14$ h, $\sigma_\gamma = 2.7 \times 10^6$ b). Since ^{135}Xe has such a large (n,γ) cross section, its inventory is depleted by absorption as well as decay. Write down the balance equations describing the inventory of ^{134}Xe and ^{135}Xe as a function of time. Find the inventory of ^{134}Xe and ^{135}Xe as a function of time for the initial condition above. If we assume there is no depletion of the initial sample (*i.e.* $\sigma_\gamma^{134}\varphi t \ll 1$), what is the saturation activity of ^{135}Xe in Ci?

35. A sample of radioactive tritium (^{3}H) is produced in a research reactor operating at low power such that the thermal flux is $\sim 10^{12} \frac{\text{n}}{\text{cm}^2\,\text{s}}$. Since the post-irradiation handling time is 4 min and activity is limited to 10 μCi, what is the maximum irradiation time for a 1 g sample of D_2O? (Data: Tritium half-life is 12.3 y, (n,γ) cross section for D is 0.52 mb).

36. The radioisotope ^{132}Te is produced in nuclear reactors as a product of fission. The ^{235}U cumulative fission yield (fraction of fission events that lead to the production of this isotope (sec. 5.4.3)) for ^{132}Te is 4.31%, and $t_{1/2} = 3.26$ d.

    (a) What is the specific activity of ^{132}Te at saturation in Ci/l? ($\varphi_o = 10^{15} \frac{\text{n}}{\text{cm}^2\,\text{s}}$ and $\Sigma_f = 0.2$ cm^{-1}) (Assume that there is no depletion of ^{132}Te by neutron activation.)

    (b) What is the time required to reach 95% of saturation?

    (c) What is the specific activity of ^{132}Te in Ci per liter 1 month after reactor shutdown? (Assume that ^{132}Te has reached saturation before shutdown.)

37. The β^- emitter ^{87}Rb ($t_{1/2} = 4.88 \times 10^{10}$ y) produces the stable daughter ^{87}Sr. Since it has a very long half-life, it is often used in the radiodating of geological samples. If a rock sample is retrieved that has a present day composition $\mathcal{N}_{^{87}\text{Rb}}/\mathcal{N}_{^{86}\text{Sr}} = 1.5$ and $\mathcal{N}_{^{87}\text{Sr}}/\mathcal{N}_{^{86}\text{Sr}} = 0.8$, and the relative abundance of the two stable strontium isotopes is $\mathcal{N}_{^{87}\text{Sr}_o}/\mathcal{N}_{^{86}\text{Sr}_o} = 0.7$, determine the age of the sample in years.

Chapter 5

Fission and Fusion

5.1 Fission Introduction

Nuclear fission is arguably the most important of all nuclear reactions. As discussed in chapter 1, much of the world's electrical power is derived by exploiting a chain of such events in a controlled way. We will briefly examine some of the basic concepts on controlling the fission reaction chain in sec. 5.5. A more complete treatment is the subject of a later course in reactor physics. Our immediate concern, in keeping with the theme of this book, is in describing the basics of the nuclear and atomic physics in the fission process and of its end products.

Credit for the discovery of fission was given to the german chemist Otto Hahn who received the 1944 Nobel prize in chemistry for his work on fission. His 1939 paper in the scientific journal, *Nature*, with Fritz Strassmann was the result of years of joint work with physicist Lise Meitner. Their interpretation of the splitting process was verified when they discovered elements lighter than uranium in their experiments. Yet it was Meitner and Frisch who, later that same year (1939), gave us the timeless terminology "nuclear fission" when they provided a theoretical description based on Bohr's liquid drop model (sec. 1.6.6.1) and detected the energetic products of the fission reaction.

Since this time of discovery, nuclear fission has enjoyed a rather spectacular history; part notorious, part sensationalized, but for the most part, an environmentally benign, safe, and extremely valuable as an almost inexhaustible source of energy with tremendous untapped potential.

5.2 Spontaneous Fission

The fission reaction is marked by the splitting or "fissioning" of a heavy nuclide into two or more lighter nuclides, sub-atomic particles, and energy. These reactions can occur spontaneously

$$\mathrm{X} \xrightarrow{\mathrm{SF}} \mathrm{Y}_1 + \mathrm{Y}_2 + y_1 + y_2 + \cdots$$

as yet another decay mode (labeled SF in the nuclide chart), although at a rather slow rate. (The partial half-life for SF in ^{238}U is $\sim 6.5 \times 10^{15}$ y.) Since there are more than two products from this nuclear decay reaction, we cannot uniquely identify the product energies from mass/energy and momentum balance. Though we can, as always, discuss the net energy released.

From the BE/A curve (Fig. 1.5), it was argued that SF is at least energetically possible for nuclides with $A > 60$. Since the liquid drop model (Eq.(1.22)) provides a reasonable nuclear model explaining the gross trends in nuclear binding, let's exploit it to make some general quantitative predictions regarding fissioning nuclei. Consider first our condition for symmetric SF

$$2\mathrm{BE}_\mathrm{Y} > \mathrm{BE}_\mathrm{X}$$

then Eq.(1.22), with A as the atomic number of X so that $A_\mathrm{Y} = A/2$, $Z_\mathrm{Y} = Z/2$, $N_\mathrm{Y} = N/2$, yields

$$a_s A^{2/3}\left(1 - 2^{1/3}\right) + \frac{a_c Z}{A^{1/3}}\left[(Z-1) - 2^{1/3}\left(\frac{Z}{2} - 1\right)\right] + \frac{a_p}{A^{3/4}}\left(2^{3/4} - 1\right) > 0$$

for even-even nuclei. When A is large, the pairing term is small with respect to the Coulomb and surface terms so that we see a direct competition between the Coulomb energy and surface tension

$$\frac{Z^2}{A} > \frac{a_s(1 - 2^{1/3})}{a_c(1 - 2^{-2/3})} \sim 15.22$$

for disintegration by SF, as we would expect for a charged, incompressible drop. If we estimate $Z = Z^* \sim A/2$ as the isobar with minimum mass (Eq.(1.24)), then the above condition reduces to the energetic possibility of SF when $A \gtrsim 61$ from the liquid drop model. The data of Fig. 1.5 suggest $A \gtrsim 85$ for SF. The difference is due to slight discrepancies between the model and the data in this region of rapidly changing curvature.

The above spontaneity condition refers to any energy release. Observations indicate that SF nuclides like ^{238}U release quite a bit more than the minimum. Indeed, the symmetric SF in ^{238}U to $A/2 = 119$ increases BE/A from about 7.57 MeV/nucleon to about 8.5 MeV/nucleon, representing a difference of $(8.5-7.59)\frac{\text{MeV}}{\text{nucleon}}\times 238$ nucleons ~ 221 MeV. From liquid drop, SF should release greater than 200 MeV when $2\text{BE}_\text{Y} - \text{BE}_\text{X} > 200$ MeV. We then predict

$$a_c(1-2^{-2/3})\frac{Z^2}{A} > a_s(2^{1/3}-1) + \frac{200}{A^{2/3}}$$

which yields as a result, again for $Z = A/2$, a SF release of > 200 MeV when $A > 175$. Asymmetric SF alters these results only slightly. The energy released in the above is just an expression of $Q_{\text{l.d.}}$ for this reaction, so that in general

$$\begin{aligned} Q_{\text{l.d.}} &= -a_s(2^{1/3}-1)A^{2/3} + a_c(1-2^{-2/3})\frac{Z^2}{A^{1/3}} \\ &= -3.38A^{2/3} + 0.222Z^2/A^{1/3} \end{aligned}$$

where $Q_{\text{l.d.}}$ indicates the liberated energy as predicted by the liquid drop model to distinguish it from that predicted from mass data, a quantity we'll later call Q_f. For ^{238}U ($A = 238, Z = 92$), $Q_{\text{l.d.}} = 173.4$ MeV, a bit lower (by about %15) than that predicted from the change in BE/A. The difference is a result of the symmetric product, ^{119}Pd in this case, being far too neutron rich to lie in the valley of stability, the only place the model is valid. Many subsequent β^- decays, contributing to Q_f, are required to produce a final stable product.

Although SF is energetically possible for $A \gtrsim 85$ and releases more than 200 MeV when $A \gtrsim 175$, only very few radionuclides participate in this mode of decay. There are two reasons. For heavy nuclides, $Z > 94$ (Pu), SF is a rapid form of decay, explaining their absence in nature. Furthermore, SF is observed for $Z > 90$ (Th) only. All lighter nuclei are stable against SF. It is only this very narrow window of radionuclides found in nature that naturally participate in SF. We are, nonetheless, fortunate for this consequence of nature. It is within and around this window that we find our nuclear fission fuels (sec. 5.3).

Within the context of the liquid drop model, stability against SF is determined by direct competition of the repulsive Coulomb force and surface tension with minor (but important) contribution from pairing. To predict a SF stability condition, envision the inverse process. Consider two charged spheres initially at $r = \infty$ with charge $Z/2$ and mass $A/2$ each. The inverse fission process requires a nuclear reaction to "fuse" the reactants. The condition for inhibition of this fusion is $B > Q_{\text{l.d.}}$, where B is the familiar Coulomb barrier, Eq.(2.16). Then, nuclei are stable against SF when

$$0.173\frac{Z^2}{A^{1/3}} > -3.38A^{2/3} + 0.222Z^2/A^{1/3}$$

Again allowing $Z \simeq A/2$, stability against SF is predicted for $A < 270$, larger than A for any known nuclide. It is, therefore, very difficult for any nuclide to overcome B and participate in SF. Those that do participate must quantum-mechanically tunnel through the barrier established by the Coulomb potential. Figure 5.1 illustrates this situation. Here, the potential energy for a system of two nuclei is depicted as a function of the distance of separation between their centers. In this hypothetical example meant to represent symmetric fission in ^{238}U, each nucleus has a mass of $A/2 = 119$ and $Z/2 = 46$. Since the nuclei are of identical composition, the distance $r = 2R$ (where R is the nuclear radius of each as per Eq.(1.13)) is that separation at which undeformed spherical nuclei just touch. At greater separation ($r \gtrsim 3R$), they exist as two individual, spherical particles mutually repelled by the Coulomb potential with the reference of potential zero at $r = \infty$. The dotted curve shows the continuation of the Coulomb potential at smaller separation. However, when $r \lesssim 2.7R$, nuclear attractive forces are influential, reducing the effect of the Coulomb potential alone. At still smaller separation, $r < 2R$, the two nuclides begin to coalesce under the increasing influence of the nuclear potential. As the two nuclei coalesce into one, surface energy is reduced. In the context of the liquid drop, energy is minimized by reducing the surface area so that one large drop is preferred over two. The potential energy then asymptotically reaches the energy of symmetric fission as $r \to 0$ (~ 173 MeV for ^{238}U).

The maximum on this curve (at $r \sim 2.0-2.5R$) is about 5.85 MeV above the fission energy and constitutes the fission barrier height. This barrier height is sometimes called E_{crit}, the critical energy for fission. In order to complete the fission process spontaneously, the fissioning nucleus must tunnel through this barrier. The successful quantum-mechanical energy redistribution required to do this is enormously improbable, explaining the tiny decay constants for SF nuclei found in nature. As such, SF is an impractical source of energy. As an example, the SF half-life of ^{238}U is $\sim 6.5 \times 10^{15}$ y, implying a specific activity of $\sim 730 \frac{\text{SF}}{\text{gd}}$. With $Q_f \sim 200$ MeV,a total ^{238}U mass of $\sim 10^{19}$ kg would be required to produce 3000 MW of fission power!

5.3 Induced Fission

In order to make fission practical as an energy source, energy exceeding E_{crit} must be supplied to the nucleus so that the Coulomb barrier may be more easily overcome rather than requiring an improbable tunneling event. Photons may provide this energy in the "photofission" process, (γ,f). It is more practical, however, to induce fission with the more easily captured neutron. As neutrons are a product of fission, a chain of neutron-induced

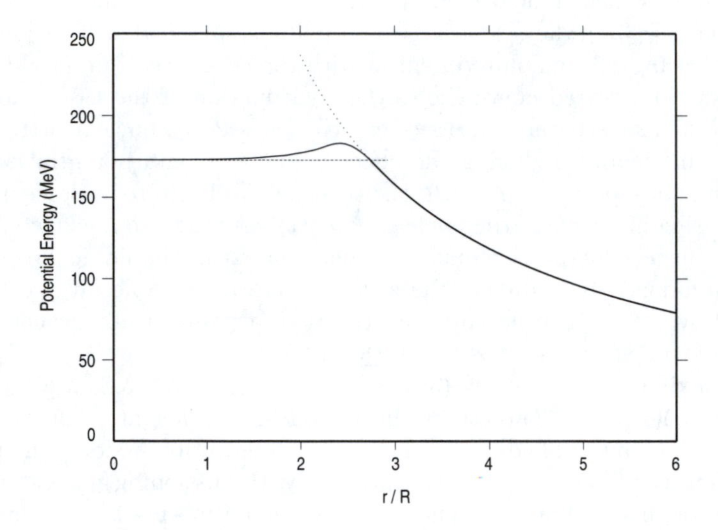

Figure 5.1: Potential energy diagram for two charged spheres with mass $A/2$ and charge $Z/2$ each, for $A = 238$ and $Z = 92$. The dotted curve represents an extrapolation of the Coulomb potential to smaller separation. The dashed line indicates the energy of symmetric fission, $Q_{\text{l.d.}}$.

fission reactions may be made possible (sec. 5.5). In addition to kinetic energy, massive particles like the neutron also bring nuclear binding energy to the compound nucleus upon absorption. It should be recalled that this single nucleon BE is just the separation energy, S_n from Eq.(1.19), which can contribute several MeV to the compound nucleus.

For neutron-induced fission

$$\mathrm{n} + \mathrm{X} \longrightarrow (\mathrm{n} + \mathrm{X})^* \longrightarrow \mathrm{Y}_1 + \mathrm{Y}_2 + y_1 + y_2 + \cdots$$

so that the Compound Nucleus (CN) excitation energy (cf: chpt. 2) is just $E^* \simeq S_n + \mathrm{KE}_n$, less a small correction for CN recoil. When $E^* \geq E_{\mathrm{crit}}$, fission is likely, although not guaranteed. Capture (n,γ) or charged particle production may compete with fission.

When $S_n \geq E_{\mathrm{crit}}$, neutrons of zero kinetic energy may induce fission. Low energy (thermal) neutron-induced fission is the basis for thermal reactors which provide most of the world's fission energy production. Neutrons from fission events are, of course, born at relatively high energies (up to several MeV, sec. 5.4.2) and must be slowed (thermalized) to low energy to enhance the likelihood of inducing further fission reactions. Such thermalization is done primarily on elastic scattering collisions as described in sec. 3.1. Nuclei that are induced to fission when S_n of the CN is above E_{crit} are called "fissile" isotopes. Otherwise, $\mathrm{KE}_n > 0$ is required, for which such nuclei are called "fissionable." An example best illustrates. The two most abundant isotopes of natural uranium are ^{235}U and ^{238}U. When the former absorbs a neutron to form the CN ^{236}U, the neutron brings $S_n = 6.54$ MeV, significantly greater than the 5.9 MeV barrier. The target nucleus, ^{235}U, is then said to be fissile. On the other hand, the last bound neutron in ^{239}U only brings 4.8 MeV in S_n, while E_{crit} is 6.2 MeV. Thus, ^{238}U is not fissile, but is fissionable. The difference between S_n and E_{crit} explains the ~ 1 MeV threshold for neutron-induced fission in ^{238}U.

Table 5.1 shows E_{crit} and S_n for several important fissioning nuclides. Note that although it is the $^{A+1}$X nucleus that splits in induced fission, it is the ^{A}X target nucleus that is referred to as fissile or fissionable. Note also that there is a general trend for odd A nuclei to be fissile in this chart. This is because of the pairing energy in the liquid drop model. Since these fissioning nuclides all have even Z and even N, the pairing energy is positive, and they have subsequently greater S_n.

Fissile nuclides can sometimes be produced from those that are not. One such example is the breeding of fissile ^{233}U from ^{232}Th in the (n,γ) reaction

$$\mathrm{n} + ^{232}\mathrm{Th} \longrightarrow \ ^{233}\mathrm{Th} + \gamma$$

and subsequent β^- decays

$$^{233}\mathrm{Th} \quad \longrightarrow \quad ^{233}\mathrm{Pa} + \beta^- + \overline{\nu} \quad (t_{1/2} = 22.3 \text{ min})$$

Table 5.1: Fissioning Nuclides

		Target Nucleus		Fissioning Nucleus	E_{crit} (MeV)	S_n (MeV)	Fissile Target?
n	+	^{232}Th	→	^{233}Th	6.4	4.8	fissionable
n	+	^{233}U	→	^{234}U	6.0	6.8	fissile
n	+	^{235}U	→	^{236}U	5.9	6.5	fissile
n	+	^{238}U	→	^{239}U	6.2	4.8	fissionable
n	+	^{239}Pu	→	^{240}Pu	5.9	6.5	fissile

$$^{233}\text{Pa} \longrightarrow {}^{233}\text{U} + \beta^- + \overline{\nu} \quad (t_{1/2} = 27 \text{ d})$$

Nuclides like ^{232}Th that can be converted to fissile nuclides by neutron absorption are called "fertile." Another important breeding reaction is shown in the following example.

Example 30: Fissile Nuclide Breeding
Show how fissile ^{239}Pu can be bred from ^{238}U.

Solution:
^{238}U absorbs neutrons in the (n,γ) reaction

$$\text{n} +^{238} \text{U} \longrightarrow {}^{239}\text{U} + \gamma$$

the product of which is β^- unstable, and the short chain

$$^{239}\text{U} \longrightarrow {}^{239}\text{Np} + \beta^- + \overline{\nu} \quad (t_{1/2} = 23.5 \text{ min})$$
$$^{239}\text{Np} \longrightarrow {}^{239}\text{Pu} + \beta^- + \overline{\nu} \quad (t_{1/2} = 2.36 \text{ d})$$

leads to the fissile ^{239}Pu.

5.4 Thermal Neutron-Induced Fission in ^{235}U

Although isotopes of all elements with $Z > 72$ (Hf) may participate in induced fission, only three (^{233}U,^{235}U,^{239}Pu) are fissile and sufficiently stable to serve as nuclear fuels. Of these three, only ^{235}U is available in nature, comprising 0.72% of natural U, making it the only useful nuclear fuel as a practical energy source without breeding.

Fission is a complex process occurring in many steps. Although the subject of a great number of studies since its discovery, fission remains not yet completely understood. What follows is a phenomenological description of the ontogeny of a fissioning ^{235}U nucleus.

Fission is a very violent nuclear reaction involving the entire nucleus. Thermal neutron-induced fission begins with the absorption of a thermal neutron to form the excited compound nucleus $^{A+1}\mathrm{X}^*$

$$\mathrm{n} + {}^{A}\mathrm{X} \longrightarrow {}^{A+1}\mathrm{X}^*$$

As mentioned in chapter 2, fission is an ideal example of a CN reaction. The projectile energy is low so that interaction with the entire target nucleus is possible, and the chance of the projectile escaping the target nucleus without appreciable energy loss is small. What's more, the target is of large mass so that it may absorb projectile energy without the average excitation energy per nucleon exceeding emission thresholds.

Decay of the excited state by fission competes only with γ decay

$$\mathrm{n} + {}^{235}\mathrm{U} \longrightarrow {}^{236}\mathrm{U}^* \longrightarrow {}^{236}\mathrm{U} + \gamma$$

for low energy neutrons. For thermal neutrons on ^{235}U, $\sigma_a = \sigma_f + \sigma_\gamma =$ 681 b and $\sigma_f = 582$ b. We then expect $\sigma_f/\sigma_a \sim 85\%$ of such CN interactions to result in the fission exit channel. The remainder are (n,γ) reactions to the ground state of the α emitter ^{236}U, an interesting nuclear reaction, but completely useless to the fission chain reaction.

Following CN formation, the $^{236}\mathrm{U}^*$ nucleus is left in a violently unstable state, undergoing large vibrations and deformations. In accord with the liquid drop model, we may paint a mental picture of the $^{236}\mathrm{U}^*$ nucleus analogous to that same electrically charged, incompressible drop now continuously elongating and contracting. The Coulomb and surface tension forces provide the driving and restoring forces, respectively. When these deformations reach a critical configuration, sometimes called the "saddle" because of its shape (Fig. 5.2), the surface force is no longer able to restore the nucleus to its original configuration. It is now energetically favorable for the deformed configuration to form two new smaller nuclei, called "primary fission fragments." The CN may persist for about 10^{-15}–10^{-14} s before reaching the critical configuration, while the moment of splitting called "scission" occurs abruptly thereafter, $\sim 10^{-20}$ s.

Beyond the scission point, the two primary fragments interact almost exclusively through the Coulomb interaction. They are repelled so strongly,

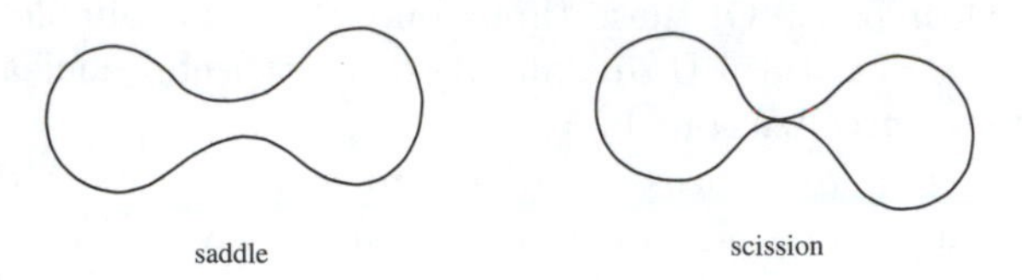

Figure 5.2: Saddle configuration and scission point.

in fact, that they incur multiple atomic ionization, leaving a net charge of $\sim +40e$ for the pair. We can then write as the net reaction to this point

$$\mathrm{n} + {}^{235}\mathrm{U} \longrightarrow \mathrm{X}_L^{+20} + \mathrm{X}_H^{+20} + 40\mathrm{e}^-$$

where the subscripts L and H refer to light and heavy primary fragments for the more probable case of asymmetric fission (sec. 5.4.3).

As they separate under Coulomb repulsion, the energetic primary fragments are in such a highly excited nuclear state from scission that neutron emission by evaporation occurs in about 10^{-17} s. Both fragments may participate to varying degrees. On average, a total of ν_p neutrons are emitted from both fragments on a single fission event. The symbol ν is commonly used in this context to refer to neutron numbers from fission. It is an unfortunate consequence of tradition that this symbol is repeated from our earlier use denoting the neutrino particle from beta decay. The context and subscript should make the distinction clear. The subscript p indicates that these neutrons are emitted "promptly" following scission and are called "prompt fission neutrons." The intent is to distinguish these from neutrons emitted on much longer time scales (up to about 1 min) called "delayed neutrons" from the slow radioactive decay of the unstable fission products. The number of delayed neutrons is given the symbol ν_d. In the thermal neutron-induced fission of ^{235}U, $\nu_p \sim 2.4$ and $\nu_d \sim 0.0158$.

Following neutron emission, primary fragments may remain in an excited nuclear state that may decay only by "prompt gamma" emission γ_p, *i.e.* there exists insufficient excitation energy for further nucleon emission. Nuclear de-excitation by γ_p occurs in about 2×10^{-14} s after prompt neutron emission.

In a solid material medium like the fuel of a nuclear reactor, the highly charged and energetic fragments eventually slow and thermalize in the surrounding media on the cumulative effect of thousands of Coulomb interactions with atomic electrons of other atoms in the material (chpt. 6). The stopping of these fragments deposits much of the energy extracted from the reactor for power applications. The fragments, being heavy and highly charged, give up kinetic energy rapidly, $\sim 10^{-12}$ s. During slowing down, they also neutralize by picking up electrons from the surrounding medium and now persist for seconds to centuries as radioactive fission products.

Both the L and H products parent their own decay chain. Since the heavy nuclides that fission require higher $N/Z(\sim 1.6)$ ratio for stability than the products (~ 1.3), all fission products are neutron rich and, hence, β^- emitters, giving rise to isobaric decay chains. In a few important cases, neutrons are emitted in decay as delayed neutrons. The delayed neutrons are very important for reactor control, despite their small numbers, since they contribute to the delicate neutron balance in a reactor only on their slow characteristic decay time. In this way, reactor neutron population and, hence, power can be controlled on this time scale (for small changes) rather than the hopelessly fast prompt neutron emission time.

The predominantly isobaric decay chains are many and long (sec. 5.4.3), and eventually result in stable final products

$$\begin{aligned} \mathrm{X}_L &\xrightarrow{\lambda} \mathrm{X}_{L'} \longrightarrow \ldots \longrightarrow \mathrm{X}_L^S \text{ (stable)} \\ \mathrm{X}_H &\xrightarrow{\lambda} \mathrm{X}_{H'} \longrightarrow \ldots \longrightarrow \mathrm{X}_H^S \text{ (stable)} \end{aligned}$$

The net fission reaction is then

$$\mathrm{n} + {}^{235}\mathrm{U} \longrightarrow \mathrm{X}_L^S + \mathrm{X}_H^S + \nu_p + \nu_d + \sum_j \gamma_{p_j} + \sum_j \gamma_{d_j} + \sum_j \left(\beta_j^- + \overline{\nu}_j\right) \quad (5.1)$$

where γ_d indicates delayed photons from decay. The numbers and types of emitted particles (fragments, neutrons, and photons) are statistical in nature (*i.e.* they vary from one fission event to another). Table 5.2 below lists the approximate range and average values for ^{235}U thermal fission, while table 5.3 lists some properties of the fission products, X_L and X_H.

The consequences of some of these properties are important. The radioactive inventory accumulated in fission products (sec. 5.4.3) can be a potentially substantial hazard. The change in volume upon fissioning implies that nuclear fuel has a potential to swell and pressurize. Finally, non-zero neutron absorption cross sections for fission products implies they are potentially parasitic to the nuclear chain reaction (sec. 5.4.4). All of these must be considered when designing a nuclear system.

Table 5.4 lists the energy release breakdown in thermal neutron-induced fission of ^{235}U. The net energy released, Q_f, is comprised

$$Q_f = \mathrm{KE}_{L+H} + \mathrm{KE}_{\mathrm{n}} + E_{\gamma_p} + \mathrm{KE}_{\beta^-} + E_{\overline{\nu}} + E_{\gamma_d} \quad (5.2)$$

where E_{γ_p} and E_{γ_d} are the prompt and delayed photon energy releases, respectively. The recoverable portion of this energy necessarily excludes

Table 5.2: Properties of ^{235}U Thermal Fission

Property	Range or Average
X_L, X_H	~center 1/3 of the periodic table
ν_p	0–8
$\langle\nu_p\rangle$	2.4
ν_d	0.00242–0.0194
$\langle\nu_d\rangle$	0.0158
KE_{L+H}	150–180 MeV
Q_f	180–210 MeV

Table 5.3: Properties of ^{235}U Thermal Fission Products, X_L and X_H

	Fission Product Property
(1)	$\sim$ 300 different nuclides
(2)	elements: ^{72}Ni to ^{163}Dy
(3)	$\sim$ 80 decay chains
(4)	initial charge $\sim +20$
(5)	initial energy 40–100 MeV
(6)	initial speed $\sim 10^9$ cm/s
(7)	range $\sim$ 1.5–3 cm (STP air), $\sim$ 5–15 μm(solids)
(8)	radioactivity: β^- emitters (with accompanying γ's)
(9)	neutron absorption cross section > 0
(10)	fragment volume > volume of ^{235}U atom

Table 5.4: Energy Release Breakdown† in ^{235}U Thermal Fission [Michaudon81]

	Particle	Energy Release per Fission (MeV)
prompt:	$\langle KE_{L+H} \rangle$	169.75
	$\langle KE_n \rangle$	4.79
	$\langle E_{\gamma_p} \rangle$	6.96
delayed:	$\langle KE_{\beta-} \rangle$	6.41
	$\langle E_{\gamma_d} \rangle$	6.23
	$\langle E_{\overline{\nu}} \langle$	8.62
	$\langle Q_f \rangle$	202.76

†Delayed neutron kinetic energy released per fission event is small compared to other releases in this table.

$E_{\overline{\nu}}$ since neutrinos interact only very weakly with matter. In addition, usually included in recoverable energy is the capture γ energies from parasitic absorption reactions of the $\nu_p + \nu_d - 1$ remaining neutrons. The net result being that the recoverable energy, $E_R \sim 200$ MeV, from thermal neutron-induced fission of ^{235}U is an excellent first approximation.

5.4.1 Conservation Principles

The conserved quantities in general CN nuclear reactions (sec. 2.4) are conserved in thermal neutron-induced fission of ^{235}U as well. As with all reactions, mass/energy and momentum are always conserved. The former quantity is to our benefit in determining energy liberated in the process only. We may not use them together, as we have done earlier, to uniquely identify product energies since the number of products in fission is more than two.

The total number of nucleons is always conserved;

$$A_L + A_H + \nu_p + \nu_d = 236$$

at all times since β^- decay does not alter the mass number. Individual nucleons are conserved among the primary fragments only. Product β^- decay does alter the Z/N ratio (as it must to move toward the valley of stability), so the final stable products will have both Z and N different from the primary fragments. Considering only the primary fragments, individual nucleons are conserved

$$\begin{aligned} Z_L + Z_H &= 92 \\ N_H + N_L + \nu_p &= 144 \end{aligned}$$

so that $A_L + A_H + \nu_p = 236$ for the primary fragments.

An example best illustrates. The primary fragment nuclides ^{90}Kr and ^{142}Ba along with four neutrons may make up the prompt products on one particular ^{235}U fission event

$$\mathrm{n} + {}^{235}\mathrm{U} \longrightarrow {}^{90}\mathrm{Kr} + {}^{142}\mathrm{Ba} + 4\mathrm{n}$$

For such a reaction, we readily verify that Z, N, and A are all independently conserved. The energy released in this process is

$$Q' = c^2 \left[m_{^{235}\mathrm{U}} - m_{^{90}\mathrm{Kr}} - m_{^{142}\mathrm{Ba}} - 3m_{\mathrm{n}}\right] = 169.5 \text{ MeV}$$

Yet, both ^{90}Kr and ^{142}Ba are unstable and parent the following decay chains

$$\begin{aligned} &{}^{90}\mathrm{Kr} \xrightarrow{\beta^-} {}^{90}\mathrm{Rb} \xrightarrow{\beta^-} {}^{90}\mathrm{Sr} \xrightarrow{\beta^-} {}^{90}\mathrm{Y} \xrightarrow{\beta^-} {}^{90}\mathrm{Zr} \text{ (stable)} \\ &{}^{142}\mathrm{Ba} \xrightarrow{\beta^-} {}^{142}\mathrm{La} \xrightarrow{\beta^-} {}^{142}_{58}\mathrm{Ce} \text{ (stable)} \end{aligned}$$

Through the series of decay events to reach the final stable products ^{90}Zr and ^{142}Ce, the neutron number has decreased and the proton number has increased by the number of β^- decay events, six. These β^- decay events contribute to the fission energy release Q_f, so

$$Q_f = c^2 \left[m_{^{235}\mathrm{U}} - m_{^{90}\mathrm{Zr}} - m_{^{142}\mathrm{Ce}} - 3m_{\mathrm{n}}\right] = 190 \text{ MeV} \tag{5.3}$$

The difference, $Q_f - Q' \sim 20.5$ MeV, is attributed to the composite Q_{β^-} in the product decay chains.

5.4.2 Prompt and Delayed Neutrons from Fission

Neutrons from fission can be lumped into two categories (sec. 5.4): (1) *prompt neutrons* — neutrons evaporated from the fission fragments immediately following scission but before coming to rest as fission products ($\sim 10^{-14}$ s after CN formation) and (2) *delayed neutrons* — neutrons emitted in the decay of fission products with half-lives up to about 1 min.

Prompt neutrons are emitted from fission with a spectrum of energies. A theoretical description of this spectrum is a very difficult problem indeed, as the physics is confounded by many complex features. For example, to determine the prompt neutron energy distribution, we would need to somehow account for the many different types of fragments emitting neutrons, the varieties of nuclear excitation levels these fragments can have, how this excitation energy varies as the fragment travels through media and interacts with other atoms, how the nuclear excitation energy gets transferred to an individual neutron, how the fragment's state of motion affects the laboratory kinetic energy of the emitted neutron, etc.

Instead, we turn to experimental determination and simple model-guided fits to the data (semi-empirical analysis), much like was done with the liquid drop model (sec. 1.6.6.1). The models, though simple, do capture some of the gross physics features. Firstly, we might assume that the fragments can be considered stationary while emitting prompt neutrons. Although this is strictly incorrect, it may lead to a reasonable first approximation. The emitted neutron then attains its kinetic energy solely from the excitation energy of the fragment. If we further propose that this energy is distributed among all nucleons in the fragment just like thermal energy is distributed among molecules of a gas, then one would expect the prompt neutron spectrum to take on a Maxwellian distribution function

$$\chi_M(\mathrm{KE_n}) = \frac{2}{\sqrt{\pi}} \left(kT_A\right)^{-3/2} \mathrm{KE_n^{1/2}} \exp\left(\frac{-\mathrm{KE_n}}{kT_A}\right)$$

characterized by the fragment's nuclear temperature T_A, used here as a free parameter. The best fit to the data fixes $kT_A \sim 1.29$ MeV. We call $\chi(\mathrm{KE_n})$ the neutron distribution function (*i.e.* $\chi(\mathrm{KE_n})\ d(\mathrm{KE_n})$ is the number of neutrons with energies in $d(\mathrm{KE_n})$ about $\mathrm{KE_n}$). The way it is expressed above, the distribution function is normalized so $\int_o^\infty \chi(\mathrm{KE_n})\ d(\mathrm{KE_n}) = 1$ and has the units of MeV^{-1}.

A slightly more sophisticated model can be developed by including fragment motion, while assuming that all fragments move with the same kinetic energy. Such a model is called the Watt spectrum[Watt52]

$$\chi_W(\mathrm{KE_n}) = a \exp\left(\frac{-\mathrm{KE_n}}{kT}\right) \sinh\left[\frac{4b\mathrm{KE_n}}{(kT)^2}\right]^{1/2}$$

where the fragment motion is incorporated in the parameter b, and

$$a = \left(\pi bkT\right)^{-1/2} \exp\left(-b/kT\right)$$

is not independent. With $b \sim 0.533$ MeV and the nuclear temperature $kT \sim 0.965$ MeV determined from fit, the multiplier becomes $a \sim 0.453\ \mathrm{MeV}^{-1}$. The two models yield quite similar forms. A comparison is shown in Fig. 5.3. The Watt spectrum is often employed in calculations since it provides a slightly better fit to the available data. It is not all together clear, however, if this result is due to the additional physics in the Watt model or a consequence of allowing an extra free parameter. The average kinetic energy per neutron is determined by integration

$$\begin{aligned}\langle \mathrm{KE_n} \rangle &= \int_o^\infty \mathrm{KE_n}\, \chi(\mathrm{KE_n})\, d(\mathrm{KE_n}) \\ &= \frac{3}{2}kT + b\end{aligned}$$

which was found by using the more general Watt model but is applicable to either model. By directly employing the Watt model parameters, we find $\langle \mathrm{KE_n} \rangle \sim 1.98$ MeV, while we can also apply this to the Maxwellian model by allowing $T \to T_A$ and $b \to 0$ to find the very similar result $\langle \mathrm{KE_n} \rangle \sim$ 1.935 MeV. The fraction of Q_f carried by fission neutrons from an average ^{235}U fission event is then $\sim \nu_p \langle \mathrm{KE_n} \rangle \sim 4.76$ MeV, as reported in table 5.4. Although prompt fission neutrons are not seen above $\sim$ 14 MeV, there is little error introduced in performing the above integration to ∞ since the distribution falls off exponentially at high energy. The most probable prompt neutron kinetic energy (*i.e.* where $d\chi/d(\mathrm{KE_n}) = 0$ at the peak of the distribution) is $\mathrm{KE_{mp}} \sim 0.73$ MeV.

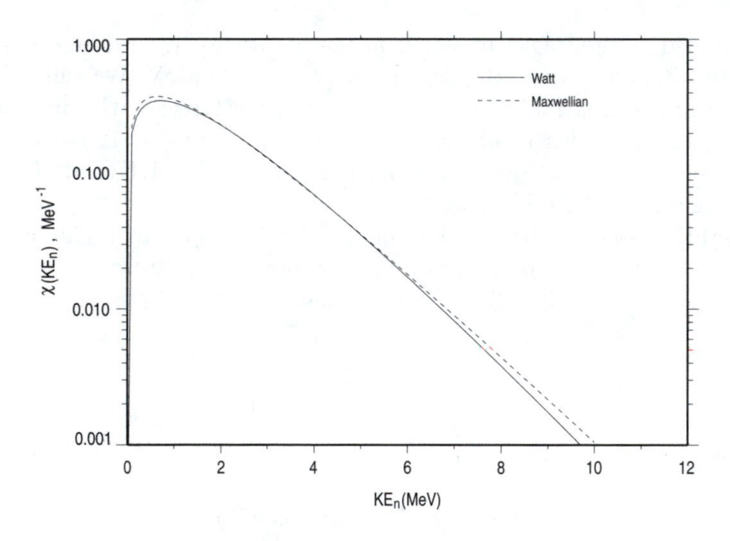

Figure 5.3: Watt and Maxwellian models of the prompt neutron spectrum from ^{235}U.

Delayed neutrons too are emitted with a distribution of energies. Their absolute energies, however, are quite a bit lower than those of prompt neutrons (maximum kinetic energies of delayed neutrons is typically $\lesssim 1$ MeV).

This is to be expected as the fission product nuclides that emit delayed neutrons have long since lost most of their original excitation energy in prompt processes. Coupled with their small numbers ($\nu_d \ll \nu_p$), delayed neutrons have little energy content to contribute to fission.

Of paramount importance, however, is the time delay in their emission and its impact on controlling the fission chain (discussed briefly in sec. 5.4). Delayed neutrons are emitted by unstable nuclides in a decay chain initiated by a product of fission. Neutron-emitting radionuclides are not produced directly in fission. The neutron-emitting fragments formed directly in fission are prompt emitters only. Neutron emission involves a strong nuclear interaction within the nucleus and, thereby, is only very short lived. Rather, neutron emitters are formed when a β^- emitting nuclide in the chain (called a neutron "precursor") decays to an excited state in the daughter product that is greater than S_n above the ground state. (Nuclear β^- decay involves the weak nuclear force and incurs much longer decay time. All fission product decay chains begin with β^- emitters.) Decay by neutron emission is then possible since the nucleus is excited to an energy level greater than that required for neutron emission. This neutron decay can occur to the ground or an excited state of the neutron-emitter's daughter product which then de-excites by γ emission. All excited levels populated by precursor β^- decay that are below S_n of the daughter product are also de-excited by γ emission.

An excellent example is the isobaric chain including the neutron precursor ^{137}I. This radionuclide is a β^- emitter with a 24.5 s half-life and decays via

$$^{137}\mathrm{I} \longrightarrow \ ^{137}\mathrm{Xe} + \beta^- + \overline{\nu}$$

with $Q_{\beta^-} \sim 5.88$ MeV. The neutron separation energy in ^{137}Xe is only 4.03 MeV. On some fraction ($\sim 7\%$) of ^{137}I decay events, β^- emission will populate an excited level in ^{137}Xe that is above S_n. This state may then rapidly decay by neutron emission to ^{136}Xe. The emission of the neutron from the excited nucleus is very rapid, almost instantaneous with the β^- emission of the precursor. Hence, the characteristic decay time for neutron emission is that of the precursor decay.

There exist many such examples of neutron precursors and there are a few dozen that occur with enough regularity to be important in the dynamics of nuclear reactors. These are usually lumped into groups with similar decay constants to aid in calculations. Table 5.5 lists the delayed group constants in a six group model[Keepin65] along with some important precursor nuclides in each group.

Table 5.5: Delayed Neutron Six Group Data for ^{235}U Thermal Fission [Keepin65]

Group #	Group $t_{1/2}$ (s)	Group Yield (n/fiss)	Nuclides
1	55.7	0.00052	^{87}Br
2	22.7	0.00346	^{88}Br, ^{137}I, ^{141}Cs
3	6.2	0.00310	^{87}Se, ^{89}Br, ^{92}Rb, ^{93}Rb, ^{134}Sb, ^{138}I
4	2.3	0.00624	^{85}As, ^{86}As, ^{88}Se, ^{90}Br, ^{92}Kr, ^{93}Kr, ^{94}Rb, ^{98}Y, ^{135}Sb, ^{139}I, ^{141}Xe, ^{142}Xe, ^{142}Cs, ^{143}Cs, ^{144}Cs
5	0.61	0.00182	^{99}Y, ^{140}I
6	0.23	0.00066	^{95}Rb, ^{96}Rb, ^{97}Rb
total		0.0158	

5.4.3 Fission Yield and Decay Chains

Of tremendous importance for fissioning systems are the types and inventory of radionuclides produced as products of the fission reaction. These are many and diverse. Some may represent an important concern as a potential hazard to man. Still others serve as probes of the nuclear physics in fission reactions. It was, in fact, the identification of ^{140}Ba that demonstrated that this neutron-induced reaction was indeed fission. In addition, holes in the periodic table (at the previously undiscovered elements $Z = 43$ (Tc) and $Z = 61$ (Pm) with no stable isotopes) were finally filled when radioisotopes of these elements were produced in fission.

Fission yield data give order to this assortment of products. Since the predominance of fission exit channels results in β^- emitting products with isobaric decay chains, the *mass* yield, $y(A)$ (sometimes called the *chain* yield) is most often reported. The chain yield is defined as the probability (usually in %) of forming a fission product with mass number A after prompt neutron emission. Since there are two products from fission, $\sum_A y(A) = 200\%$. (A small fraction of all fission events result in the emission of three nuclei, so called "ternary" fission. The dominant mode of ternary fission involves the emission of the very stable α particle in about 1/500 of all ^{235}U thermal neutron-induced fission events. Other light nuclei from ^{2}H to ^{10}Be are emitted with much lesser frequency. Since the fraction is low, we'll ignore ternary fission here.) A plot of $y(A)$ vs. A for thermal neutron-induced fission in ^{235}U is show in Fig. 5.4. Obvious from this figure is the relative rarity for symmetric fission. Much more likely are fission products with mass ratio of about $A_H/A_L \sim 3/2$. The most probable products are near $A_H \sim 140$ and $A_L \sim 94$, having chain yields on the order of 6.6%. Since each half of this distribution (on either side of $A = 119$) is symmetric, the most probable light and heavy product is also the average. We can then write for the "average" thermal neutron-induced fission event in ^{235}U

$$\langle \mathrm{n} + ^{235}\mathrm{U}\rangle \longrightarrow \ ^{94}\mathrm{X}_L + ^{140}\mathrm{X}_H + 2.4\mathrm{n}$$

accompanied by the release of 200 MeV. This is the average behavior only. No individual fission event can proceed this way.

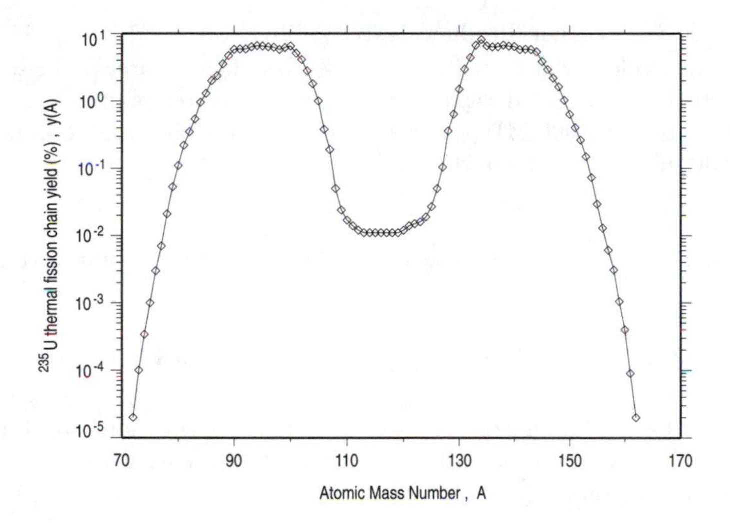

Figure 5.4: Fission chain yield curve for thermal neutron-induced fission of ^{235}U. The data were taken from [IAEA69].

Requiring more information, we can ask for the yield of individual nuclides directly from fission[Wahl88]. This *independent yield*, $y(A, Z)$, represents the probability (again, most usually in %) for formation of fission product nuclide ^A_ZX immediately following prompt neutron emission but before β^- decay (if it is not stable). The chain yield is the sum of independent yields

$$y(A) = \sum_j y(A, Z_j)$$

for fixed A (ignoring small departures from isobaric chains for delayed neutron emitters). Consider, for example, the $A = 90$ decay chain beginning with ^{90}Kr

$$\begin{array}{ccccccccccc} ^{90}\mathrm{Kr} & \xrightarrow{\beta^-} & ^{90}\mathrm{Rb} & \xrightarrow{\beta^-} & ^{90}\mathrm{Sr} & \xrightarrow{\beta^-} & ^{90}\mathrm{Y} & \xrightarrow{\beta^-} & ^{90}\mathrm{Zr} & & \text{(stable)} \\ \uparrow & & \uparrow & & \uparrow & & \uparrow & & \uparrow & & \\ \text{fission} & & \text{fission} & & \text{fission} & & \text{fission} & & \text{fission} & & \end{array}$$

Each component of the chain may be produced in fission (but no two on the same event since this would violate our conservation rules of sec. 5.4.1) with different independent yields. This is an excellent example of an open system, production and decay problem as introduced in sec. 4.5.

Example 31: Fission Product Activity: ^{90}Kr

The independent yield for ^{90}Kr is $\sim 5\%$. If it has no precursors in the $A = 90$ chain and β^- decays with $t_{1/2} = 32.3$ s, how many Ci of ^{90}Kr are present in a 3000 $\mathrm{MW_{th}}$ power reactor after 10 h of operation, and after 30 min following shutdown?

Solution:

A fission reactor that generates 3000 $\mathrm{MW_{th}}$ has a continuous fission rate of

$$R_f \sim 3000 \times 10^6 \text{ (J/s)} \frac{1 \text{ MeV}}{1.6 \times 10^{-13} \text{ J}} \frac{1 \text{ fiss}}{200 \text{ MeV}} \sim 9.4 \times 10^{19} \text{ (fiss/s)}$$

Since its half-life is short, ^{90}Kr activity reaches saturation after only a few minutes. With no precursors, the equilibrium (saturation) activity is the production rate

$$A_\infty(^{90}\mathrm{Kr}) = y(90, 36) R_f \sim 1.3 \times 10^8 \text{ Ci}$$

an enormous activity. But because of its short half-life, this product decays quickly after reactor shutdown. Since there is no precursor, the decay is a simple exponential which after 30 min following shutdown becomes

$$A(30 \text{ min}) = A_\infty e^{-\lambda T} \sim 2.18 \text{ nCi}$$

The independent yields are bounded

$$0 \leq y(A, Z) \leq 100\%$$

by definition. We find that for thermal neutron-induced fission in ^{235}U, $y(A, Z) \leq 7\%$ for any product. Displaying the yield data isotopically, *i.e.* as a function of A for fixed Z, we find there is a maximum yield for each element

$$\begin{array}{cccccccc} y(A_1, Z) & < & y(A_2, Z) & < & \cdots & & & \\ \cdots & < & y(A_{p-1}, Z) & < & y(A_p, Z) & > & y(A_{p+1}, Z) & > & \cdots \\ & & & & & & \cdots & > & y(A_n, Z) \end{array}$$

where $y(A_p, Z)$ is the independent yield for the most probable among all products of element Z.

Since charge is conserved in fission ($Z_L + Z_H = 92$), chains are complementary. (Identifying $\nu_p(A)$ also fixes the complementary isotope.) An example of complementary chains are shown below

$$\begin{array}{ccccccc} \longrightarrow & {}^{93}_{36}\mathrm{Kr} & \longrightarrow & {}^{93}_{37}\mathrm{Rb} & \longrightarrow & {}^{93}_{38}\mathrm{Sr} & \longrightarrow \\ & \ddots & & \vdots & & \cdot^{\cdot^{\cdot}} & \\ & & \ddots & \vdots & \cdot^{\cdot^{\cdot}} & & \\ & & & \cdot & & & \\ & & \cdot^{\cdot^{\cdot}} & \vdots & \ddots & & \\ & \cdot^{\cdot^{\cdot}} & & \vdots & & \ddots & \\ \longrightarrow & {}^{140}_{54}\mathrm{Xe} & \longrightarrow & {}^{140}_{55}\mathrm{Cs} & \longrightarrow & {}^{140}_{56}\mathrm{Ba} & \longrightarrow \end{array}$$

where the dotted lines are intended to indicate complements among the products. Complementary isotopes must have identical independent yields.

There are many ways to present the yield data. We have already seen two; the independent yield, and the chain yield, the former being more general and containing all the yield information. Often, we are interested in the inventory of a fission product nuclide in the middle of a decay chain. This particular nuclide may be fed directly from fission, as well as from decay of its precursor. In many cases of interest, the precursors, being several β^- decay steps away from stability, will have much shorter half-lives than the product in question. For such cases, the important yield is not that of the independent product, rather it is that of the product and all its precursors. Such a yield is called a "cumulative" yield, $y_c(A, Z)$, equal to the probability (in %) of formation for a fission product with atomic mass number A and atomic number Z, and all its radioactive precursors after prompt neutron emission, so

$$y_c(A, Z) = \sum_{Z_j \leq Z} y(A, Z_j)$$

Here, the summation is over all $Z_j \leq Z$ since these are the nuclides that β^- decay to ${}^{A}_{Z}$X.

An important nuclide in this regard is ^{90}Sr in the same $A = 90$ chain discussed above. Since its $t_{1/2} = 28.8$ y while that of its precursors, ^{90}Kr and ^{90}Rb, are 32.3 s and 2.6 min, respectively, the important yield is the cumulative yield for ^{90}Sr. The precursors quickly decay to ^{90}Sr, so on the time scale of more than a few minutes, essentially all precursor atoms are transmuted to ^{90}Sr. The independent yield for ^{90}Sr is $< 10^{-3}\%$, yet its cumulative yield is $\sim 5.8\%$. This is very close to the chain yield for $A = 90$ and only slightly greater than the independent yield for ^{90}Kr. These are general trends as we can interpret the cumulative yield as a partial chain yield and the independent yields, for other than the first isobar in the chain, are generally small.

Example 32: Fission Product Activity: ^{90}Sr
Because of its long half-life and propensity to settle in bone tissue with long biological half-life, ^{90}Sr is a potential biological hazard. Find the total ^{90}Sr activity in the reactor of example 31 after 1 y of operation and again 1 y after shutdown.

Solution:
Since 1 y is much greater than the $t_{1/2}$ of ^{90}Kr and ^{90}Rb

$$A(^{90}\mathrm{Sr}) \sim y_c(90, 38) R_f \left(1 - e^{-\lambda_{^{90}\mathrm{Sr}} t}\right)$$

For our 3000 MW$_{\mathrm{th}}$ reactor we have $R_f \sim 9.4 \times 10^{19}$ fiss/s and $y_c(90, 38) \sim 0.058$, the total activity of ^{90}Sr after 1 y operation is $\sim 3.5 \times 10^6$ Ci. One year following shutdown, this decays only to $\sim 3.4 \times 10^6$ Ci.

5.4.4 Fission Product Poisoning

Certain fission products, among the multitude of possibilities, have a high affinity for neutrons (*i.e.* large σ_a). This makes them especially important in nuclear reactors as they can rob neutrons that might otherwise be destined to induce fission reactions.

By far, the most important of these fission product "poisons" is ^{135}Xe, with a thermal neutron absorption cross section of $\sigma_a \simeq 2.7 \times 10^6$ b. ^{135}Xe is produced both as a β^- decay product of ^{135}I and directly in fission. The

$A = 135$ fission decay chain is

$$^{135}\text{Te} \xrightarrow{\beta^-} {}^{135}\text{I} \xrightarrow{\beta^-} {}^{135}\text{Xe} \xrightarrow{\beta^-} {}^{135}\text{Cs} \xrightarrow{\beta^-} {}^{135}\text{Ba (stable)}$$

Since the $t_{1/2}$ of ^{135}Te is very short ($\sim$ 19 s) compared with that of ^{135}I and ^{135}Xe, 6.57 h and 9.1 h, respectively, we can safely assume that all ^{135}I is produced directly in fission with a cumulative yield of $y_{c_\text{I}} \sim 0.0639$. However, since both ^{135}I and ^{135}Xe have half-lives on the same order, we must follow the time dependence of both radionuclides. The rate equation for ^{135}I becomes

$$\frac{dN_\text{I}}{dt} = y_{c_\text{I}}\Sigma_f\varphi - \lambda_\text{I}N_\text{I}$$

where the production function $y_{c_\text{I}}\Sigma_f\varphi$ is considered a local constant under conditions of no fissile fuel depletion (*i.e.* Σ_f = const.) and steady flux ($\varphi(t) = \varphi$). Here, we have written this expression as a function of the local ^{135}I concentration (atom density, N_I) rather than the total inventory, $\mathcal{N}_\text{I}$ as we had done earlier (sec. 4.5), since we realize that in a real nuclear system like a reactor, the flux (and, hence, production rate and product concentration) will vary spatially.

Since ^{135}Xe can be produced directly in fission as well as β^- decay of the precursor ^{135}I, there are two production terms in the rate equation for ^{135}Xe concentration, N_X. As well, ^{135}Xe is lost both through decay and depletion by neutron absorption to ^{136}Xe, so that

$$\frac{dN_\text{X}}{dt} = y_\text{X}\Sigma_f\varphi + \lambda_\text{I}N_\text{I} - \lambda_\text{X}N_\text{X} - \sigma_{a_\text{X}}\varphi N_\text{X}$$

where $y_\text{X} \sim 0.00237$ is the independent fission yield of ^{135}Xe. With no initial concentration ($N_\text{I}(0) = N_\text{X}(0) = 0$), the solutions to these two rate equations become

$$\begin{aligned} N_\text{I}(t) &= \frac{y_{c_\text{I}}\Sigma_f\varphi}{\lambda_\text{I}}\left(1 - e^{-\lambda_\text{I}t}\right) \qquad (5.4) \\ N_\text{X}(t) &= \frac{(y_{c_\text{I}} + y_\text{X})\Sigma_f\varphi}{\lambda_\text{X} + \sigma_{a_\text{X}}\varphi}\left[1 - e^{-(\lambda_\text{X}+\sigma_{a_\text{X}}\varphi)t}\right] \\ &\quad + \frac{y_{c_\text{I}}\Sigma_f\varphi}{\lambda_\text{X} - \lambda_\text{I} + \sigma_{a_\text{X}}\varphi}\left[e^{-(\lambda_\text{X}+\sigma_{a_\text{X}}\varphi)t} - e^{-\lambda_\text{I}t}\right] \end{aligned}$$

At equilibrium (*i.e.* $t \to \infty$),

$$\begin{aligned} N_{\text{I}_\infty} &= \frac{y_{c_\text{I}}\Sigma_f\varphi}{\lambda_\text{I}} \qquad (5.5) \\ N_{\text{X}_\infty} &= \frac{(y_{c_\text{I}} + y_\text{X})\Sigma_f\varphi}{\lambda_\text{X} + \sigma_{a_\text{X}}\varphi} \end{aligned}$$

so that the competition for neutrons between poison and fissile material

$$\frac{\Sigma_{a\mathrm{X}_\infty}}{\Sigma_f} = \frac{y_{c_\mathrm{I}} + y_\mathrm{X}}{1 + \lambda_\mathrm{X}/(\sigma_{a\mathrm{X}}\varphi)}$$

never goes beyond the chain yield of ^{135}Xe. After this reactor shutdown, however, ^{135}Xe concentration continues to build from ^{135}I decay without removal of ^{135}Xe from absorption. ^{135}I and ^{135}Xe are eventually depleted by decay, yet for intermediate times, N_X may reach as much as an order of magnitude (or more) above N_{X_∞} depending on the value of φ before shutdown. Using the equilibrium concentrations N_{I_∞} and N_{X_∞} as initial conditions for the shutdown problem ($\varphi \to 0$), we find

$$N_\mathrm{X}(T) = N_{\mathrm{X}_\infty} e^{-\lambda_\mathrm{X} T} + \frac{\lambda_\mathrm{I} N_{\mathrm{I}_\infty}}{\lambda_\mathrm{X} - \lambda_\mathrm{I}} \left[e^{-\lambda_\mathrm{I} T} - e^{-\lambda_\mathrm{X} T}\right] \tag{5.6}$$

for all times T after shutdown, a simple binary build-up and decay problem with non-zero initial concentration (sec. 4.4.2). This concentration peaks at $T = T_\mathrm{max}$

$$T_\mathrm{max} = \frac{1}{\lambda_\mathrm{X} - \lambda_\mathrm{I}} \ln\left[\frac{\lambda_\mathrm{X}}{\lambda_\mathrm{I}}\left(1 - \frac{y_{c_\mathrm{I}} + y_\mathrm{X}}{y_{c_\mathrm{I}}} \frac{\lambda_\mathrm{X} - \lambda_\mathrm{I}}{\lambda_\mathrm{X} + \sigma_{a\mathrm{X}}\varphi}\right)\right] \tag{5.7}$$

The maximum concentration is most easily found by setting $dN_\mathrm{X}/dt = 0$ at the peak so that $N_\mathrm{X}|_\mathrm{max} = \frac{\lambda_\mathrm{I}}{\lambda_\mathrm{X}} N_\mathrm{I}(T_\mathrm{max})$, then

$$\frac{N_\mathrm{X}|_\mathrm{max}}{N_{\mathrm{X}_\infty}} = \frac{\lambda_\mathrm{I} N_{\mathrm{I}_\infty}}{\lambda_\mathrm{X} N_{\mathrm{X}_\infty}} \left[\frac{\lambda_\mathrm{X}}{\lambda_\mathrm{I}}\left(1 - \frac{y_{c_\mathrm{I}} + y_\mathrm{X}}{y_{c_\mathrm{I}}} \frac{\lambda_\mathrm{X} - \lambda_\mathrm{I}}{\lambda_\mathrm{X} + \sigma_{a\mathrm{X}}\varphi}\right)\right]^{\frac{\lambda_\mathrm{I}}{\lambda_\mathrm{I} - \lambda_\mathrm{X}}} \tag{5.8}$$

Figure 5.5 shows an example ^{135}Xe concentration buildup after reactor shutdown normalized to N_{X_∞} for three cases, $\varphi = 5 \times 10^{13}$ cm^{-2} s^{-1}, 10^{14} cm^{-2} s^{-1}, and 2×10^{14} cm^{-2} s^{-1}. For the $\varphi = 10^{14}$ cm^{-2} s^{-1} case, $T_\mathrm{max} \sim 10.143$ h, $N_{\mathrm{I}_\infty}/N_{\mathrm{X}_\infty} \sim 9.58$, and $N_\mathrm{X}|_\mathrm{max}/N_{\mathrm{X}_\infty} \sim 4.56$. Because of this ^{135}Xe concentration buildup, some reactors may be prevented from restarting immediately. Several hours of reactor "dead time" may result while waiting for the ^{135}Xe concentration to decay.

5.4.5 Conversion Constants

It is illustrative to compare the energy density of fissile material with that of other energy sources. As a point of comparison, let's consider a 3000 MW$_\mathrm{th}$ ($\sim$ 1000 MW electric) power plant. This is the scale of a typical commercial power facility. At 200 MeV per fission in ^{235}U, we need $R_f \sim 9.4 \times 10^{19}$ fiss/s to amass this power (cf: example 31). Since each fission event involves a

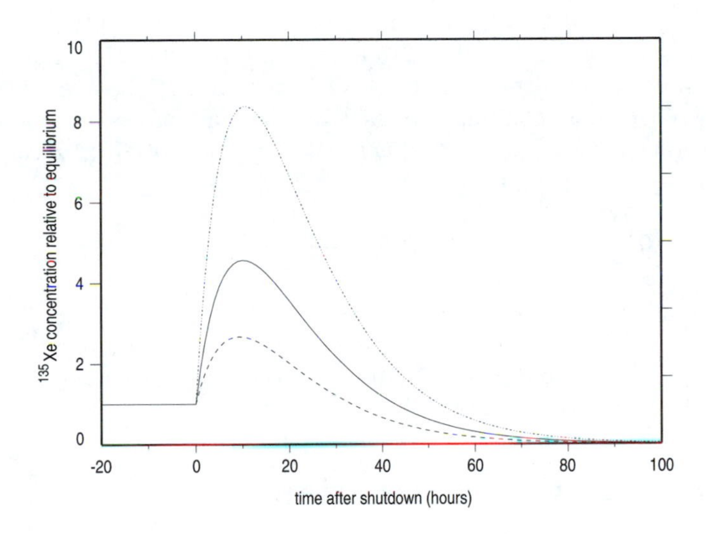

Figure 5.5: ^{135}Xe concentration after shutdown of a reactor at $t = 0$ that was originally operating at a steady flux of $\varphi = 5 \times 10^{13}$ cm^{-2} s^{-1} (dash curve), $\varphi = 10^{14}$ cm^{-2} s^{-1} (solid curve), and $\varphi = 2 \times 10^{14}$ cm^{-2} s^{-1} (dot curve), normalized to N_{X_∞}.

single ^{235}U atom which then can no longer participate in energy production, one atom of ^{235}U is "consumed" per fission. In order to produce 3000 MW_{th} continuously for one day, we must consume

$$\begin{aligned} 3000\ \text{MWd} &= 9.4 \times 10^{19} \frac{\text{fiss}}{\text{s}} \times (8.64 \times 10^4\ \text{s}) \times 235.0439 \frac{\text{g}}{\text{mole}} \\ &\quad \times \frac{1\ \text{mole}}{0.6022 \times 10^{24}\ \text{at}} \times \frac{1\ \text{kg}}{1000\ \text{g}} \\ &= 3.17\ \text{kg} \end{aligned}$$

of ^{235}U. Note that 3.17 kg is not destroyed in this consumption. We are converting only 200 MeV/c^2 $\sim$ 0.215 amu of the original 235 amu nuclear mass per atom to energy, about 0.1%. The remainder (the overwhelming majority) of the mass continues to reside as the mass of fission products.

To liberate the same energy (3000 MWd) with coal ($\sim$ 12.2 GJ/ton), we would need to consume

$$3000 \times 10^6 \frac{\text{J}}{\text{s}} \times (8.64 \times 10^4\ \text{s}) \times \frac{1\ \text{ton coal}}{12.2 \times 10^9\ \text{J}} = 21,246\ \text{ton coal}$$

Burning crude oil ($\sim$ 6 GJ/barrel) requires the consumption of about 43,200 barrels for the same energy production. Even in detonating TNT ($\sim$ 4.2 GJ/ton), we would need 61,700 ton to release the same quantity of energy in fissioning only 3.17 kg of ^{235}U.

5.5 Fission Chain Reaction

Recall that among the products of fission are several neutrons (sec. 5.3). Should these go on to induce subsequent fission events in other fissile or fissionable nuclei, a chain of such events is possible. A *nuclear reactor* is a device designed to produce the chain reaction in a controlled manner. In this way, useful power can be produced continuously.

The key to harnessing nuclear power for peaceful purposes is control. Since more than one neutron is emitted on average, the potential exists for a fission chain to multiply and grow without bound. Such uncontrolled chains have potentially serious consequences. The parameter nuclear engineers use to assess control of the nuclear core in a reactor is the *multiplication factor*, k. Consider the state of a nuclear reactor at any instant in time. This system is populated by some large but finite number of neutrons that originated in fission events. Let's call this population of neutrons a *generation*. Some of these neutrons proceed to induce fission reactions that produce the next generation of neutrons. The multiplication factor is then defined as

$$k = \frac{\text{number of neutrons in one generation}}{\text{number of neutrons in the preceding generation}}$$

It is only when this factor is identically equal to unity that we can have a stable, self-sustained chain reaction. This situation is called the *critical* condition. Under these conditions, our reactor experiences a constant rate of fission reactions and, hence, power production. When k is greater than unity, the reactor is *super-critical* and neutron production, fission rate, and power production all increase. If unchecked by some means of reducing k, a super-critical system will continue to multiply neutrons and power. Such a system is bounded only by the eventual self-dismantling of the system. When $k < 1$, our system is *sub-critical*, and power, fission rate, and neutron production will decay and eventually die out.

In defining the critical condition, it was not necessary to specify any absolute quantity. The state of criticality of the reactor is independent of neutron flux and power level. The system may be critical at low power, high power, and everywhere in between. Of course, to change power level, the system must be made sub- or super-critical.

The responsibility of a nuclear engineer is to design a nuclear system which is capable of maintaining the detailed level of neutron balance required for a critical reactor. The formalism and methods necessary for this task comprise the subject of reactor physics. The reader is directed to references at the chapter's end for more advanced discussion. Mechanisms that dictate the fate of individual neutrons are important in maintaining neutron inventory and, hence, the condition of criticality. Obviously, to maintain $k = 1$, neutrons must be generated in the system at the same rate at which they are removed. Neutrons are generated in fission reactions and are removed either by absorption reactions with nuclei comprising the reactor fuel, moderator, or internal structure; or they escape the nuclear core (leak out) of the reactor.

Detailed neutron balance then depends on the size and composition of the system. Consider first a reactor fabricated from a cube of pure ^{235}U. If this block is made very small, most of the neutrons generated in fission within the block will leak from the system, and the system is sub-critical. Neutrons may in fact leak very quickly if the size of the system, a, is $\ll 1/\Sigma$ (where a is the length of our cube side). As a is increased, however, the probability a neutron will leak from the surface is reduced. When the size is increased so that $a \gg 1/\Sigma$, leakage of neutrons occurs only near the outer surface of the system. The contribution to leakage will indeed increase as a^2, yet neutron production will increase as a^3, so that the fraction of neutrons leaking decreases as $1/a$. Production will catch up and, eventually, compensate leakage, at which point the system will be critical. The size of the critical system, $a = a_c$, is called the *critical dimension.* The mass of fuel at this condition is the *critical mass.*

In the scenario above, the reactor composition was specified (pure ^{235}U) and the critical dimensions were determined. One may proceed in just the opposite way by specifying the dimensions of the system and determining the critical composition. Consider a very large pool of water containing some small concentration of ^{235}U atoms in homogeneous solution. To simplify matters, imagine that this pool is so large that neutron leakage is insignificant. Then, the balance is determined by fission and absorption rates only. The water in this system serves to moderate neutrons so that they are more likely to induce fission, as well as providing a medium for solution. In the course of moderation, however, some neutrons are absorbed in water (sec. 3.2) and cannot contribute to generating neutrons. When the ^{235}U concentration is low, neutron absorption in water will dominate and the system will be sub-critical. By increasing the fuel (^{235}U) concentration, however, neutron absorption in ^{235}U will increase, thereby increasing the fission rate. Since a low concentration system is sub-critical and a very large dimension system with pure ^{235}U is super-critical (previous scenario), there must be some intermediate concentration where the system is just critical. Then, as the ^{235}U concentration is increased, there will be a *critical concentration* where neutron generation will exactly balance the net absorption rate.

Of course, we realize that real nuclear systems are far more complex than in the idealized examples above. The mechanisms dictating detailed neutron balance are much more involved. We've already seen the potential of parasitic absorption from fission product poisons (sec. 5.4.4). Temperature changes in the reactor can alter absorption rates and, thereby, affect criticality. Likewise, moderator density changes and voids can influence the average Σ_a and Σ_s, and affect criticality. These often subtle influences must be considered by the nuclear engineer when designing a reactor and a means of control.

As we briefly mentioned in an earlier discussion (sec. 5.4.2), delayed neutrons are instrumental in allowing practical control of a reactor. Were one able to guarantee $k = 1$ always, there would be no need for concern. Yet, because of the need to make changes in the reactor power level and because of slight and unavoidable variations in the system conditions, k will always move away from exact unity, even if only infinitesimally. In a reactor without delayed neutrons, the response of the system to small changes (or perturbations) occurs on a time scale determined by the average time required for prompt fission neutrons to be absorbed, the *prompt neutron lifetime*. The response time may then be quite short, $\sim 10^{-4}$–10^{-1} s, depending on the reactor composition and size of the perturbation. Considering delayed neutrons, however, the situation is quite different. These neutrons contribute to the detailed balance just as do prompt neutrons. The delayed neutrons, though, are emitted to the system on the β^- decay time of the precursors. The response of the system is then much slower (seconds to hundreds of seconds), and the system is now controllable by human or programmed intervention.

5.6 Fusion Reactions

The other method, of course, to extract energy stored in the nucleus is to approach the peak of the BE/A curve (Fig. 1.5) from the low mass end. Assembling light nuclei to form larger, more tightly bound states liberates energy. Such reactions are termed "fusion," as we imagine the two reactants in this process to fuse together to form the larger product. Though the product nucleus is larger than either of the reactants, the net result must be to reduce rest mass.

Consider the fusion reaction combining a deuterium and tritium nucleus

$$\mathrm{D} + \mathrm{T} \longrightarrow {}^{4}\mathrm{He} + \mathrm{n}$$

to form helium. This is considered the most important of all fusion reactions since it has the highest cross section (at low reactant temperature, $\lesssim 100$ keV) and, thereby, may be the most accessible to terrestrial efforts in developing a controlled thermonuclear fusion reactor. In the D-T reaction, $Q = 17.59$ MeV is released in the form of kinetic energy of the two products. A straightforward application of mass/energy and momentum balance for this two-product reaction assuming stationary reactants yields, from Eq.(2.8), $\mathrm{KE_n} = 14.05$ MeV and $\mathrm{KE_{He}} = 3.54$ MeV. Energetic neutrons from this reaction are sometimes used for nuclear physics or radiation effects on materials experiments. They are also problematic for nuclear fusion reactors employing the D-T reaction. Such energetic neutrons are highly penetrating, making shielding of this radiation from man and sensitive equipment quite challenging. As well, these neutrons have the eventual fate of being absorbed somewhere in the fusion reactor chamber or surrounding structure, often leading to the production of radionuclides in these materials through neutron activation (sec. 4.5).

Some other important fusion reactions of light nuclides are listed below in table 5.6. Where there are two products, it is a trivial exercise to determine the product kinetic energies from the techniques developed in chapter 2. The D-D product branching is very close to 50%. The latter two fusion reactions in table 5.6 involving, ^{6}Li and ^{11}B, are classified as "aneutronic" (without neutrons). Neither they nor any of their side reactions (reactions involving products) produce neutrons. This is considered a strong advantage in controlled fusion, thereby eliminating the radioactive inventory associated with fusion neutron activation. These aneutronic fuels are, however, considered advanced fuels since they have a much lower cross section for fusion than the D-T or D-D reactions. D-^{3}He is also considered an advanced fuel reaction since its fusion cross section is some two orders of magnitude below that of D-T at 10 keV, a typical fusion reactant temperature. D-^{3}He is not an aneutronic fuel, however, since in any fusion system burning D-^{3}He, there will occur a significant rate of D-D reactions.

Table 5.6: Light Nuclide Fusion Reactions

Reaction			Q (MeV)
$\mathrm{D + D}$	$\longrightarrow$	$\mathrm{T + p}$	4.03
	$\longrightarrow$	$^3\mathrm{He + n}$	3.27
$\mathrm{T + T}$	$\longrightarrow$	$^4\mathrm{He + 2n}$	11.33
$\mathrm{D +} {}^3\mathrm{He}$	$\longrightarrow$	$^4\mathrm{He + p}$	18.35
$\mathrm{p +} {}^6\mathrm{Li}$	$\longrightarrow$	$^4\mathrm{He} + {}^3\mathrm{He}$	4.02
$\mathrm{p +} {}^{11}\mathrm{B}$	$\longrightarrow$	$3(^4\mathrm{He})$	8.68

The reactions

$$\mathrm{D + D} \longrightarrow {}^4\mathrm{He}$$

and

$$\mathrm{H + H} \longrightarrow {}^2\mathrm{He}$$

are not allowed by momentum conservation. The latter is also not allowed owing to the instability of ^{2}He. The reaction

$$\mathrm{D + D} \longrightarrow {}^4\mathrm{He} + \gamma$$

is rarely observed. The Q-value for this reaction is 28.5 MeV, well above the S_n or S_p for ^{4}He. Much more likely are the D-D fusion reactions in table 5.6.

The tritium fuel for the D-T or T-T fusion reactions must be produced artificially. Tritium is a β^- emitter with $t_{1/2} = 12.3$ y and is, therefore, not available in nature. The neutron-induced reactions in Li

$$\begin{aligned} \mathrm{n +} {}^6\mathrm{Li} &\longrightarrow {}^4\mathrm{He + T}\,, \quad Q = 4.78\,\mathrm{MeV} \\ \mathrm{n +} {}^7\mathrm{Li} &\longrightarrow {}^4\mathrm{He + T + n}\,, \quad Q = -2.47\,\mathrm{MeV} \end{aligned}$$

are available as "breeding" reactions for T. Fusion reactor designs often exploit neutrons from the D-T or D-D reactions for this purpose.

5.6.1 Energy Release from Deuterium

In a fusion reactor fueled with deuterium, the equally likely heavy products, T and ^{3}He, readily fuse with additional deuterium. On average, four deuterium nuclei are consumed in the net process

$$4\text{D} \longrightarrow \text{T} + \text{p} + ^3\text{He} + \text{n}\,, \quad Q = 7.3 \text{ MeV}$$

and the reactions T(d,n)^{4}He and ^{3}He(d,p)^{4}He, with greater fusion cross section, readily follow. The net consumption

$$\boxed{6\text{D} \longrightarrow 2(^4\text{He}) + 2\text{n} + 2\text{H}} \tag{5.9}$$

results, with a net energy release of 43.24 MeV. The higher probability D-T and D-^{3}He reactions are said to "catalyze" the D-D fuel by supplying energy to the system and elevating the fuel temperature to increase the D-D cross section and rate of reactions. The catalyzed D-D reaction, then, liberates about 7.2 MeV per deuteron or about 3.6 MeV/amu. By comparison, ^{235}U fission releases about 200 MeV/235 amu $\sim$ 0.85 MeV/amu. On a per mass basis, catalyzed D-D fusion represents about 4.2 times the energy density of fission fuels. The energy density of D-T fuel is comparable at $\sim$ 3.5 MeV/amu. The 3.17 kg of ^{235}U that we had consumed to produce 3000 MWd of energy (sec. 5.4.5) is the equivalent of about 0.75 kg of deuterium or D-T mixed fuel.

5.6.2 Solar Fusion

It has long been established that fusion reactions power the sun and, in fact, all stars. Prior to this realization, attempts had been made to attribute the $\sim 4 \times 10^{26}$ W continuous solar power to chemical combustion or gravitational energy, the latter being released as the sun was postulated to shrink. (The size of our sun is not changing and is not expected to for billions of years.) Fossil records on earth indicate that the sun is an extremely stable nuclear furnace with constant output for more than 10^9 y. Yet, all proposed non-nuclear fuels would have been exhausted in a few thousand years to a few million years at best.

The fuel for stellar fusion is the most abundant of all materials in the universe, hydrogen. More than 90% of all atomic and nuclear matter in the universe is ^{1}H. Only a tiny fraction ($\lesssim$ 1%) of the remainder is comprised of matter other than ^{4}He. These concentrations are primordial. Stellar nucleosynthesis accounts for only a tiny fraction of all heavy nuclei. No doubt this is an important fraction, however. All planets, our earth, and even ourselves are the products of such rare materials born in stars long dead. We owe not only our origin but continual existence to stars, as all life-giving energy originates there.

Stellar fusion is an involved, multi-stage process. New stars are comprised only of hydrogen ions, which only very slowly fuse via

$$\mathrm{H}^+ + \mathrm{H}^+ \longrightarrow {}^2\mathrm{H}^+ + \beta^+ + \nu\,, \quad Q = 1.442 \text{ MeV}$$

or occasionally through the very rare ($\sim 0.25\%$) three body process

$$\mathrm{H}^+ + \mathrm{H}^+ + \mathrm{e}^- \longrightarrow {}^2\mathrm{H} + \nu$$

with the same energy release. In order to find Q for these reactions using atomic masses, we can imagine adding two electrons on both sides of the above reaction equations. We then have for the more probable route

$$\mathrm{H} + \mathrm{H} \longrightarrow {}^2\mathrm{H} + \beta^+ + \nu$$

maintaining electric charge. As with all computations of this sort (sec. 1.6.4), we make error on the order of the atomic binding energy, small on the order of Q. The free ionic state, accompanied by an equal number of free electrons which is dictated by the hot interior of the sun, is called the "plasma" state of matter. Fusing systems, like the interior of stars, need to be maintained at elevated temperature (about 1.5×10^7 K for our sun) to overcome the strong Coulomb repulsion (barrier) for these reactions (sec. 5.6.3).

The tremendous temperatures and densities ($\sim 10^{26}$ protons/cm^3) in our sun's interior are maintained against expansion by the gravitational potential afforded by the sun's enormous mass ($\sim 2\times10^{30}$ kg). Despite these terrific densities and temperatures, proton-proton fusion proceeds rather slowly. In round figures, the p-p reaction rate density in the solar core is $\sim 10^8$ cm^{-3} s^{-1}, implying a reaction probability per proton of only $\sim 10^{-18}$/s. The mean lifetime of a solar proton at the core is then on the order of 3×10^{10} y. This slow p-p rate controls the rate of solar fuel burn and is sometimes called the "bottleneck" in the fusion process to ^{4}He, as it is by far the slowest step.

Since the deuteron density is so low, D-D reactions are extremely unlikely. Rather, the product deuteron much more readily fuses with the multitude of available protons

$$\mathrm{H} + {}^2\mathrm{H} \longrightarrow {}^3\mathrm{He} + \gamma\,, \quad Q = 5.49 \text{ MeV}$$

This is a comparably fast reaction so that deuterons are, essentially, immediately converted to ^{3}He. Since ^{4}Li is unstable against breakup to H + ^{3}He, the fusion of H + ^{3}He does not proceed. Instead, ^{3}He nuclei wander through the solar interior until happening upon another ^{3}He nucleus. At this interaction, there is a relatively high probability of forming ^{4}He in the fusion reaction

$$^{3}\text{He} + {}^{3}\text{He} \longrightarrow {}^{4}\text{He} + 2\text{H} + \gamma\,, \quad Q = 12.86\ \text{MeV}$$

The net process then consumes four protons to produce a single ^{4}He nucleus

$$\boxed{4\text{H} \longrightarrow {}^{4}\text{He} + 2\gamma + 2\beta^{+} + 2\nu} \qquad (5.10)$$

and releases 26.72 MeV. A certain small percentage of ^{3}He interactions can proceed via another path

$$^{3}\text{He} + {}^{4}\text{He} \longrightarrow {}^{7}\text{Be} + \gamma$$

which produces

$$\begin{aligned} {}^{7}\text{Be} + \text{e}^{-} &\longrightarrow {}^{7}\text{Li} + \nu \\ {}^{7}\text{Li} + \text{H} &\longrightarrow 2({}^{4}\text{He}) \end{aligned}$$

or

$$\begin{aligned} {}^{7}\text{Be} + \text{H} &\longrightarrow {}^{8}\text{B} + \gamma \\ {}^{8}\text{B} &\longrightarrow {}^{8}\text{Be} + \beta^{+} + \nu \\ {}^{8}\text{Be} &\longrightarrow 2({}^{4}\text{He}) \end{aligned}$$

All branches lead to ^{4}He production and the same net energy release per ^{4}He nucleus formed.

When heavier nuclei are present, ^{4}He can be produced in catalyzed reactions. The "carbon" or CNO cycle is a good example:

$$\begin{aligned} \text{H} + {}^{12}\text{C} &\longrightarrow {}^{13}\text{N} + \gamma \\ {}^{13}\text{N} &\longrightarrow {}^{13}\text{C} + \beta^{+} + \nu \\ \text{H} + {}^{13}\text{C} &\longrightarrow {}^{14}\text{N} + \gamma \\ \text{H} + {}^{14}\text{N} &\longrightarrow {}^{15}\text{O} + \gamma \\ {}^{15}\text{O} &\longrightarrow {}^{15}\text{N} + \beta^{+} + \nu \\ \text{H} + {}^{15}\text{N} &\longrightarrow {}^{12}\text{C} + {}^{4}\text{He} \end{aligned}$$

The net result in the CNO chain is

$$\boxed{4\text{H} \longrightarrow {}^{4}\text{He} + 3\gamma + 2\beta^{+} + 2\nu} \qquad (5.11)$$

There is no net accumulation of ^{12}C in this chain; it acts only as a catalyst in the eventual formation of ^{4}He. Some small fraction ($\lesssim 0.1\%$) of the intermediary ^{15}N may produce the nuclides ^{16}O, ^{17}O, and ^{17}F.

The above processes, p-p and CNO, account for almost all (with CNO only about 1%) of the energy production in young stars. Once most of the proton fuel has been exhausted, helium fusion may occur. Since ^{8}Be is unstable to 2(^{4}He), the reaction

$$^{4}\mathrm{He} + {}^{4}\mathrm{He} \longrightarrow {}^{8}\mathrm{Be}$$

yields no net consumption of helium. Instead, the three body process

$$3(^{4}\mathrm{He}) \longrightarrow {}^{12}\mathrm{C}$$

consumes helium and begins the path to heavier nuclei. The helium fusion reaction has higher Coulomb barrier and, thus, requires an older, hotter star. Once ^{12}C is formed, then α-heavy particle reactions may proceed

$$\begin{aligned}
^{4}\mathrm{He} + {}^{12}\mathrm{C} &\longrightarrow {}^{16}\mathrm{O} + \gamma\,, \quad Q = 7.16\ \mathrm{MeV} \\
^{4}\mathrm{He} + {}^{16}\mathrm{O} &\longrightarrow {}^{20}\mathrm{Ne} + \gamma\,, \quad Q = 4.73\ \mathrm{MeV} \\
^{4}\mathrm{He} + {}^{20}\mathrm{Ne} &\longrightarrow {}^{24}\mathrm{Mg} + \gamma\,, \quad Q = 9.31\ \mathrm{MeV}
\end{aligned}$$

Stars become hotter and grow in size as these reaction occur. After their ^{4}He fuel is used up, stars may shrink, raising core temperatures, and fuse still heavier isotopes

$$\begin{aligned}
^{12}\mathrm{C} + {}^{12}\mathrm{C} &\longrightarrow \begin{cases} ^{20}\mathrm{Ne} + {}^{4}\mathrm{He} \\ ^{23}\mathrm{Na} + \mathrm{p} \end{cases} \\
^{16}\mathrm{O} + {}^{16}\mathrm{O} &\longrightarrow \begin{cases} ^{28}\mathrm{Si} + {}^{4}\mathrm{He} \\ ^{31}\mathrm{P} + \mathrm{p} \end{cases}
\end{aligned}$$

Fusion energy production ends with $A = 56$ nuclei (*i.e.* ^{56}Ni, ^{56}Co, ^{56}Fe). Heavier nuclei are produced in gravitational collapse.

5.6.3 Coulomb Barrier and Fusion Reaction Rates

The obstacle to fusion reactions in the laboratory is the reactant mutual repulsion, the Coulomb barrier (sec. 2.5). For the charged, light fuels of fusion, the Coulomb barrier is a substantial hindrance. In D-T, for example, deuterons and tritons are repelled by a potential of $B \sim 444$ keV, and it is almost 1 MeV for D-^{3}He. Despite the high barrier energy, tunneling allows a non-zero cross section at energies below B. Deuterium ions accelerated on a tritium target have a peak fusion cross section of about 5 b at 100 keV, but this falls off by about three orders of magnitude at 10 keV.

The accelerator approach to fusion is an important means of producing energetic neutrons, but falls short of practical energy production. To see this, we can estimate the average energy output per unit energy input in accelerating a deuteron beam

$$pE_{\mathrm{DT}}/E_b$$

where $E_{\mathrm{DT}} = 17.59$ MeV, E_b is the beam energy of 100 keV, and p is the interaction probability. This quantity must be greater than one to produce net energy. To be optimistic, let's assume the best of all conditions, *i.e.* solid tritium target ($\rho_{\mathrm{T}} \sim 0.2$ g/cm^3), maximum cross section ($\sigma \sim 5$ b at 100 keV), and $p \sim \Sigma_{\mathrm{DT}}\Delta x$. The latter assumes constant interaction probability over the deuteron's path ($\sim 6.7\ \mu$m) in the solid tritium target, a very optimistic assumption (sec. 6.1.2). Under these conditions, $N_{\mathrm{T}} \sim 4 \times 10^{23}$ cm^{-3} so that $\Sigma_{\mathrm{DT}} \sim 0.2$ cm^{-1} and $p \sim 1.34 \times 10^{-4}$. Even considering these very favorable conditions, the ratio pE_{DT}/E_b is only 0.024. Net energy production is impossible this way.

Thermal systems (Maxwellian particle distributions at temperature T, sec. 5.4.2) overcome these limitations by allowing very hot mixtures of fuels to thermally "cook" in a crucible called a reaction chamber. We then pay only the initial heating penalty. Once the fuel is heated from room temperature, fusion reactions provide the thermal energy to maintain the fuel at elevated temperature. (*i.e.* Energy extracted from fusion reactions can make up for heat transferred out of the system and the energy required to heat additional fuel introduced to compensate losses from leakage and burnup.) In order to make this practical, thermal energies sufficient to yield high fusion cross section are needed (1–20 keV, about (10–200) $\times 10^6$ K). Fluids at these temperatures attain the "plasma" state of matter where a large fraction of the fuel atoms are ionized by inelastic (ionization) collisions. Such a system need not have thermal energy, kT, at the fusion cross section peak to attain high reaction rates. Particles in the high energy tail of a distribution with lower T somewhat compensate. To calculate the fusion reaction rate correctly for a thermal system, it is necessary to average Eq.(3.18) over the energy dependent cross section and particle number density to find

$$\mathcal{R} = N_{\mathrm{D}} N_{\mathrm{T}} \langle \sigma v \rangle \tag{5.12}$$

where v is the relative velocity of the interacting particles. Values of the fusion reaction rate parameter, $\langle \sigma v \rangle$, vs. fuel temperature in keV for several fusion reactions are shown in Fig. 5.6.

To maintain reacting fluids under the conditions for fusion requires extreme measures. The plasma at the sun's core is maintained in this state by virtue of its enormity. The gravitational potential of its tremendous mass contains the burning fuel for p-p fusion against huge internal pressures,

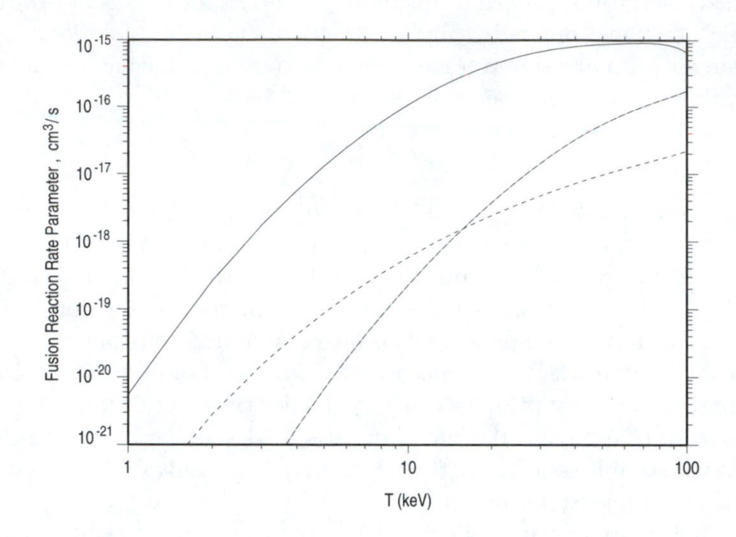

Figure 5.6: Fusion reaction rate parameter, $\langle \sigma v \rangle$ (cm^3/s), vs. fuel temperature (keV) for several important fusion reactions: D-T (solid line), D-D (dashed line), and D-^{3}He (dot-dash line).

Example 33: Thermonuclear Fusion Reaction Rate
Calculate the fusion reaction rate for D-T fuel with $N_{\mathrm{D}} = N_{\mathrm{T}} = 10^{14}$ cm^{-3} under the following conditions: (a) 10 keV deuteron beam on a cold tritium plasma and (b) thermal D-T fuel at 10 keV.

Solution:
(a) A deuteron at $v = v_{\mathrm{D}} = \sqrt{2E_b/m_{\mathrm{D}}} \sim 10^8$ cm/s is incident on a cold tritium plasma. The tritium reservoir being cold simply means that $kT_{\mathrm{T}} \ll E_b$. Then, the interaction energy is just E_b for which the fusion cross section is $\sigma_{\mathrm{DT}}(10 \text{ keV}) \sim 2$ mb. Then

$$\mathcal{R} = (10^{14})^2 2 \times 10^{-3} \times 10^{-24}(10^8) \sim 2 \times 10^9 \text{ fus/(cm}^3 \text{ s)}$$

(b) When these fuels react in a thermal system, we find $\langle \sigma v \rangle_{\mathrm{DT}} \sim 10^{-16}$ cm^3/s, for which $\mathcal{R} \sim 10^{12}$ fus/(cm^3 s), about 500 times greater.

$p = NkT \sim 2(10^{32}\ \mathrm{m}^{-3})1.38 \times 10^{-23}\ \mathrm{J/K}(15 \times 10^6\ \mathrm{K}) \sim 4.14 \times 10^{16}\ \mathrm{Pa} \sim$ 414 billion earth atmospheres.

Assembling such gravitational containment on earth is an obvious impossibility. Yet, we are attempting to replicate a fusion environment on a small scale. Thermonuclear weapons reach these conditions for a fleeting instant by rapidly compressing fusion fuels with fission explosives as initiators. This is an uncontrolled release of fusion power. In a controlled fashion, fusion has not yet produced net energy in the lab. This eventuality is the dream of many scientists worldwide, though we may be getting close. A technique conceived as a miniature mock-up of a thermonuclear detonation, called inertial confinement fusion (ICF), is expected to reach energy "break-even" and beyond within the next decade in the National Ignition Facility (NIF) device planned to be built in Livermore, CA. In ICF, tiny frozen D-T pellets on the order of 1 mm in diameter are compressed to fusion conditions by lasers, ion or electron beams. Repetitive pulsing of such devices 1–10 times per second is required to produce the continuous electrical energy output of $\sim$ 1000 MW. Energy *break-even* is the condition whereby equal quantities of energy are generated in fusion reactions as is applied to the fuel to initiate and sustain the burn. *Ignition* is the condition wherein a fraction of the fusion product energy generated in fusion reactions is used to sustain burn conditions, so that after an initial initiation input of energy, there is no longer required any input of external energy.

Sustaining burning plasma conditions in a steady system is equally challenging, if not more. Fuels at the required temperature ($\sim$ 10 keV) are not quite compatible with solid materials. The question thus arises how to house such hot fusion fuels. From what do we make the "crucible"? The concern is not so much that plasmas may melt or vaporize solid containing walls. Though they are at extremes of temperature, the particle densities in plasmas are many orders below solid or even STP gas density so that comparatively little heat is transferred to the chamber walls. Rather, maintaining a hot plasma in contact with cold (300 K) surfaces, has the effect of cooling the high thermal conductivity plasma and quenching the fusion burn. Instead of confining with solid walls, plasmas, being made up of free ions and electrons, are good conductors of electricity and may be influenced and contained by magnetic fields. Such an approach is classified "magnetic confinement fusion" (MCF). Throughout the world, there are several devices of this variety showing progress in various aspects of magnetic containment of plasmas and approach to fusion conditions. The best performance thus far is found in a toroidal device called the "tokamak", an abbreviation of the Russian for toroidal chamber with magnets. Toroidal devices are effective because plasma particles can follow magnetic field lines in a closed toroidal (racetrack) configuration, increasing the difficulty for these particles to leave the chamber interior and reach the chamber walls.

Record fusion conditions were recently (1997) achieved in the Joint European Torus (JET) tokamak device in England. More than 12 MW of fusion power and 11 MJ of fusion energy were produced in a single experiment with only 24 MW of input power. The latter implies a 50% fusion to input power ratio. All three results are records to date of this writing. Though impressive, these results are still far from break-even conditions, let alone energy production. A new, still larger, device ITER (International Thermonuclear Experimental Reactor), designed to take the next step toward fusion energy production, is now in the engineering design phase.

5.6.4 The Promise and Challenge of Fusion

Controlled thermonuclear power holds great promise for our world's future energy needs. It is an energy source with tremendous energy density, as we have seen. But, by far, fusion's greatest advantages are that it is a clean energy source with a very large reservoir of readily available fuel. It is a clean source of power in so far as it burns no fossil fuels, thereby producing no polluting effluent. What's more, the reaction by-products (product particles) are none other than benign ^{4}He in the case of D-T fusion and ^{3}He, T, and H in the case of D-D fusion. Though the radioisotope tritium is present, the inventory and associated potential hazard are trivial in comparison to that in spent fission fuels. Fusion neutron activation of the reactor chamber and surrounding structure may accumulate to a rather substantial radionuclide inventory, but one which is trapped at the reactor site and not in danger of remote contamination.

The practically inexhaustible fuel reserves, however, places fusion as the most important and highly sought after energy source in the history of mankind. There is, in fact, enough deuterium in the world's oceans to supply all the planet's energy needs for billions of years. A simple calculation illustrates the potential. Deuterium is abundant as 0.015% of all hydrogen isotopes. The oceans cover about 71% of the earth's surface at an average depth of about 12,500 feet ($\sim$ 3.8 km). Since this depth, d, is much less than the earth's radius, $R_e = 6.37 \times 10^3$ km, we can estimate the earth's water volume

$$V_{\mathrm{H_2O}} \sim 0.71\left(4\pi R_e^2\right) d \sim 1.37 \times 10^{18}\ \mathrm{m}^3$$

In this volume of water, there are

$$\mathcal{N}_{\mathrm{D}} = 2(0.00015)\mathcal{N}_{\mathrm{H_2O}} = 1.375 \times 10^{43}\ \text{atoms D}$$

This number of deuterons burned in a catalyzed D-D reactor (7.2 MeV/D) represents a net energy of $\sim 4.8 \times 10^{30}$ J after a 30% efficient thermal to electric cycle. For the year 1995, about 4.7×10^{19} J of electrical energy was consumed worldwide (sec. 1.2.1). Catalyzed D-D burning of seawater can then supply this constant rate for more than 100 billion years. Even allowing for a very liberal growth in electric energy demand and complete elimination of all fossil fuel burning still does not detract from the conclusion. There is more fusion fuel available than earth inhabitants will ever need.

In spite of the optimistic potential, extracting net energy from a burning plasma has been elusive. The reasons are varied and many. As this is not a text on plasma physics or fusion technology, we'll not delve into the specifics. Rather, it is more appropriate to point out some of the very difficult challenges inherent in such a task.

We have already mentioned the Coulomb barrier and that it necessitates high temperature fuel in the plasma state to reach fusion conditions. The plasma state is very tenuous and notorious for being wildly unstable. It is not difficult to see why. Heating a fluid to fusion temperatures requires a tremendous quantity of energy. This burden increases with plasma density. But more importantly, high density plasmas are difficult to contain. Thermonuclear plasmas are intentionally driven (heated) far from thermal equilibrium with their surroundings. At higher densities and pressures, acoustic instability is inevitable. Even without instability, containing the plasma kinetic pressure by the reactor chamber and support structure is no trivial task. At reactor conditions of $\sim 10^{15}$ cm^{-3} and 10 keV, the kinetic pressure becomes $\sim$ 16 atm. Raising the density to near atmospheric gas densities ($\sim 10^{19}$ cm^{-3}) increases the pressure to in excess of 10^5 atm! The constraint of low particle density necessitates the complexity of vacuum conditions, but does provide the advantage of keeping the plasma thermal content rather low and avoiding any possibility of an explosion hazard. In addition, plasmas are electrically conducting fluids. Although this provides for magnetic confinement, plasma electrodynamics introduces many additional modes of motion and instability. Meeting all of the constraints is no easy task. Plasmas have many, many modes of characteristic motion.

Plasmas must be contained for a sufficiently long time to allow a reasonable probability for fusion reactions. An extension of the problem in example 33 will illustrate. The D-T fusion cross section for 10 keV (10^8 cm/s) deuterons into a cold tritium plasma is about 2 mb. The D-T fusion mean free path for such particles at 10^{14} cm^{-3} is

$$\lambda_{\text{mfp}} = (N_{\text{T}}\sigma)^{-1} \sim 5 \times 10^{12} \text{ cm}$$

These particles then need to spend an average of $t \sim \lambda_{\text{mfp}}/v \sim 5 \times 10^4$ s ($\sim$ 14 h) in the fusion device before being consumed in fusion. The problem is reduced to about 100 s and 10^{10} cm by employing thermal distributions of D-T at 10 keV. As well, not every particle in this system needs to fuse to produce an net energy gain. Yet, this simple calculation does point to a difficulty. It is quite a daunting task to contain such energetic particles in a reasonable sized system for this path length before it leaks out due to Coulomb interactions with other plasma particles. (This type of interaction allows plasma particles to escape the grip of the confining magnetic field.) Even for a closed device like a large tokamak (perhaps with radius of as much as 3 m), our 10 keV deuteron must make $\sim \frac{10^{10}\ \mathrm{cm}}{2\pi R} \sim 5.3 \times 10^6$ transits before fusing.

Further problems abound. The challenges of energy containment in plasma systems is often severely compromised by photon radiation. Hot plasmas consist of energetic charged particles which emit photon radiation when colliding with other charged particles in the plasma. This radiation is called "bremsstrahlung," German for breaking radiation (sec. 6.1.3), most often in the X-ray range of the electromagnetic spectrum for fusion plasmas. Controlled thermonuclear plasmas are much too rarefied to be in thermal equilibrium with this radiation. The photons are then lost from the plasma. In addition, impurity particles, gas molecules leaking into the evacuated reaction chamber or metal atoms evolved from the chamber inner walls, also contribute to radiation loss. These atoms may achieve states of ionization other than fully stripped, thereby allowing atomic transitions to radiate power out of the plasma. Such radiators are often quite efficient energy loss mechanisms and may prevent some plasma system from reaching fusion conditions. They do sometimes have an advantage, however. Accident scenarios involving runaway fusion reactions and vessel melt down are eliminated from possibility because both radiation from the ingress and evolution of impurity particles and the reduction of reaction rate parameter at high temperatures. A fusion power reactor can never melt down or explode.

The hurdles in front of fusion, though high, are gradually being overcome. One day soon, perhaps in our lifetime, we may realize the dream of harnessing practically limitless fusion power for peaceful purposes.

References

[Dolan82] Dolan, T. J., *Fusion Research: Vol.1, Principles*, Pergamon Press, NY, 1982.

[Duderstadt76] Duderstadt, J. J., Hamilton, L. J., *Nuclear Reactor Analysis*, John Wiley & Sons, NY, 1976.

[Eisberg85] Eisberg, R., and Resnick, R., *Quantum Physics of Atoms, Molecules, Solids, Nuclei, and Particles*, John Wiley & Sons, NY, 1985.

[Evans55] Evans, R. D., *The Atomic Nucleus*, McGraw Hill, Inc., NY, 1955.

[Firestone96] Firestone, R. B., Shirley, V. S., Baglin, C. M., Chu, S. Y. F., and Zipkin, J., *Table of Isotopes*, 8^{ed}, John Wiley & Sons, Inc., NY, 1996.

[Foukal] Foukal, P. V., *Solar Astrophysics*, John Wiley & Sons, Inc., NY, 1990.

[GE96] General Electric Co., *Nuclides and Isotopes*, 15ed., 1996.

[Glasstone60] Glasstone, S., Loveberg, R. H., *Controlled Thermonuclear Reactions*, Van Nostrand, NY, 1960.

[IAEA69] *Physics and Chemistry of Fission*, Proceedings of the Second IAEA Symposium on the Physics and Chemistry of Fission, IAEA, Vienna, 1969.

[Kaplan63] Kaplan, I., *Nuclear Physics*, Addison-Wesley, Inc., Reading, 1963.

[Keepin65] Keepin, G. R., *Physics of Nuclear Kinetics*, Addison Wessley, Reading, MA, 1965.

[Krane88] Krane, K. S., *Introductory Nuclear Physics*, John Wiley & Sons, NY, 1988.

[Lamarsh66] Lamarsh, J. R., *Introduction to Nuclear Reactor Theory*, Addison-Wesley, Reading, MA, 1966.

[Lamarsh83] Lamarsh, J. R., *Introduction to Nuclear Engineering*, 2^{ed}, Addison-Wesley Publishing Co. Inc., Reading, MA, 1983.

[Michaudon81] Michaudon, A., *Nuclear Fission and Neutron-Induced Fission Cross-Sections*, Pergamon Press, NY, 1981.

[Vandenbosch73] Vandenbosch, R., and Huizenga, J. R., *Nuclear Fission*, Academic Press, Inc., NY, 1973.

[Wagemans91] Wagemans, C., *The Nuclear Fission Process*, CRC Press, Boca Raton, FL, 1991.

[Wahl88] Wahl, A. C., *Nuclear-Charge Distribution and Delayed-Neutron Yields for Thermal-Neutron-Induced Fission of* ^{235}U, ^{233}U, *and* ^{239}Pu *and for Spontaneous Fission of* ^{252}Cf, Atomic Data and Nuclear Data Tables, **39** (1988) 1.

[Watt52] Watt, B. E., *Energy Spectrum of Neutrons from Thermal Fission of* ^{235}U, Phys. Rev., **87** (1952) 1037.

Problems and Questions

1. Explain what is meant by "fissile," "fertile," and "fissionable" nuclei. Give an example of each.

2. Write down and explain the multi-step fission process.

3. Examine the effect of nuclear pairing on SF stability by calculating $Q_{\text{l.d.}}$ including the pairing energy from the liquid drop model. Calculate $Q_{\text{l.d.}}$ for N, Z: (a) even-even, (b) odd-odd, and (c) odd-even, in all cases keeping $A \sim 200$.

4. The following nuclei have induced fission critical energies as listed below. From these data, determine if the target nuclei are fissile or fissionable.

		Target Nucleus		Fissioning Nucleus	E_{crit} (MeV)
n	+	^{230}Th	→	^{231}Th	6.2
n	+	^{237}Np	→	^{238}Np	6.0
n	+	^{241}Am	→	^{242}Am	6.4
n	+	^{242}Pu	→	^{243}Pu	5.8
n	+	^{244}Cm	→	^{245}Cm	6.2
n	+	^{248}Cm	→	^{249}Cm	5.5

5. For the thermal neutron-induced fission reaction

$$\text{n} + {}^{235}\text{U} \longrightarrow {}^{84}\text{Br} + {}^{148}\text{La} + \nu_p \text{n}$$

show that nucleon (N, Z, A) conservation holds for the primary products listed above and determine ν_p. Show the complete decay chains to stable fission products. Determine the total fission energy release (Q_f).

6. Consider a 3000 MW_{th} power reactor operating at steady conditions for one year. Find expression for the ^{90}Kr and ^{90}Sr activity during operations and after shutdown. How much activity of each (in Ci) are present after 10 min and 1 y reactor operation? How much activity of each (in Ci) are present 30 min after the reactor has shut down?

7. One thermal neutron-induced fission event in ^{235}U leads to the prompt production of four neutrons and the primary fission product ^{121}Ag. Find the identity of the other primary fission product. What is the Q' value in this primary reaction (*i.e.* neglecting beta decay)? If the primary fragments carry 150 MeV of the liberated energy, estimate the kinetic energy shared by the prompt fission neutrons.

8. Consider the following reaction:

$$\mathrm{n} + {}^{235}\mathrm{U} \longrightarrow {}^{90}\mathrm{Kr} + {}^{142}\mathrm{Ba} + 4\mathrm{n}$$

Determine the decay chains for the primary fragments in the above fission reaction. Determine the Q' and Q_f values.

9. Integrate the distribution functions to find the average and most probable energies in the Maxwellian and Watt neutron distributions. Also, show that the Watt distribution goes over to the Maxwellian distribution in the limit $b \to 0$.

10. The delayed neutron precursor ^{137}I has an independent fission yield of $\sim$ 3.0% in ^{235}U thermal fission. If its decay branches to neutron emission 8% of the time, estimate the delayed neutron emission rate for this product as a function of time for a 3000 $\mathrm{MW_{th}}$ reactor. What is the delayed neutron kinetic energy from this reaction if $\mathrm{KE}_{\beta^-}|_{\max} \sim 1.2$ MeV?

11. The fission product poison ^{149}Sm is produced as the stable end product of the $A = 149$ chain in ^{235}U thermal fission.

$$\begin{array}{ccccc} {}^{149}\mathrm{Nd} & \xrightarrow{\beta^-} & {}^{149}\mathrm{Pm} & \xrightarrow{\beta^-} & {}^{149}\mathrm{Sm}\ \text{(stable)} \\ \uparrow & & & & \\ \text{fission} & & & & \end{array}$$

Determine an expression for the ^{149}Sm inventory after reactor startup. If the independent yield for ^{149}Nd is 1.1% and $\sigma_a = 58,700$ b for ^{149}Sm, find the equilibrium concentration of ^{149}Sm when $\Sigma_f \sim 0.2\ \mathrm{cm}^{-1}$. Find an expression for the ^{149}Sm concentration 5 months after the reactor is shutdown.

12. How many thermal neutron-induced fission events per second in ^{235}U are required to sustain power to one 100 W light bulb? What mass of ^{235}U is required to power this bulb for one day? Repeat for DT fusion and coal.

13. A 60 W light bulb is operated, on average, for 10 h a day for 52 weeks. How many thermal neutron-induced fission events in ^{235}U are required to keep this light bulb lit for this period? How many grams of ^{235}U would be needed to keep this light bulb lit for this period of time? How many grams of coal would be needed instead to power the bulb for the same time?

14. Along the new corridor of Interstate-85 through Burlington, NC, there are approximately 100 street lights rated at 500 W each. If these lights

come on for an average of 10 h every day, determine how much ^{235}U will be needed to be fissioned by a thermal neutron reactor to keep the lights lit for 10 y. How many tons of coal would be needed to replace the ^{235}U fission power? If D-D fusion were an option, how many grams of deuterium would be needed to replace the ^{235}U? Assume the conversion to electricity is 30% for all of these fuels.

15. What energy is required to overcome the Coulomb barrier of the p-p fusion reaction? For an initially distant proton at rest, what mass stellar object is required to accelerate the proton to the barrier energy (found above) as the proton reaches the star surface? Assume the star is comprised of constant density matter at $\rho \sim 1400$ kg/m^3.

16. Estimate the D-D fusion energy stored in an eight ounce glass of ordinary drinking water. For how long would this energy store power a 1 kW toaster?

17. What plasma density ($N = N_D = N_T$) would be required to generate 3000 MW$_{th}$ in a 10 m^3 fusion reactor at 5 keV? What plasma temperature is required for the same size and density reactor fueled with D-^{3}He? Calculate the plasma kinetic pressure in atmospheres for both cases.

Chapter 6

Interaction of Radiation with Matter

In what has preceded, we have discussed in some detail the nature of interactions involving the atomic nucleus. Whether in binary interactions like induced fission or single reactant events like radioactive decay, of primary interest has been the kinematic properties of such processes. With rare exception, little attention has been paid to the fate and consequences of the energetic reaction products. Yet in the *real* world (our macroscopic, every day existence), these are of great concern.

In this chapter, we will explore the nature of interactions among particles of nuclear radiation with material media. The physical description of these interactions is of paramount importance since it gives us the fundamental tools necessary to quantify the affect of radiation on all material objects including ourselves. Such quantitative description of material damage allows us to identify precautionary measures, construct protective barriers, and establish tolerance limits in a rational way. Furthermore, since particles of nuclear radiation cannot be detected directly, it is only through their interactions with matter that we may identify their existence and determine their properties. As well, these energetic particles have the potential, *en masse*, to affect material damage and alter macroscopic properties of materials. One excellent example is the radiation embrittlement of fission reactor vessel steel. The continuous bombardment on this structure by fission neutrons may, over time, accumulate to significant atomic displacement so as to substantially alter the vessel ductility, making it more susceptible to fracture.

Though there are many possible interaction mechanisms involving the many possible radiation particles (we'll discuss several in detail shortly), there are a couple of commonalities. Firstly, all particles of nuclear radiation possess initial energy much greater than the thermal energy of the material medium with which they interact. These particles are initially far from thermodynamic equilibrium with their surroundings and interact in such a way as to deposit their incident energy in the medium, continuously tending toward thermalization. Eventually, all of the particle's initial energy appears as either latent heat or raises the thermodynamic temperature, depending on the state of the system. We have already seen an example of this in our discussions of neutron slowing down (chpt. 3). Secondly, all nuclear radiation is "ionizing." That is, it possesses sufficient initial energy so that it may directly ionize ordinary matter, or it may produce such ionization by its secondary radiations. As an example of the former, a β^- particle is ionizing since it has initial kinetic energy much greater than the ionization potential, I, of all ordinary materials, and interacts strongly with atomic electrons. When such a particle collides with an atomic electron, the electron may be "ejected" from its bound state. The latter case is a bit more subtle. We consider, for example, thermal neutrons to be ionizing even though they possess (on average) insufficient kinetic energy to ionize matter. They may, however, produce very energetic charged particles or photons in absorption reactions, and it is these products that perform ionization directly.

The discussion of radiation interactions is in keeping with the established theme of this text in so far as being a kinematic development of binary interactions. We must stray, however, from the strictly nuclear context. Radiation interactions (excluding neutron interactions) are primarily atomic, *i.e.* in the energy range of interest, roughly 0.01–10 MeV, ionizing radiation acts predominantly with the atomic electrons in ordinary matter.

It facilitates the discussion to group radiation interactions into two categories, those that involve light and heavy charged particles, and those in which electrically neutral particles participate. Neutrons and γ-rays fall into this latter category. (We should also lump hard X-rays into this group. Though not nuclear in origin, hard X-rays are ionizing and overlap the γ-ray spectrum.) These particles are distinguished by their interaction with individual atomic electrons or nuclei. As we have seen for neutrons (chpt.3), this feature results in a constant interaction probability per unit length so that the uncollided flux is diminished exponentially and without loss of energy. This is in stark contrast to the interaction character of charged particles which interact via the long range Coulomb force with many electrons and nuclei simultaneously. The attenuation of charged particles is not exponential in general, and there is continuous loss of kinetic energy along the particle's trajectory. We'll begin our discussion of radiation interactions with a close look at classical Coulomb scattering and charged particle stopping in matter, before investigating γ-ray interaction mechanisms.

6.1 Charged Particle Interactions

Charged particles interact through the Coulomb force with the particles (nuclei and electrons) that comprise matter. Of our immediate concern is the stopping in matter of the charged particles originating in nuclear processes, the products of decay and nuclear reactions; $\beta^{\pm}$, e^- (fast electrons), p, α, ... (light ions), and A^+ fission fragments (heavy ions). Though greatly disparate in rest mass, the fundamental interaction mechanism is substantially similar. Let's begin by developing the classical Coulomb scattering cross section so that we may discuss energy exchange in such interactions before elaborating on charged particle stopping in matter in the following sections.

6.1.1 Coulomb Scattering

Consider the situation depicted in Fig. 6.1. Incident charged particle m_x with charge Z_x approaches target m_{X}, Z_{X} at lab velocity $\mathbf{v}_x$. The completely general problem of this sort is quite difficult indeed. Let's make several assumptions for simplicity and analytic tractability. Though potentially limiting in strict application, these will not detract from the basic physics and our conclusions about charged particle interactions. Firstly, we consider projectile m_x to interact with only one target m_{X} at a time and only through the Coulomb force. Secondly, the projectile is initially of sufficient energy to be considered "fast" or "swift." That is, its kinetic energy is much greater than the energy of thermal agitation in the medium and much greater than the atomic ionization potential, I. The latter allows us to consider the interaction as occurring among "free" (unbound) particles. Such interactions are elastic, and as we know, elastic interactions conserve the kinetic energy in the system so that whatever energy is lost by the projectile is gained by the target. The fast projectile assumption also allows the target to be considered effectively stationary in the lab system. We must, however, restrict the projectile speed so that $v_x \ll c$, the non-relativistic approximation. This third assumption usually poses no difficulty until treating electron interactions, a limitation that does not detract from our conclusions regarding the nature of such interactions and one that will be overcome later (sec. 6.1.3).

To facilitate analysis, let's fourthly assume that the target remains stationary throughout the interaction, *i.e.* X is of infinite mass. In this limit, the lab (L) and center of momentum (C) systems are identical. For the case of repulsion (like sign interaction, ala α particle scattering on a heavy nucleus), this becomes the Rutherford scattering problem. Rutherford, in fact, employed the results of the model we are about to develop to explain the large angle scattering of ^{214}Po α particles on thin foil targets of gold,

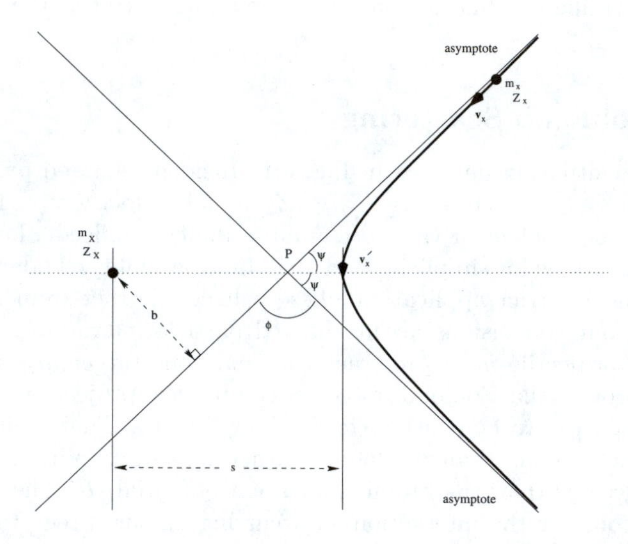

Figure 6.1: Repulsive Coulomb scattering event between projectile m_x, Z_x with initial speed v_{x_o} and a stationary target m_X, Z_X. The thin solid lines are asymptote to the projectile trajectory and define the impact parameter b. The distance of closest approach s is also shown.

silver, copper, and aluminum. Since the experimental results were consistent with only the Coulomb scattering interpretation of energetic charged particles on a small, dense atomic nucleus, Rutherford is thus credited with introducing this standard model of the atom.

The total energy in such a system can be written as the sum of time or trajectory dependent projectile kinetic and potential energies

$$\mathrm{KE}_x + P_x = \frac{1}{2} m_x v_x^2(t) + \frac{e^2 Z_x Z_{\mathrm{X}}}{4\pi\epsilon_o r(t)}$$

where $r(t)$ is the line-of-sight distance between X and x at all times t. The initially distant (non-interacting or free) projectile with speed v_{x_o} possesses energy that is entirely kinetic. As the total energy of the system must be conserved throughout the interaction, we express a balance between the initial projectile kinetic energy and the shared kinetic and potential energy at the turning point

$$\frac{1}{2} m_x v_{x_o}^2 = \frac{1}{2} m_x (v_x')^2 + \frac{e^2 Z_x Z_{\mathrm{X}}}{4\pi\epsilon_o s}$$

where s is the distance of closest approach and v_x' is the projectile speed at that position (Fig. 6.1). On the approach portion of the projectile trajectory, it is being continuously slowed by the repulsive interaction with X, while on departure it is continuously accelerated. Since X is infinitely massive, the final speed for x is also v_{x_o}, though it has been deflected through lab scattering angle

$$\phi = \pi - 2\psi$$

from its original trajectory. At the turning point, the projectile no longer possesses any linear momentum in the direction toward the target. Such a trajectory can be shown to be hyperbolic[Evans55]. For the special case where the impact parameter, b, is identically zero (*i.e.* a head-on collision), then $v_x' = 0$ also, and the closest approach distance is minimized to

$$\delta = s|_{\mathrm{min}} = \frac{e^2 Z_x Z_{\mathrm{X}}}{4\pi\epsilon_o E_o}$$

called the "collision diameter." Here, $E_o = \frac{1}{2} m_x v_{x_o}^2$ is the initial projectile kinetic energy in arriving to within δ of X. For nuclear reactions, when $\delta = R$ (the nuclear radius), E_o is the familiar Coulomb barrier, B (sec. 2.5).

Rearranging the energy equation for the more general case, we have

$$1 = \left(\frac{v_x'}{v_{x_o}}\right)^2 + \frac{\delta}{s}$$

Conservation of angular momentum about X provides another expression for $v_x' = f(v_{x_o})$

$$m_x v_{x_o} b = m_x v_x' s$$

so that combined with energy conservation, we have

$$b^2 = s\,(s - \delta) \tag{6.1}$$

The approach distance s can be related to the scattering angle ϕ by the equations of a hyperbolic curve. Assigning the length a' to be the distance from X to point P and a as the distance from P to the turning point, then

$$a' = a\left(1 + \tan^2\psi\right)^{1/2} = a \sec\psi$$

Since $b = a' \sin\psi$, we find

$$s = a + a' = b\frac{1 + \cos\psi}{\sin\psi} = b \cot\frac{\psi}{2}$$

Finally, utilizing Eq.(6.1), we obtain the result

$$\boxed{b = \frac{1}{2}\delta \cot\frac{\phi}{2}} \tag{6.2}$$

The meaning of the collision diameter is thus revealed. When $b = \delta/2$, then the scattering angle is $\phi = \pi/2$.

We now have the tools to calculate the relative interaction probability for such events, *i.e.* the interaction cross section. Since the interaction probability is strongly dependent on scattering angle, let's define

$\sigma(\mathbf{\Omega})\, d\Omega$ = cross section for scattering into differential solid angle $d\Omega$ about $\mathbf{\Omega}$

where $\mathbf{\Omega} = \mathbf{v}/|\mathbf{v}|$ is the unit direction vector after the interaction. Such quantities are known as differential cross sections. There are many such entities that may be differential in several independent variables. We'll not dwell on the definitions of these or their properties except to note that the differential cross section above is integrally related to the cross section through

$$\sigma = \int_{\Omega} \sigma(\mathbf{\Omega})\, d\Omega$$

It is important for our purposes that we carry the angular information, however, since there is a one-to-one correspondence between scattering angle and energy exchange.

Since the differential cross section represents that fraction of scattering events which scatter projectiles into solid angle $d\Omega$ about $\mathbf{\Omega}$ (Fig. 6.2), it follows that

$$\sigma(\mathbf{\Omega})\, d\Omega = -2\pi b\, db$$

where the factor 2π comes from integration over the azimuth, Φ, (*i.e.* $d\Omega = d\Phi \sin\phi\, d\phi$) and exploiting azimuthal symmetry. The factor -1 comes from the inverse dependence of b on ϕ. From Eq.(6.2), it follows that

$$db = -\frac{1}{4}\delta \csc^2\frac{\phi}{2}\, d\phi$$

from which we can find directly

$$\sigma(\mathbf{\Omega})\, d\Omega = \frac{\pi\delta^2}{4}\left(\frac{1}{2}\sin\phi\right)\frac{d\phi}{\sin^4\frac{\phi}{2}}$$

But since $d\Omega = 2\pi \sin\phi\, d\phi$ also, we finally arrive at

$$\boxed{\sigma(\mathbf{\Omega})\, d\Omega = \frac{\delta^2}{16}\frac{d\Omega}{\sin^4\frac{\phi}{2}}} \tag{6.3}$$

the classical Rutherford scattering cross section. We may also choose to write this in a form that is explicit in particle properties by employing the definition of δ

$$\sigma(\mathbf{\Omega})\, d\Omega = \left[\frac{e^2 Z_x Z_{\mathrm{X}}}{8\pi\epsilon_o m_x v_{x_o}^2}\frac{1}{\sin^2\frac{\phi}{2}}\right]^2 d\Omega$$

A more general result, allowing the recoil of a finite mass target, is found by replacing the projectile mass with the reduced mass of the system

$$\mu = \frac{m_x m_{\mathrm{X}}}{m_x + m_{\mathrm{X}}}$$

and the lab scattering angle with the C system angle ϕ_c, so that

$$\boxed{\sigma(\mathbf{\Omega})\, d\Omega = \left[\frac{e^2 Z_x Z_{\mathrm{X}}}{16\pi\epsilon_o(\mu/m_x)E_o}\frac{1}{\sin^2\frac{\phi_c}{2}}\right]^2 d\Omega} \tag{6.4}$$

where, again, $E_o = \frac{1}{2}m_x v_{x_o}^2$ is the projectile initial kinetic energy in the lab system.

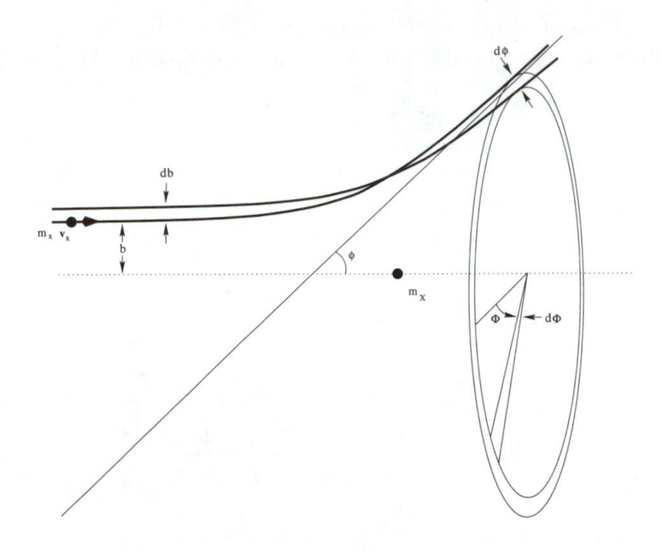

Figure 6.2: Coulomb scattering interaction redrawn from Fig. 6.1 to show the differential scattering angle dependence on impact parameter.

Before continuing, it should be mentioned that Eq.(6.4) is in agreement with the exact quantum-mechanical result only when the normalized de Broglie wavelength of the projectile is small compared to the collision diameter

$$\frac{\lambda_{\text{deB}}}{2\pi} \ll \delta$$

which reduces to

$$\frac{v_{x_o}}{c} \ll \frac{2Z_x Z_\text{X}}{137}$$

or

$$E_o \ll \left(\frac{2Z_x Z_\text{X}}{137}\right)^2 m_x c^2$$

This condition is easily satisfied, requiring only that $E_o \ll 85$ GeV for α particles incident on gold nuclei as per the Rutherford problem.

Since we are interested in energy loss (slowing down) of the projectile, it is much more instructive to express the differential scattering cross section as a function of energy transfer rather than scattering angle. In keeping with our binary collision notation of chapter 2, we express the energy exchange (T) in the lab system when the target is initially at rest as equal to the energy given the target

$$T = E_\text{Y} = E_x - E_y$$

Utilizing expressions (2.37) and (2.39), we immediately find

$$T = 4\frac{m_x m_X}{(m_x + m_X)^2} E_o \sin^2\frac{\phi_c}{2} = 4\frac{\mu^2}{m_x m_X} E_o \sin^2\frac{\phi_c}{2} \tag{6.5}$$

The energy transfer is maximized for $\phi_c = \pi$, head-on collisions, where

$$T_{\max} = 4\frac{m_x m_X}{(m_x + m_X)^2} E_o = 4\frac{\mu^2}{m_x m_X} E_o \tag{6.6}$$

Using the expression above for T (Eq.(6.5)), we can write

$$dT = 2\frac{\mu^2}{m_x m_X} E_o \sin\phi_c \, d\phi_c$$

so that the differential Coulomb scattering cross section (Eq.(6.4)) can be re-written with the aid of $\sigma(T)\,dT = \sigma(\mathbf{\Omega})\,d\Omega$

$$\boxed{\sigma(T)\,dT = \frac{e^4 Z_x^2 Z_X^2}{16\pi\epsilon_o^2 E_o}\left(\frac{m_x}{m_X}\right)\frac{dT}{T^2}} \tag{6.7}$$

From this result, it is evident that energetic charged particles more readily deposit energy in material media on Coulomb interactions with atomic electrons, *i.e.* the ratio m_x/m_X is maximized for target electrons for any species of incident particle.

Two limiting cases appear: like-particle interactions, and heavy projectile on light target interactions. In like-particle interactions (*i.e.* e-e), one finds $T_{\max} = E_o$, the projectile deposits all its original kinetic energy in a single head-on collision, exactly as we saw for neutrons (sec. 3.1), showing that this is a general result, independent of the nature of the interaction. In the opposite limit, the energetic projectile is much more massive than the target, *i.e.* α-e for which $T_{\max} = 4(m_e/m_\alpha)E_o$, and only a tiny fraction of the initial α kinetic energy is lost to the atomic electron. The maximum speed given to a single electron in a head-on interaction is

$$v_e|_{\max} = \left(2\frac{T_{\max}}{m_e}\right)^{1/2} = 2v_\alpha$$

This is a familiar classical result. It is exactly the speed, for example, that would be attained by any light object struck by a much heavier one as represented in the frame of the light object before the interaction. (Test this by considering the recoil speed of a rubber ball thrown against a brick wall assuming an elastic collision.)

Example 34: Heavy Charged Particle Interactions with Electrons
Consider a single Coulomb interaction between a 5 MeV α particle and an electron. The maximum energy transferred in for such a collision is

$$T_{\max} = 4\frac{m_e}{m_\alpha}5 \text{ (MeV)} \sim 2.7 \text{ keV}$$

about 1/2000 of the initial α kinetic energy. The implication is that many thousands of such interactions are required to slow the incident particle of radiation to thermal energies.

6.1.2 Light and Heavy Ion Slowing Down

The passage of ions through matter is largely inhibited by Coulomb interactions with atomic electrons as shown in the previous section. In so far as this is the dominant mechanism determining ion passage through matter, all ions can be treated similarly, at least to first order. All ions are considered "heavy charged particles" with respect to their atomic electron scattering target, though it will be useful to discuss subtleties between the passage of "light ions" (like the familiar particles of radiation, p, d, α, ...) and "heavy ions" (like fission fragments).

Let's first consider some implications from the results in example 34. It is clear that since only a tiny fraction of the incident ion's kinetic energy is lost in a single interaction, many thousands of such interactions are required to bring this particle to rest. Targets that stop these energetic ions are called "thick absorbers" as they are thick enough to completely stop the incident particle and the incident kinetic energy is absorbed in the target. From example 34, we can estimate that at least 2000 of the maximum energy exchange interactions would be required to absorb all the initial α energy. In reality, this represents a terribly gross underestimate. Most interactions do not exchange $T_{\max}$. The average energy transferred, $\langle T \rangle$, is quite a bit less. As well, $\langle T \rangle$ decreases with E so that less energy is exchanged on average as the incident particle slows. Consequently, many tens of thousands of Coulomb interactions are required in the stopping of an energetic ion in a thick absorber. Furthermore, since the energy transfer necessary to ionize ordinary materials is on the order of a few tens of eV, many thousands of primary ionizations are created along the path of ion travel. The products of ionization events produced directly by the incident particle are called "primary ionization pairs." Energetic electrons created in this process are often of sufficient energy to create *secondary ionization.* The fundamental principle of radiation detection is based on sensing the presence of these "free" charges produced by the passage of energetic particles of radiation.

The majority of interactions between a fast ion and an atomic electron are in fact glancing collisions, deflecting the ion trajectory an insignificantly tiny amount. (Even a head-on collision with an electron changes the ion momentum by a infinitesimal fraction.) Ion trajectories are, therefore, nearly straight paths through matter until stopped. The length of this path, referred to as the "range" (R), is only a function of projectile mass, charge, and energy for a given target material. Hence, all radiation particles of the same type and energy, have nearly identical range. An attenuation curve for energetic ions passing through an arbitrary thick absorber is depicted in Fig. 6.3 as the projectile number density normalized to that at incidence (n_o) from the left. Though ions continuously lose energy along their trajectory, the density incident on the absorber surface is attenuated only near $x \sim R$.

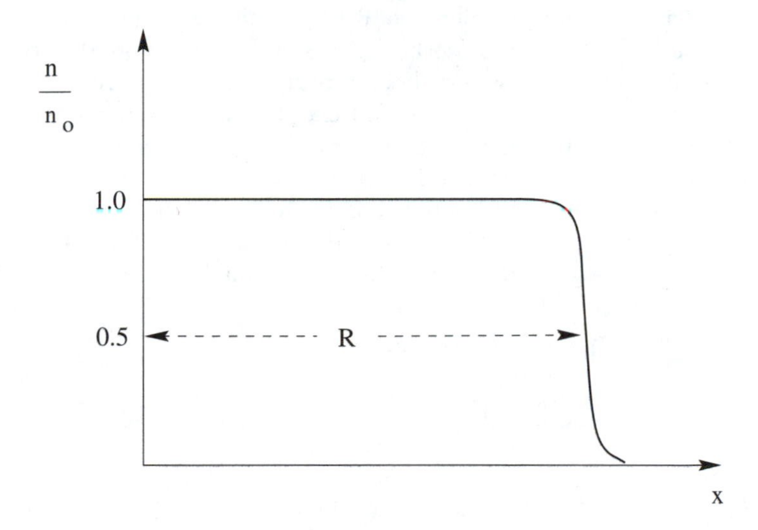

Figure 6.3: Conceptualization of fast ion attenuation in a thick absorber. Ions, incident at left in a thick absorber, continuously lose energy as they traverse the medium. Since their path is straight, however, there is little loss of particles until R.

An estimate of ion range can be obtained from the Coulomb scattering formalism as derived in the previous section. Since $\sigma(T)\ dT$ from Eq.(6.7) is the cross section for ion energy exchange in dT about T to a single electron, and NZ is the electron density in the absorber, then

$$NZ\,dx \int_T \sigma(T)\,dT \quad = \quad \text{interaction probability in path } dx$$

Because our incident ion undergoes many Coulomb interactions in traversing dx, the net energy loss $(-dE)$ in dx is well represented by the average energy transfer

$$\langle T \rangle = \frac{\int_T T\sigma(T)\,dT}{\int_T \sigma(T)\,dT} \tag{6.8}$$

so that

$$-dE = \langle T \rangle NZ\,dx \int_T \sigma(T)\,dT$$

We then identify the quantity $-dE/dx$ as the linear rate of energy loss by the incident ion, often called the "stopping power" of the absorber. Since the majority of the transferred energy is carried by struck electrons which themselves deposit energy locally (sec. 6.1.3), $-dE/dx$ is also the "linear energy transfer" (LET) to the absorber. For completeness, note that the interactions that are responsible for charged particle stopping are inherently *inelastic*, involving atomic ionization and excitation. Though we describe the energy exchange via elastic Coulomb interaction theory, energy transferred in the form of electromagnetic radiation (atomic recombination, deexcitation line radiation, and bremsstrahlung X-rays) is inescapable. This may be a significant contribution to electron stopping (sec. 6.1.3), but is insignificant in ion stopping. There is little error made, then, in assuming all ion kinetic energy is deposited locally.

We can now write for $-dE/dx$

$$-\frac{dE}{dx} = NZ\,dx \int_{T_{\rm min}}^{T_{\rm max}} T\sigma(T)\,dT$$

where $T_{\rm max} = 4(\frac{m_e}{m_x})E$ and $T_{\rm min}$ is the minimum energy transfer allowed. Assigning $T_{\rm min}$ to zero is not permissible as the Coulomb cross section diverges there. This is overcome by realizing that arbitrarily small energy transfers are not allowed by the quantum mechanics. Only those energy transfers which raise electrons to higher quantum levels are permitted. Calculating the average of minimum energy transfers (I) for many electron atoms is a hopeless task. These values are found experimentally, and can be estimated by the empirical expression[Marmier69]

$$I(\text{eV}) \sim 9.1Z\left(1 + 1.9/Z^{2/3}\right) \tag{6.9}$$

for an absorber of atomic number Z. The following mixture law can be used for non-elemental absorbers

$$\ln I = \frac{\sum_j f_j Z_j \ln I_j}{\sum_j f_j Z_j}$$

where f_j is the number of atoms of type j in a molecule of the mixture.

Example 35: Average Atomic Ionization and Excitation Energy
Calculate the average atomic excitation and ionization energy I, for ordinary water.

Solution:
According to expression (6.9), $I_{\rm H} \sim 26.4$ eV and $I_{\rm O} \sim 107.4$ eV. Then, by the mixture law

$$\ln I_{\rm H_2O} = \frac{2(\ln 26.4) + 8(\ln 107.4)}{10} = 4.396$$

Hence, $I_{\rm H_2O} \sim 81.1$ eV.

With I as the lower integration limit, the stopping power becomes

$$-\frac{dE}{dx} = \frac{NZe^4 Z_x^2}{8\pi\epsilon_o^2 E}\frac{m_x}{m_e}\ln\left(\frac{4m_e E}{m_x I}\right) \tag{6.10}$$

also known as the "Bethe equation" in the non-relativistic limit. The strict energy range of applicability is $\frac{m_x}{4m_e} I < E < m_x c^2$, though Eq.(6.10) fails at low energy primarily because of electron pickup (charge exchange) by the slowing ion, thus altering the effective Z_x along the path. Expression (6.10) has also been multiplied by a factor of 2 to bring it into agreement with the exact quantum mechanical result[Evans55]. The identification of $T_{\rm min} = I$ also allows the explicit evaluation of the average energy transfer, Eq.(6.8)

$$\langle T \rangle = \frac{T_{\rm min} T_{\rm max}}{T_{\rm max} - T_{\rm min}} \ln\left(T_{\rm max}/T_{\rm min}\right) = \frac{4m_e I E}{4m_e E - m_x I} \ln\left(\frac{4m_e E}{m_x I}\right) \tag{6.11}$$

so that except at low energy, $\langle T \rangle \sim \ln E$. A 5 MeV α particle in water ($I_{\rm H_2O} \sim 81.1$ eV, example 35) has an initial average energy exchange $\langle T \rangle \sim$ 294 eV, to be compared with $T_{\rm max} \sim 2.7$ keV (example 34).

Despite its quantitative limitations, the parametric dependencies appearing in Eq.(6.10) are important and confirmed by experiment. The stopping power is strongly dependent on projectile charge as Z_x^2, and inversely on energy so that stopping becomes more rapid at the end of the path. The latter is illustrated in a hypothetical stopping power curve for an average incident particle on a thick absorber shown in Fig. 6.4. Such curves are sometimes called "specific ionization" curves since the projectile creates a number of ionization pairs/mm $\propto -dE/dx$. The $1/E$ dependence in $-dE/dx$ is apparent in the initial portion of Fig. 6.4. At the end of the ion trajectory, however, $-dE/dx$ turns over and abruptly drops to zero as the ion slows and acquires charge, making $Z_x \rightarrow 0$.

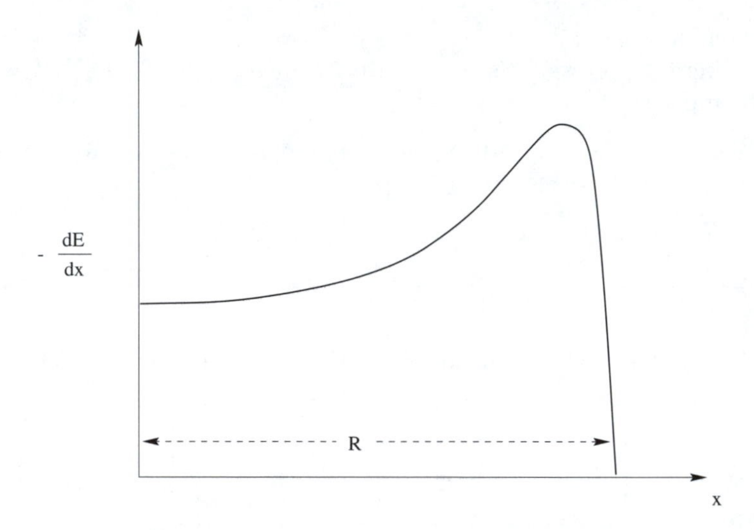

Figure 6.4: General behavior of the stopping power expression along charged projectile path in a thick absorber for heavy charged particles incident from the left at $x = 0$. The gradual increase in $-dE/dx$ is indicative of of the $1/E$ dependence in the stopping power. Charge neutralization and eventual stopping are responsible for the sharp drop in $-dE/dx$ at the end of the particle trajectory.

For a given particle of radiation (*i.e.* fixed Z_x, m_x, E), we expect from Eq.(6.10)

$$-\frac{dE}{dx} \propto NZ \propto \frac{\rho}{A} Z \propto \rho$$

taking Z/A to be nearly constant and ignoring the weak Z dependence in the logarithmic factor, indicating higher density materials are more effective absorbers.

Since ions pass through matter in nearly straight trajectories, the stopping power expression can be used to estimate range

$$R = \int_{E_o}^{0} \left(-\frac{dE}{dx} \right)^{-1} dE$$

Yet, an analytic treatment is of dubious value by the inability of Eq.(6.10) to quantify the true stopping power at low energy. An analytic expression for R has only crude quantitative merit. Instead, fast charged particle ranges may be found either experimentally or by computational modeling. An empirical expression for α particles in air (at STP) is given in [RHH70]

$$R(\text{cm}) = \begin{cases} 0.56E(\text{MeV}) & , E < 4 \text{ MeV} \\ 1.24E(\text{MeV}) - 2.62 & , 4 \leq E < 8 \text{ MeV} \end{cases} \tag{6.12}$$

Examining parametric dependencies in R provides connection formulae for range in different materials and for different species of projectile. As we have found $-dE/dx \sim \rho$, we should expect $R \sim 1/\rho$ for a given projectile specie at fixed initial energy in different materials. Indeed, the inverse density dependence is observed, yet a more precise empirical result was obtained by Bragg and Kleeman indicating $R \sim \sqrt{A}/\rho$. The range for a given ion in material b reference to that in material a is then given by the Bragg-Kleeman rule

$$\frac{R_b}{R_a} = \frac{\rho_a}{\rho_b} \sqrt{\frac{A_b}{A_a}} \tag{6.13}$$

A connection for different ion species in the same absorber material (*i.e.* NZ = const.) is found by employing the explicit form of the stopping power expression (Eq.(6.10)) to the range integral. The range for particle of type 2 relative to that of type 1 in the same material, ignoring the weak logarithmic dependence, is

$$\frac{R_2}{R_1} = \frac{m_2}{m_1} \frac{Z_1^2}{Z_2^2} \tag{6.14}$$

when E/m of the projectile is held constant (*i.e.* two different projectiles are compared at the same speed). Example 36 illustrates the connection formulae.

The picture of swift ions from a monoenergetic source all having identical range and possessing identical energy loss as a function of position is somewhat idealized. The statistical nature of the Coulomb interaction process produces slight variations in range even for monoenergetic particles.

Example 36: Light Ion Range
What is the range of a 1 MeV α particle in air at STP? In aluminum? What is the range of a 1 MeV proton in STP air?

Solution:
In air at 1 MeV, $R = 0.56$ cm from the empirical expression, Eq.(6.12). At STP, $\rho_{\text{air}} \sim 0.001293$ g/cm^3 and the density of aluminum is $\rho_{\text{Al}} \sim 2.7$ g/cm^3. From the materials connection formula (Eq.(6.13)), we have

$$\begin{aligned} R_{\alpha_{\text{Al}}} &= R_{\alpha_{\text{air}}} \frac{\rho_{\text{air}}}{\rho_{\text{Al}}} \sqrt{\frac{A_{\text{Al}}}{A_{\text{air}}}} \\ &= (0.56 \text{ cm}) \frac{0.001293}{2.7} \sqrt{\frac{27}{14}} \\ &= 3.72 \times 10^{-4} \text{ cm} = 3.72\ \mu m \end{aligned}$$

Compared at the same speed as a 1 MeV proton

$$E_\alpha = \frac{m_\alpha}{m_p} E_p \sim 4 \text{ MeV}$$

and the α range at 4 MeV is $R_\alpha(4 \text{ MeV}) = 2.24$ cm in STP air. Then, for the proton range, we use the connection expression (6.14)

$$\begin{aligned} R_p(1 \text{ MeV}) &= R_\alpha(4 \text{ MeV}) \frac{m_p}{m_\alpha} \frac{Z_\alpha^2}{Z_p^2} \\ &= 2.24(\text{cm}) \frac{1}{4} \frac{2^2}{1} = 2.24 \text{ cm} \end{aligned}$$

Typically this variation, called "range straggling," is only a few percent of the range. The attenuation curve (Fig. 6.3) illustrates this effect. In the absence of range straggling, all monoenergetic incident particles terminate their trajectories at exactly the same distance. The attenuation curve would then have perfectly square corners. Instead, these are rounded by range straggling. Some particles in the ensemble have slightly shorter range, some slightly longer. The range is then formally defined for a collection of particles as that distance at which the number density of particles drops to 1/2 its original value. Likewise, the stopping power curve (Fig. 6.4) for an ensemble of particles would show a rounded peak and tail at the curve end due to straggling. Such curves are called "Bragg curves" or "Bragg ionization curves."

Heavy ions like fission fragments also deposit energy in absorbers via Coulomb collisions with atomic electrons. However, the methods developed above are of little practical use and may serve only as guidance regarding the underlying physics. Most responsible for the departure from the simple theoretical description is the tendency for highly charged fission fragments to recombine by absorbing electrons as they slow. To a somewhat lesser extent, elastic collisions with nuclei, especially at the end of the trajectory, also confound the theoretical description. To illustrate the magnitude of charge changing, consider the following situation. On average, fission fragments are observed to have range similar to that of ~ 4 MeV α particles. For an average fission fragment with $A \sim 117$, $E_A \sim 85$ MeV, and initial net charge $Z_A \sim 20$, our formalism developed for light ions suggests that we should compare with same speed α particles at $E_\alpha = (m_\alpha/m_A)E_A \sim 2.9$ MeV. We should then expect

$$R_A(85\text{ MeV}) = R_\alpha(2.9\text{ MeV})\frac{m_A}{m_\alpha}\frac{Z_\alpha^2}{\langle Z_a^2\rangle}$$

where $\langle Z_a^2\rangle$ represents the square of the atomic charge averaged over the fragment trajectory. Since $R_A(85\text{ MeV}) \sim R_\alpha(4\text{ MeV})$ and $R_\alpha(2.9\text{ MeV}) \sim 0.7R_\alpha(4\text{ MeV})$, we predict

$$\langle Z_a^2\rangle^{1/2} \sim 9$$

far from the initial net fission fragment charge.

A more contemporary approach to the slowing down problem involves numerical solution to the range expression based on the Bethe equation (Eq.(6.10)), and empirical data particularly at low energy. The *continuous slowing down approximation* (CSDA) is such a model in which energetic charged particles are assumed to lose energy in matter continuously via Coulomb interactions. The local value of stopping power is evaluated at the particle's instantaneous energy. Tables of CSDA ranges vs. particle energy for many common materials and elements can be found in [Janni82] for protons, [ICRU93] for protons and α particles, and [Hubert90] for heavy charged particles. [Shultis96] provides constants for fits to CSDA model ranges for electrons and protons in a few common materials. The empirical expressions and connection formulae given above agree with the CSDA predictions to within $\lesssim 3\%$ for protons in air at 1–2 MeV for example, but may differ by 30–70% or more outside their narrow energy range of applicability. In calculations requiring a high degree of accuracy over a wide range of energies, CSDA is preferred.

6.1.3 Fast Electron Slowing Down

Fast electrons too interact with matter via the Coulomb force, suffering inelastic collisions with atomic electrons, and elastic and inelastic collisions with nuclei. Sufficient similarity exists in the slowing down of all charged particles that the formalism developed in prior sections of this chapter may be exploited. There are, however, important differences to be noted between the stopping of heavy charged particles and electrons, namely (1) fast electrons from nuclear decay are relativistic, (2) $\beta^{\pm}$ decay electrons are not monoenergetic, (3) electrons are light projectiles and are either identical in mass or much lighter than their electron or nucleus scattering target, respectively, and (4) the electromagnetic radiation that is emitted when all charged particles are accelerated is appreciable in the energy loss of electrons in matter. The latter constitutes an inelastic interaction between swift electrons and atomic nuclei.

The implications of these differences are abundant. Most obviously, these confounding factors make electron energy loss and penetration difficult to describe analytically. For inelastic interactions with atomic electrons, the fully relativistic stopping power expression first provided by Moller [Moller32] is

$$\begin{aligned} -\left.\frac{dE}{dx}\right|_c &= \frac{NZe^4}{4\pi\epsilon_o^2 m_e v^2}\left[\ln\left(\frac{2m_e c^2}{I}\right) + \ln\left((\gamma-1)(\gamma+1)^{1/2}\right)\right. \\ &- \left.\left(\frac{3}{2}+\frac{1}{\gamma}-\frac{1}{2\gamma^2}\right)\ln 2 + \frac{1}{16}\left(1-\frac{2}{\gamma}+\frac{9}{\gamma^2}\right)\right] \end{aligned} \tag{6.15}$$

where $\gamma = (1 - v^2/c^2)^{-1/2}$ is the relativistic factor, m_e is the electron rest mass, and the subscript c indicates that this is the collisional contribution to stopping power only. In the non-relativistic limit, $v \ll c$, this becomes

$$-\left.\frac{dE}{dx}\right|_c = \frac{NZe^4}{4\pi\epsilon_o^2 m_e v^2}\left[\ln\left(\frac{2m_e v^2}{I}\right) - \frac{5}{2}\ln 2 + \frac{1}{2}\right] \tag{6.16}$$

which differs from the Bethe form (Eq.(6.10)) by only the numerical factor $-\frac{5}{2}\ln 2 + \frac{1}{2} \sim 1.23$. When considering multiple interactions combining the effects of inelastic electrons and elastic nucleus scattering, it is often an acceptable approximation to replace Z with $Z^2 + Z$ in Eq.(6.15).

The emission of electromagnetic radiation as the electron slows cannot be ignored here. Classical electrodynamics tells us that radiation must accompany any charged particle acceleration. In the context of fast charged particle slowing down in matter, this radiation is called "bremsstrahlung," German for *breaking radiation.* Though present, bremsstrahlung radiation contributes negligibly to ion stopping, as velocity changes are tiny in individual interactions with electrons. As electrons penetrate material media, however, they experience large angle deflections where momentum changes are on the order of p, resulting in appreciable radiation emission. The radiative contribution to the electron stopping power in the highly relativistic limit is

$$- \left. \frac{dE}{dx} \right|_r = \frac{NZe^4E}{137(4\pi\epsilon_o)^2 m_e^2 c^4} \left[4 \ln \left(\frac{2E}{m_e c^2} \right) - \frac{4}{3} \right] \tag{6.17}$$

where $E = \mathrm{KE} + m_e c^2$ is the total energy. The increased importance for light particles is evident in the m_e^{-2} dependence. The total stopping power is then the sum of the collisional and radiative contributions

$$- \frac{dE}{dx} = - \left. \frac{dE}{dx} \right|_c - \left. \frac{dE}{dx} \right|_r \tag{6.18}$$

For relativistic electrons, the relative contribution of the two loss mechanisms becomes

$$\frac{dE/dx|_r}{dE/dx|_c} \sim \frac{Z}{137(4\pi)} \frac{E}{m_e c^2} \sim \frac{Z}{1722} \frac{E}{m_e c^2}$$

so that an increasing importance of radiation loss appears at high energy for interactions with high Z materials. As an example, a 9.5 MeV electron stopping in uranium ($Z = 92$) has roughly equal radiative and electron scattering loss.

Torturous electron trajectories (Fig. 6.5) result from the substantial momentum change on individual interactions when electrons interact in matter. This can also include "backscatter" where incident fast electrons appear to be reflected from the surface toward the direction from which they came. Consequently, the range of fast electrons is much less than the distance of travel and is quite impossible to obtain analytically. Laboratory or numerical experiments are required. (CSDA range calculations are performed for fast electrons as well. Some results are available in [ICRU84].) Figure 6.6 illustrates a hypothetical attenuation or "transmission" curve for monoenergetic fast electrons incident on a homogeneous material from vacuum at $x = 0$. The fall off of electron density with distance, $n(x)$, is noticeably different from that of ions (Fig. 6.3). Whereas nearly all ions persist to R, there is considerable and continuous electron depletion with distance. The range for such particles is identified with the extrapolation of the linear, central portion of the transmission curve to zero intensity, called the "extrapolated" range, R_e.

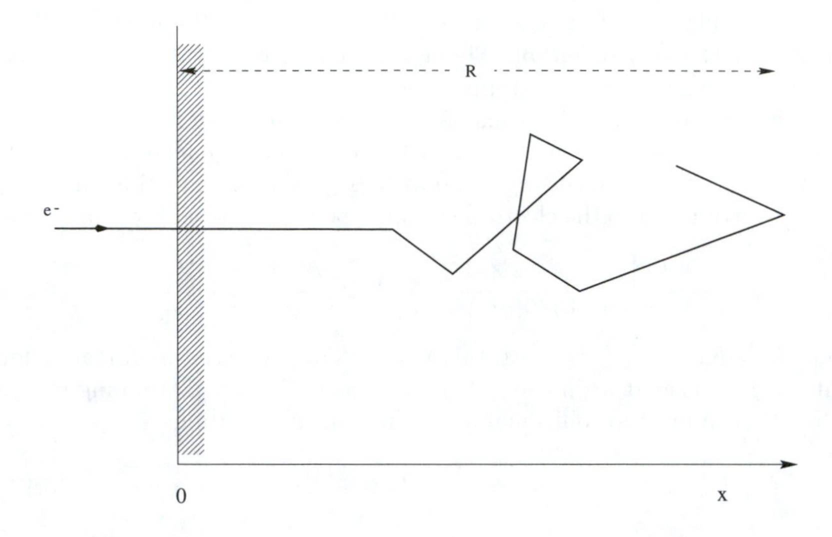

Figure 6.5: Idealized trajectory for an energetic electron entering a material from a vacuum interface at $x = 0$. Omitted are the effects of numerous small angle scattering events with atomic electrons which appear as slight curvature in the fast electron trajectory. Rather, what is shown is the effect of relatively few large angle, small energy transfer interactions with nuclei. Examples of computer simulated electron trajectories can be found in Figs. 9.5 and 9.6 of [Shultis96].

A transmission curve for $\beta^{\pm}$ spectrum electrons (Fig. 6.7) has different signature, *i.e.* nearly exponential fall off with distance. This fortunate result is entirely a coincidence of the $\beta^{\pm}$ spectrum convolution with electron slowing down. (The reader should note that the exponential attenuation of $\beta^{\pm}$ is neither rigorous nor completely general, and is presented to show trends only.) The range for this ensemble of particles is identified as the maximum distance of penetration, that distance at which the detected electron density falls to the background level of detection (dashed line in Fig. 6.7). The maximum range, R_m, is found to be experimentally identical to R_e when the monoenergetic (KE_o) source has $\text{KE}_o = \text{KE}_{\beta^{\pm}}|_{\max}$. This consequence is attributed to the multiple collision nature of electron slowing down. After a few large angle collisions, all angular information is lost, *i.e.* the electron population becomes isotropic. Further electron transport

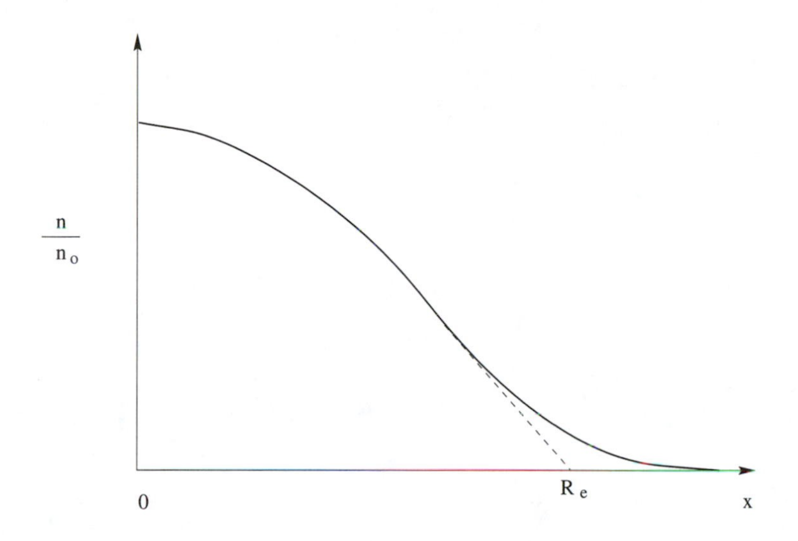

Figure 6.6: Representative transmission curve for monoenergetic electrons. In contrast to ion stopping, attenuation and energy reduction are continuous. R_e indicates the "extrapolated" range.

can then be described as diffusive. As well, the kinetic energy distribution becomes randomized after a few collisions so that the initial energy spectrum becomes unimportant.

A widely employed empirical range-energy relation for fast electrons is given by Katz and Penfold[Katz52]

$$R_m = \begin{cases} 412E^{1.265-0.0954\ln E} & ,0.01 \leq E \leq 3\ \text{MeV} \\ 530E - 106 & ,1 \leq E \leq 20\ \text{MeV} \end{cases} \tag{6.19}$$

where $E = \text{KE}_{\beta^\pm}|_{\text{max}}$ for $\beta^\pm$ spectrum electrons, and $E = \text{KE}_o$ for monoenergetic electrons, E in MeV. Here R_m is expressed as a density × thickness product, $R_m(\text{mg/cm}^2) = \rho(\text{mg/cm}^3) \times R(\text{cm})$, where R is the linear range. It proves very convenient to do so. Since R is found to depend on density as $1/\rho$ across various materials for the same incident electron energy, then R_m is a constant across materials and only dependent on energy.

Before closing the discussion of electron interactions, two additional points need to be mentioned. Firstly, the description of electron slowing down provided above applies equally well to positrons, as well as negatrons, the only difference being in the ultimate fate of the positron. When the positron slows, after many interactions, it will annihilate with a free electron to produce annihilation photons (sec. 6.2.3.2). Secondly, fast electrons

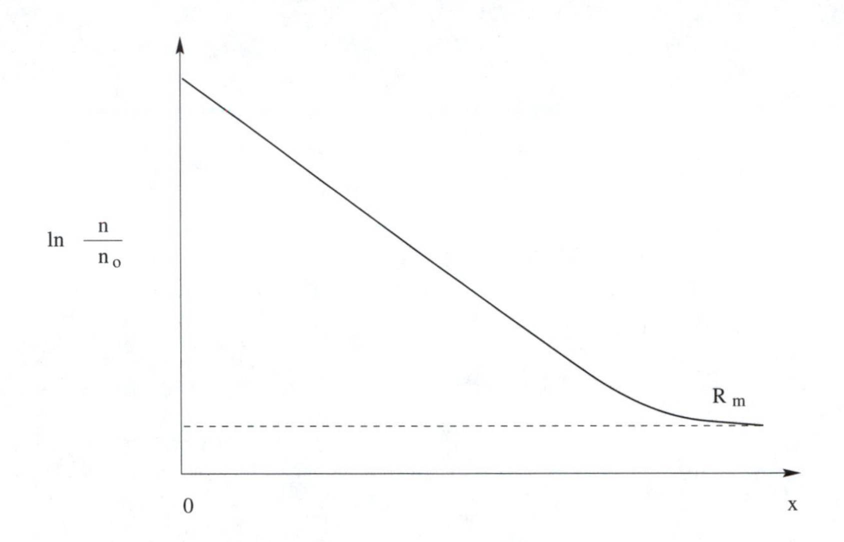

Figure 6.7: Hypothetical transmission curve for $\beta^{\pm}$ spectrum electrons. The dashed line represents the background level of detection for electrons in a transmission experiment. The nearly exponential attenuation of electron density is a consequence of the convolution of $\beta^{\pm}$ spectra with electron slowing down. It is not, however, a rigorous or completely general result.

Example 37: Electron Range/Energy Relationships

For 1 MeV β^- particles, we find $R_m = \rho R \simeq 400\,\text{mg/cm}^2$ ($\sim 0.4\,\text{g/cm}^2$) from Eq.(6.19). Since this is a constant for all materials, we determine the linear range $R = R_m/\rho$. The linear range for these fast electrons in various materials is listed in the table below.

Material	Density (g/cm^3)	Linear Range, R (cm)
Air	0.001293	310
H_2O	1.0	0.4
Al	2.7	0.148
Fe	7.87	0.051
Pb	11.34	0.0353

may possess speed greater than the phase velocity of light in the medium. In such a situation, electromagnetic radiation called Čerenkov[Čerenkov37] radiation is emitted. As this is a very soft interaction, it does not contribute to electron energy loss in any appreciable way. The mechanism does, however, provide an explanation of the faint bluish glow seen surrounding the core of pool nuclear reactors that is produced by the Čerenkov radiation of Compton (sec. 6.2.2) scattered electrons from fission γ's.

6.2 Photon Interaction Mechanisms

Photon interaction mechanisms, regardless of the photon origin, are dictated solely by photon energy and target atom properties. There are some dozen processes possible[Evans55] by which photons may interact with nuclei and electrons. In the energy range encountered in nuclear transitions, perhaps $0.01 \leq E_\gamma \leq 10$ MeV, only three mechanisms are important in attenuating photons; the photoelectric effect, the Compton effect, and pair production. The energy range of dominance for these (sec. 6.3) is a function of the atomic number of the target atom, though it is always the case that the photoelectric effect is dominant at low energy, Compton at intermediate energy, and pair at high energy. In Pb ($Z = 82$) for example, below ~ 0.7 MeV the photoelectric effect is most important, while above ~ 3.5 MeV pair production is more probable, and Compton dominates in between.

6.2.1 The Photoelectric Effect

The photoelectric effect has been discussed previously (sec. 1.4.1) in the context of the corpuscular nature of photons. Here, we examine the kinematics of the process. The photoelectric interaction is a photon absorption process wherein an incident photon is completely absorbed by a target atom to eject an energetic electron from its bound state. As we shall show shortly, momentum conservation requires that the initial state of the electron be bound. The photoelectric effect cannot occur with a free electron. The incident photon, instead, interacts with the entire atom invoking atom recoil to conserve momentum. The ejected photoelectron suffers the fate of any energetic electron in matter (sec. 6.1.3). In passing, we note that energetic photons may interact with nuclei directly in a similar fashion and eject nucleons in the nuclear photoeffect (sec. 2.2) or "photodisintegration." This interaction, however, appears with very low probability in comparison and, thereby, contributes negligibly to photon attenuation.

Symbolically, let's describe the photoelectric process this way

$$\gamma + {}^A\mathrm{X} \longrightarrow {}^A\mathrm{X}^+ + \mathrm{e}^-$$

leaving the recoiling atom in a singly ionized state which will later recombine and produce characteristic X-rays. The photoelectric absorption process conserves mass/energy such that

$$E_\gamma = \mathrm{KE}_e + \mathrm{BE}_e + \mathrm{KE}_A \tag{6.20}$$

The most tightly bound atomic electrons have the highest probability of photoelectric absorption with about 80% in the K shell, so long as $E_\gamma > \mathrm{BE}_e^{\mathrm{K}}$.

The following expressions, which include partial screening of inner shell electrons, may be used to approximate the electron binding energy for K, L, and M shell electrons in high Z atoms[Marmier69]

$$\begin{aligned} \mathrm{BE}_e^{\mathrm{K}} &= R_y(Z-1)^2 \\ \mathrm{BE}_e^{\mathrm{L}} &= \frac{1}{4}R_y(Z-5)^2 \\ \mathrm{BE}_e^{\mathrm{M}} &= \frac{1}{9}R_y(Z-13)^2 \end{aligned} \tag{6.21}$$

where R_y is the Rydberg constant in eV (*i.e.* $R_y = chR_\infty = 13.6$ eV, cf: table 1.7). The average BE_e per electron can be approximated

$$\langle \mathrm{BE}_e \rangle \sim \frac{\mathrm{BE}_e|_{\mathrm{total}}}{Z} \sim 15.73 Z^{4/3} \quad , \quad \mathrm{eV}$$

again, best for high Z atoms.

Returning to mass/energy and momentum conservation

$$\mathbf{p}_\gamma = \mathbf{p}_e + \mathbf{p}_A$$

let's first examine the recoil energy. The magnitude of atomic recoil can be estimated by allowing the approximation, $\mathbf{p}_e \sim 0$, *i.e.* the recoil atom carries all the momentum of the incident photon

$$p_\gamma \sim p_A$$

which, upon squaring, implies

$$\mathrm{KE}_A \sim \frac{E_\gamma^2}{2m_Ac^2} \tag{6.22}$$

a tiny quantity and $\ll E_\gamma$. Then, in the limit that all incident photon momentum is carried by the recoil atom, the photoelectron kinetic energy is written

$$\mathrm{KE}_e = E_\gamma \left(1 - \frac{E_\gamma}{2m_Ac^2}\right) - \mathrm{BE}_e$$

Table 6.1: K Shell Binding Energy and Maximum Atom Recoil Energy for Several Target Atoms Interacting by the Photoelectric Effect with a 1 MeV Photon

nuclide	$\mathrm{BE}_e^{\mathrm{K}}$ (keV)	KE_A (keV)
^{27}Al	1.96	0.02
^{56}Fe	8.5	0.0096
^{208}Pb	89.2	0.0026
^{238}U	112.6	0.00255

Table 6.1 illustrates the magnitude of electron binding energy and atom recoil energy for the photoelectric process in several possible absorbers. These calculations demonstrate that the recoil energy is always negligible, whereas BE_e may become significant at high Z.

If instead we were to presume the atom absent in the photoelectric process ($\mathrm{BE}_e \to 0$), energy conservation then implies

$$\mathrm{KE}_e = E_\gamma$$

yet, momentum conservation, $p_e = p_\gamma$, requires

$$\mathrm{KE}_e \stackrel{?}{=} \frac{E_\gamma^2}{2m_e c^2}$$

Both relations cannot be satisfied simultaneously. This contradiction implies that this process is not allowed. The photoelectric effect cannot occur without the presence of a third body to conserve momentum.

Since E_γ must be greater than the electron binding energy, photoelectric absorption is a threshold reaction. Let's consider the effect of recoil on the threshold energy. At threshold, $E_\gamma = E_\gamma^{\mathrm{t}}$ is the minimum quantum of photon energy required for photoelectron ejection, *i.e.* $\mathrm{KE}_e \to 0$, so

$$E_\gamma^{\mathrm{t}} = \mathrm{KE}_A + \mathrm{BE}_e$$

Employing the maximum recoil energy for KE_A (Eq.(6.22)) as conservative, we find the quadratic expression

$$(E_\gamma^{\mathrm{t}})^2 - 2m_A c^2 E_\gamma^{\mathrm{t}} + 2m_A c^2 \mathrm{BE}_e = 0$$

which has the solution

$$E_\gamma^{\mathrm{t}} = m_A c^2 \left[1 \pm \left(1 - \frac{2\mathrm{BE}_e}{m_A c^2}\right)^{1/2}\right]$$

Since $\mathrm{BE}_e \ll m_A c^2$, a binomial expansion may be used. Choice of the lower root as minimum provides

$$E_\gamma^{\mathrm{t}} \sim \mathrm{BE}_e$$

so that recoil is shown to have a negligible effect on the threshold energy.

6.2.1.1 Photoelectric Cross Section

Let's consider some general properties of the cross section for photoelectric absorption so that we can examine its magnitude and the dependence of E_γ and Z. We write

$$\sigma_{pe}^{j} = f(E_\gamma, Z)$$

as the cross section (cm^2/atom) for photoelectric interaction among photons of energy E_γ on a j shell electron (j =K, L, M, ...) in atoms of atomic number Z. Since the photoelectric effect is a threshold reaction, $\sigma_{pe}^{j} = 0$ when $E_\gamma < \mathrm{BE}_e^j$. As well, we have shown that it is necessary that the interacting electron be a bound state. Since electron binding decreases with shell distance from the nucleus, then $\sigma_{pe}^{\mathrm{K}} > \sigma_{pe}^{\mathrm{L}} > \sigma_{pe}^{\mathrm{M}} \cdots$ per electron at fixed E_γ and Z.

The abrupt change in σ_{pe}^{j} at BE_e^j is called the shell effect or the "edge" effect. Figure 6.8 illustrates K, L, and M absorption edges in Pb. The highest energy edge is that belonging to the K-shell at $\sim$ 88 keV. The groupings of edges at successively lower energies are those of the L and M shells, respectively. Each of these groupings have several distinct edges corresponding in number to the "multiplicity" of the respective shell = $2l+1$ where $l = 0, 1, 2, \ldots$ for the K, L, M, ... shells. The three L edges appear at 13.1, 15.3, and 15.9 keV, while there are five M edges between 2.4–3.9 keV.

The presence of sharp absorption edges and relativistic electron effects at high energies makes analytic treatment difficult. At intermediate energies, a crude analytic approximation of σ_{pe}^{K} is[Heitler54]

$$\sigma_{pe}^{\mathrm{K}} = 32(m_e c^2)^{7/2} \frac{\alpha^4 Z^5}{E_\gamma^{7/2}} \sigma_{\mathrm{Th}}$$

where $\alpha = \frac{e^2}{2\varepsilon_o hc} \simeq \frac{1}{137}$ is the "fine structure constant" and $\sigma_{\mathrm{Th}} \sim 0.665$ b/e is the Thomson cross section (sec. 6.2.2.1). Substituting physical constants yields

$$\sigma_{pe}^{\mathrm{K}} \simeq 6 \times 10^{-8} \frac{Z^5}{(E_\gamma / m_e c^2)^{7/2}}, \quad \text{b/atom} \tag{6.23}$$

illustrating very strong Z and E_γ dependence. As a very rough rule of thumb, the contributions from other electron shells increases the photoelectric absorption cross section by $\sigma_{pe} \sim \frac{5}{4}\sigma_{pe}$. These expressions should

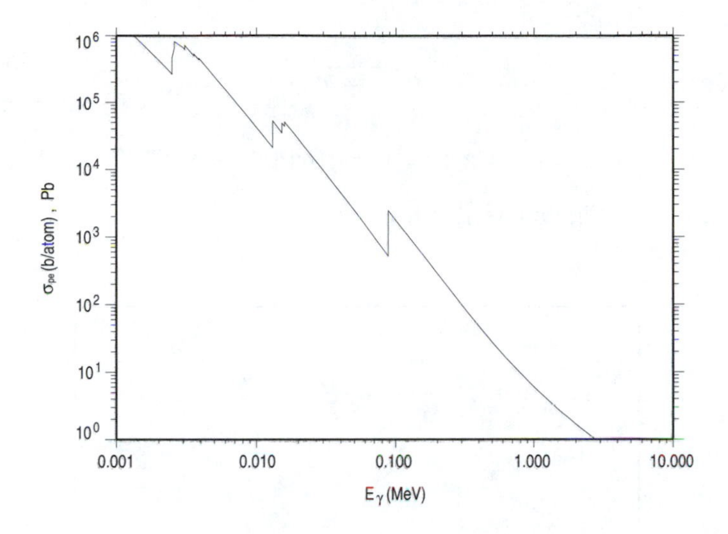

Figure 6.8: Photoelectric absorption cross section for lead. The sharp discontinuities appear at atomic shell (K, L, M, ...) absorption energies.

be used carefully as they are approximate only and fail in the relativistic limit ($E_\gamma/m_e c^2 \gg 1$) and near absorption edges.

6.2.1.2 Auger Effect

Following the ejection of a K-shell electron in the photoelectric effect, the atom is left with a K-shell vacancy. Since $\mathrm{BE}_e^{\mathrm{K}} > \mathrm{BE}_e^{\mathrm{L}}$, a radiative transition may fill this void. This is marked by the emission of a characteristic X-ray in a process called *fluorescence.* When $\mathrm{BE}_e^{\mathrm{K}} - \mathrm{BE}_e^{\mathrm{L}} > \mathrm{BE}_e^{\mathrm{L}}$, however, another possibility exists. This state may decay by a non-radiative transition that emits an L-shell electron with $\mathrm{KE}_e = \mathrm{BE}_e^{\mathrm{K}} - 2\mathrm{BE}_e^{\mathrm{L}}$. This effect, named after its french discoverer, is called the "Auger effect" (pronounced ō zhā′) and the ejected electron is called an "Auger electron." The terms "internal photoelectric effect" and "autoionization" are sometimes used. Auger electrons are also seen following electron capture (sec. 4.2.2.5) and internal conversion (sec. 4.2.2.8) since these decay modes leave the atom with a low lying electronic vacancy.

The competition between K-shell fluorescence and the emission Auger electron is quantified by the empirical expression for the K-shell fluorescence yield[Marmier69], y_f^{K}

$$y_f^{\mathrm{K}} = \frac{a}{1+a} \tag{6.24}$$

where

$$a = \left(-6.4 + 3.4Z - 0.000103Z^3\right)^4 \times 10^{-8}$$

and illustrated in Fig. 6.9 as y_f^{K} vs. Z.

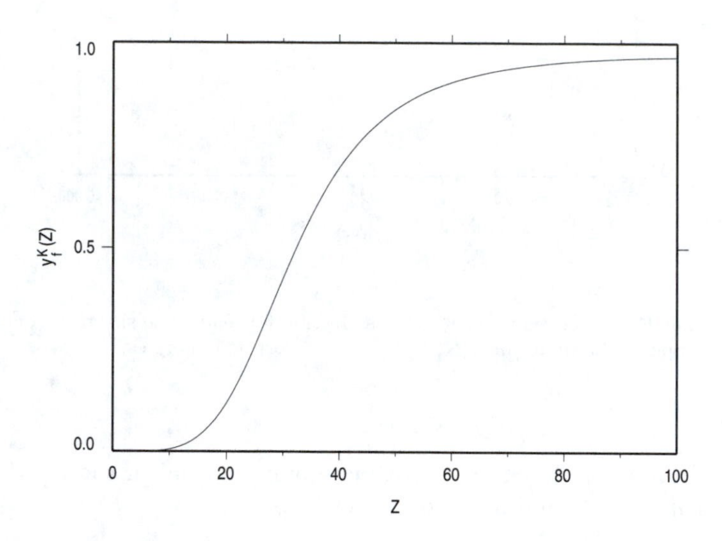

Figure 6.9: K-shell fluorescence yield, y_f^{K}, vs. Z.

6.2.2 The Compton Effect

At intermediate energies, a Compton collision is more likely than photoelectric absorption. In the Compton effect, an incident photon scatters incoherently with an atomic electron, imparting energy to the recoiling electron and producing a secondary or scattered photon, γ', of different energy and, hence, frequency. Symbolically, we may write

$$\gamma +^{A}\mathrm{X} \longrightarrow {}^{A}\mathrm{X}^{+} + \mathrm{e} + \gamma'$$

for Compton scattering, though here the atom ${}^{A}\mathrm{X}$ serves merely as a reservoir of electrons and is not required to allow this reaction as it was in the photoelectric effect. To show this, we again estimate the upper limit for the recoil energy of A from $p_A \leq p_\gamma$ so that

$$\mathrm{KE}_A \leq \frac{E_\gamma^2}{2m_A c^2}$$

which is, again, extremely tiny. The difference here is that there exists an additional particle, the scattered photon γ', to participate in momentum balance. Without inconsistency, we may allow $\mathrm{KE}_A, \mathrm{BE}_e$, and p_A to be zero, *i.e.* the electron need not be bound to an atom.

To negligible error, except where E_γ approaches BE_e at very low E_γ (a region of little interest since photoelectric absorption dominates there), we can assume the electron is free and write mass/energy and linear momentum balance

$$\begin{aligned} E_\gamma &= \mathrm{KE}_e + E_{\gamma'} \\ \mathbf{p}_\gamma &= \mathbf{p}_e + \mathbf{p}_{\gamma'} \end{aligned} \tag{6.25}$$

a two-particle scattering problem, as illustrated in Fig. 6.10. We must, however, treat the scattered electron relativistically since energy transfers on the order of $m_e c^2$ or greater are common. As this process is azimuthally symmetric, it will suffice to treat the problem as two-dimensional. In the incident photon direction then, we have

$$p_\gamma = p_e \cos\phi_e + p_{\gamma'} \cos\theta \tag{6.26}$$

and in the direction perpendicular to $\mathbf{p}_\gamma$

$$0 = p_e \sin\phi_e - p_{\gamma'} \sin\theta \tag{6.27}$$

Eliminating ϕ_e between these by squaring yields

$$p_\gamma^2 + p_{\gamma'}^2 - 2p_\gamma p_{\gamma'} \cos\theta = p_e^2$$

which, with the relativistic momentum relation (Eq.(1.12)), becomes

$$E_\gamma^2 + E_{\gamma'}^2 - 2E_\gamma E_{\gamma'} \cos\theta = \mathrm{KE}_e^2 + 2m_e c^2 \mathrm{KE}_e$$

Employing energy conservation $\mathrm{KE}_e = E_\gamma - E_{\gamma'}$, we can finally express $E_{\gamma'}, \mathrm{KE}_e = f(E_\gamma, \theta)$

$$\boxed{E_{\gamma'} = \frac{E_\gamma}{1 + \frac{E_\gamma}{m_e c^2}(1 - \cos\theta)}} \tag{6.28}$$

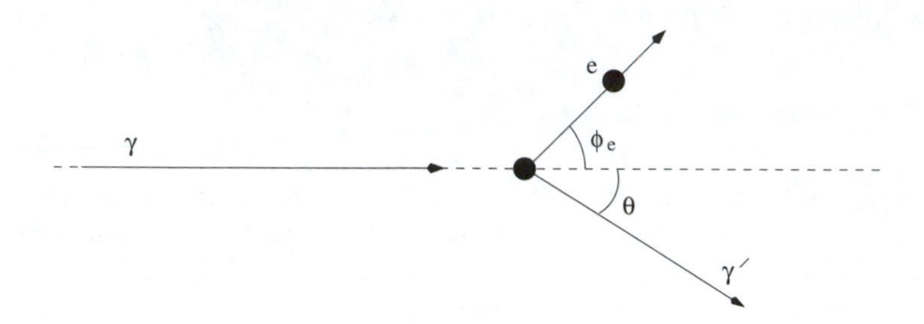

Figure 6.10: Notation for Compton scattering interactions. Incident photon, γ, scatters incoherently on a free electron, e, imparting momentum and energy shared with scattered photon, γ'.

and

$$\boxed{\mathrm{KE}_e = E_\gamma \left[1 - \frac{1}{1 + \frac{E_\gamma}{m_e c^2}(1 - \cos\theta)} \right]} \tag{6.29}$$

We should further employ the momentum conservation expressions (6.26) and (6.27) in ratio to reveal

$$\cot\phi_e = \frac{p_e \cos\phi_e}{p_e \sin\phi_e} = \frac{p_\gamma - p_{\gamma'}\cos\theta}{p_{\gamma'}\sin\theta}$$

which, after some algebra and trigonometric identity, reduces to

$$\cot\phi_e = \left[1 + \frac{E_\gamma}{m_e c^2} \right] \tan\frac{\theta}{2} \tag{6.30}$$

so that, although $0 \leq \theta \leq \pi$, the Compton electron scattering angle is restricted to $0 \leq \phi_e \leq \pi/2$, *i.e.* Compton electrons are never scattered backward.

The photon scattering angle ranges over all polar angles $0 \leq \theta \leq \pi$, resulting in a continuum of scattered photon energies from a minimum of

$$E_{\gamma'}\big|_{\min} = \frac{E_\gamma}{1 + 2E_\gamma/m_e c^2}$$

at $\theta = \pi$, called "backscatter" since the Compton scattered photon trajectory is back along the incident photon path, to

$$E_{\gamma'}\big|_{\max} = E_\gamma$$

at $\theta = 0$. The backscatter photon has the high energy limit

$$\lim_{\frac{E_\gamma}{m_e c^2} \to \infty} E_{\gamma'}|_{\min} = \frac{m_e c^2}{2} \sim 0.256 \text{ MeV}$$

The Compton scattered electron must also exhibit an energy continuum called the "Compton continuum" from

$$\text{KE}_e|_{\min} = 0$$

at $\theta = 0$ to

$$\text{KE}_e|_{\max} = E_\gamma \left[\frac{2E_\gamma / m_e c^2}{1 + 2E_\gamma / m_e c^2} \right]$$

at $\theta = \pi$. The high energy boundary to this electron continuum at $\text{KE}_e|_{\max}$ is referred to as the "Compton edge" energy. Note that the special case $\theta = 0$ corresponds to $E_{\gamma'}|_{\max} = E_\gamma$ and $\text{KE}_e|_{\min} = 0$. The product state in this limit is indistinguishable from the reactant state, *i.e.* there is no evidence of an interaction when $\theta \to 0$. The Compton scattering kinematics are summarized in Figs. 6.11 and 6.12 where we plot $E_{\gamma'}/E_\gamma$ vs. θ and KE_e/E_γ vs. ϕ_e, respectively, for $E_\gamma/m_e c^2 = 0.5, 1.0, 2.0$, and 10.0.

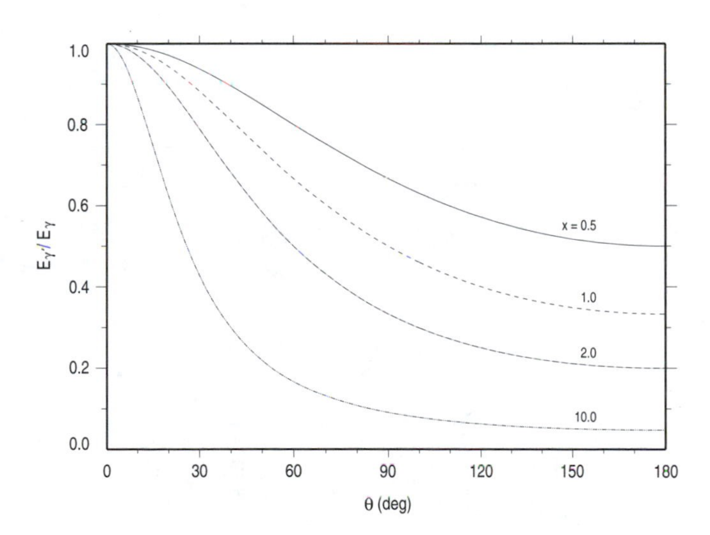

Figure 6.11: Normalized Compton scattered photon energy, $E_{\gamma'}/E_\gamma$ as a function of photon scattering angle θ for $x = E_\gamma/m_e c^2 = 0.5, 1.0, 2.0$, and 10.0.

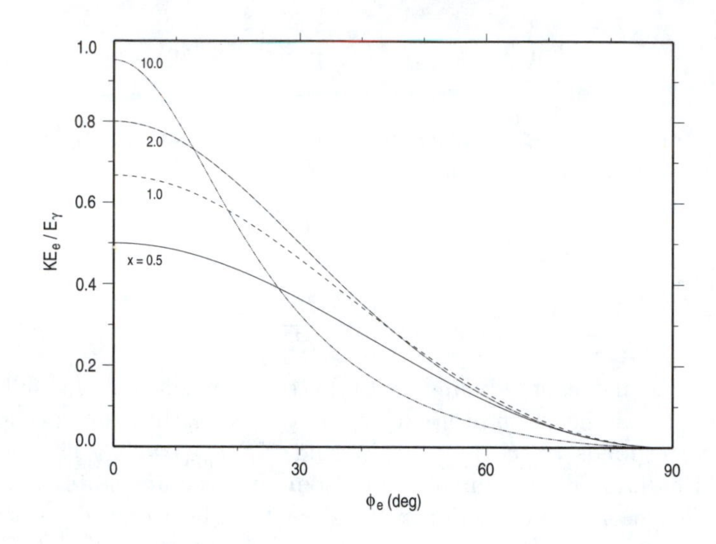

Figure 6.12: Normalized Compton scattered electron energy, KE_e/E_γ as a function of electron scattering angle ϕ_e for $x = E_\gamma/m_e c^2 = 0.5, 1.0, 2.0$, and 10.0.

Example 38: Compton Scattering
Find the backscatter and Compton edge energies for $E_\gamma = 0.5, 1.0$, and 10.0 MeV.

Solution:
From Eqs.(6.28) and (6.29), we find backscatter and Compton edge, respectively, at $\theta = \pi$.

E_γ (MeV)	Backscatter $E_{\gamma'}\|_{\mathrm{min}}$ (MeV)	Compton edge $\mathrm{KE}_e\|_{\mathrm{max}}$ (MeV)
0.5	0.169	0.331
1.0	0.2035	0.7965
10.0	0.249	9.751

6.2.2.1 Thomson Cross Section

In the low energy limit $E_\gamma \ll m_e c^2$, photon interactions with electrons may be described as elastic, coherent scattering. For such processes, theoretically described by Thomson and called "Thomson scattering," a low energy incident photon is absorbed by an atomic electron causing forced, resonant oscillations. The oscillating electron re-emits a quantum of electromagnetic radiation of the same frequency as the incident photon.

Restricting interactions to the energy range $\mathrm{BE}_e \ll E_\gamma \ll m_e c^2$, so that the oscillating electron can be considered a non-relativistic and "quasi-free" particle, the differential Thomson scattering cross section is obtained

$$\sigma_{\mathrm{Th}}(\theta)\, d\Omega = \frac{1}{2}\left(\frac{e^2}{4\pi\varepsilon_o m_e c^2}\right)^2 \left[1+\cos^2\theta\right]\, d\Omega \qquad (6.31)$$

where θ is the scattered photon emission angle relative to the incident photon direction of propagation. The result is strongly forward and reverse peaked emission. Since there is predicted no reduction in photon energy, there is no energy deposition in this process and no energy extracted from the incident photon beam.

Integration over solid angle yields the total Thomson cross section

$$\sigma_{\mathrm{Th}} = \int_\Omega \sigma_{\mathrm{Th}}(\theta)\, d\Omega = \frac{8}{3}\pi r_e^2 \sim 0.665 \text{ b/electron} \qquad (6.32)$$

where $r_e = \frac{e^2}{4\pi\varepsilon_o m_e c^2} \sim 2.818 \times 10^{-15}$ m and is called the "classical electron radius." This terminology is rather unfortunate though; r_e has absolutely nothing to do with the dimensions of the electron. Instead, r_e is just that distance from a corpuscle possessing electric charge e at which the electrostatic potential energy $\frac{e^2}{4\pi\varepsilon_o r_e}$ is identical with the electron rest energy $m_e c^2$. We can interpret r_e in the context of Thomson scattering as giving the "effective" collision radius.

Interestingly, r_e is connected with other important atomic length scales. We immediately notice that

$$r_e = \frac{1}{2\pi}\alpha\lambda_C$$

where the fine structure constant $\alpha = \frac{e^2}{2\varepsilon_o hc} \sim 1/137$ again appears, and $\lambda_C = \frac{h}{m_e c}$ is the "Compton wavelength," a wavelength numerically equal to that of photons with $E_\gamma = m_e c^2$ but physically represents one half the maximum wavelength shift in Compton scattering, *i.e.* $\Delta\lambda|_{\max} = (\lambda' - \lambda)_{\max} = 2\lambda_C$. As well, we readily notice

$$r_e = \alpha^2 a_o$$

where $a_o = \frac{\varepsilon_o h^2}{\pi m_e e^2}$ is the first Bohr radius.

Returning to the Thomson cross section, we recognize that there appears no explicit energy dependence. As well, since σ_{Th} represents an interaction probability with a single electron, the interaction probability for an atom with atomic number Z should be $Z\sigma_{\text{Th}}$. The simple Z dependence in the atomic cross section and the lack of energy dependence made Thomson scattering an ideal technique for determining atomic numbers of elements.

Thomson scattering on heavier targets like protons is possible, though at greatly reduced probability by the inverse square mass dependence in r_e. The subject of Thomson scattering also provides some historical perspective. It was, in fact, departure from the predicted Thomson interaction signature, *i.e.* incoherent scattering, that lead to the discovery of the Compton effect. In addition, some useful terminology has been developed which we'll see again in our discussion of the Compton cross section below.

At still lower energies, $E_\gamma \lesssim \text{BE}_e$, the electrons of an atom may act collectively to scatter radiation as a collective sum of individual electron contributions in what is referred to as *Rayleigh scattering.* The radiated power in Rayleigh scattering behaves as λ^{-4} which provides explanation for such interesting phenomena as the blue color of the atmosphere and the red sunset (cf: [Jackson75] for a detailed discussion), yet Rayleigh scattering is most often of little consequence in nuclear radiation stopping since there is little effect on photon direction or energy, and the photoelectric effect is almost always dominant in this energy region ([Shultis96], sec. 3.4.4).

6.2.2.2 Klein-Nishina Cross Section for Compton Scattering

The direct, analytic determination of the Compton effect cross section is a rather involved task as it requires a relativistic treatment of the quantum wave mechanics. The result was first obtained by Klein and Nishina[Klein29]

$$\sigma_C(\theta)\, d\Omega = \frac{1}{2} r_e^2 \left(\frac{E_{\gamma'}}{E_\gamma}\right)^2 \left[\frac{E_\gamma}{E_{\gamma'}} + \frac{E_{\gamma'}}{E_\gamma} - \sin^2\theta\right] d\Omega \tag{6.33}$$

as the differential Compton scattering cross section per electron for unpolarized incident photons. A similar expression is obtained for polarized photons. The former, however, is most usually encountered in situations of nuclear decay or energy production, *i.e.* nuclear decay and fission γs are emitted at random orientation and are, hence, unpolarized. On the other hand, high energy photons from particle accelerators, for example, may be emitted with well oriented polarization vectors. Refer to [Evans55] and [Marmier69] for further discussion of the Compton cross section for polarized photons.

By using the Compton scattered photon energy expression (Eq.(6.28)), we can write the Compton cross section as an explicit function of E_γ and

θ alone

$$\sigma_C(\theta)\, d\Omega = \frac{1}{2} r_e^2 \frac{(1+x+x^2)(1+\cos^2\theta) - x\cos\theta(1+2x+\cos^2\theta)}{[1+x(1-\cos\theta)]^3}\, d\Omega \tag{6.34}$$

where $x = E_\gamma/m_e c^2$, the incident photon energy normalized to electron rest mass. In the low energy limit, we find

$$\lim_{E_\gamma \to 0} \simeq \frac{1}{2} r_e^2 \left(1+\cos^2\theta\right)\, d\Omega$$

just the Thomson cross section, Eq.(6.31), as we expect. At high energy, $x \to \infty$, the highest order (x^2) dominates the numerator of Eq. 6.34, but with order x^3 in the denominator, we find $\sigma_C(\theta) \sim 1/x \sim 1/E_\gamma$ at constant θ. The overall behavior is summarized in Fig. 6.13, where $\sigma_C(\theta)/r_e^2$ is shown as a function of θ for $x = E_\gamma/m_e c^2 = 0$ (Thomson scattering), $1/2, 1$, and 10.

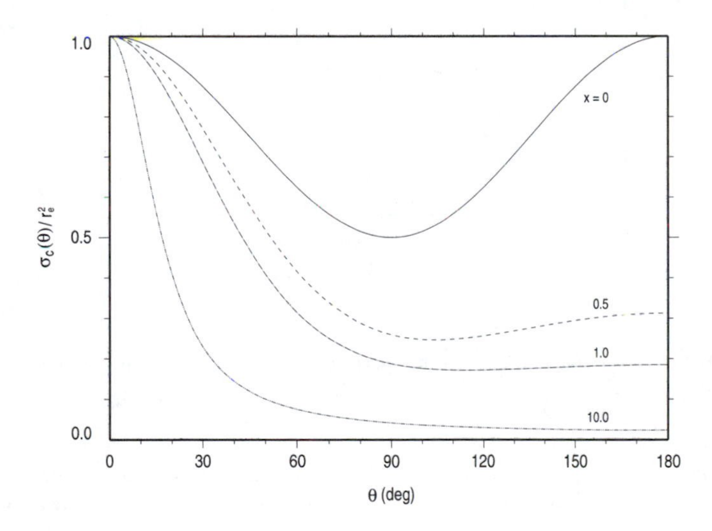

Figure 6.13: Normalized Klein-Nishina cross section, $\sigma_C(\theta)/r_e^2$ as a function of θ for $x = E_\gamma/m_e c^2 = 0$ (Thomson scattering), $1/2, 1$, and 10.

The Compton scattered electron kinetic energy spectrum is also predicted from this formalism. The related quantity

$$\sigma_C(\mathrm{KE}_e) = \sigma_C(\theta) \frac{d\theta}{d\mathrm{KE}_e}$$

is the cross section to find Compton scattered electrons at energy KE_e. This quantity is often important in experimental situations where Compton interaction events are observed in radiation detectors by the signature of the electron energy distribution. Figure 6.14 shows the Compton scattered electron energy distribution as a plot of $\sigma_C(\mathrm{KE}_e)/r_e^2$ as a function of KE_e for $x = 1/2, 1$, and 2. Obvious here is the strongly forward and reverse peaked scattering, with the tendency toward forward peaking as E_γ increases. Moreover, the KE_e distribution flattens as KE_e increases. The sharp cut off appears at the Compton edge (cf: example 38).

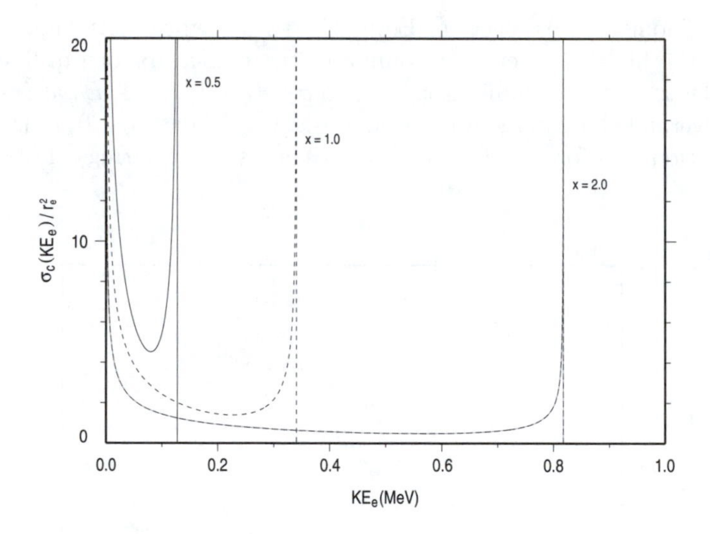

Figure 6.14: Normalized Compton electron energy cross section, $\sigma_C(\mathrm{KE}_e)/r_e^2$ as a function of KE_e for $x = E_\gamma/m_ec^2 = 1/2, 1$, and 2.

At low energy, $E_\gamma \lesssim \mathrm{BE}_e$, the "free" electron assumption breaks down, and binding energy effects must be included in the Compton (incoherent) cross section[Hubbell82, Storm70]. These corrections are included in the photon attenuation data presented in sec. 6.3. Again, however, photoelectric interactions dominate photon attenuation in this energy region so that the inclusion of electron binding effects in incoherent scattering is of only very minor significance and is often ignored in calculations.

6.2.3 Pair Production

At still higher energy, the production of negatron, positron pairs is favored. In pair production, the incident photon is completely absorbed (disappears), creating the energetic pair, e^- and e^+. Particle, anti-particle pairs are formed together to conserve charge when a neutral photon is converted. (We'll write the symbols e^- and e^+ for the pair to distinguish them from the nuclear decay particles, $\beta^\pm$.) The resulting massive particles are provided kinetic energy in the interaction. Their fate resides in classical slowing down via Coulomb interactions (sec. 6.1.3). Upon coming to rest, though, the ultimate demise of e^+ is annihilation (sec. 6.2.3.2).

Complicating the kinematics is the requirement for a third massive body for momentum conservation. Writing the pair production interaction symbolically

$$\gamma + \mathrm{X} \longrightarrow \mathrm{X} + \mathrm{e}^- + \mathrm{e}^+$$

where X is an arbitrary, charged, non-zero rest mass third body. It is in the Coulomb field of X that the pure, electromagnetic energy of the photon is converted to the rest mass of the pair. As atomic nuclei and electron comprise the majority of matter, it is in their presence that we consider pair production.

Since rest mass is created from energy, pair formation is an endoergic process which must have a threshold. Let's first determine the pair production threshold, E_γ^{t}, for an atomic nucleus target, $m_{\mathrm{X}} \sim A$. Mass/energy and momentum conservation for this process take the form

$$\begin{aligned} E_\gamma &= 2m_ec^2 + \mathrm{KE}_{e^-} + \mathrm{KE}_{e^+} + \mathrm{KE}_A \\ \mathbf{p}_\gamma &= \mathbf{p}_{e^-} + \mathbf{p}_{e^+} + \mathbf{p}_A \end{aligned} \tag{6.35}$$

However, since the recoil nucleus is of much greater mass than the electron pair products, A serves to carry nearly all the incident photon momentum. We may write without serious error

$$p_\gamma \sim p_A$$

from which we estimate the recoil energy

$$\mathrm{KE}_A \sim \frac{E_\gamma^2}{2m_Ac^2}$$

At threshold, $E_\gamma = E_\gamma^{\mathrm{t}}$ and $\mathrm{KE}_{e^-} = \mathrm{KE}_{e^+} = 0$. The incident photon is just energetic enough to form the electron pair without imparting additional energy in the form of $\mathrm{KE}_{e^\pm}$. We then have for mass energy conservation

$$E_\gamma^{\mathrm{t}} = \mathrm{KE}_A + 2m_ec^2 = \frac{(E_\gamma^{\mathrm{t}})^2}{2m_Ac^2} + 2m_ec^2$$

an expression quadratic in E_γ^{t}, for which the solution is

$$E_\gamma^{\mathrm{t}} = m_Ac^2\left[1 \pm \left(1 - 4\frac{m_e}{m_A}\right)^{1/2}\right]$$

As the ratio m_e/m_A is always $\ll 1$, this expression may be reduced to the first order in a binomial expansion. Choosing the lower root as the smallest allowable, we finally find

$$E_\gamma^{\mathrm{t}} \simeq 2m_ec^2 = 1.022\ \mathrm{MeV}$$

as anticipated. This is so only because $\mathrm{KE}_A \ll E_\gamma$. As an example, let's estimate the recoil energy for an $A = 100$ nucleus participating in pair production with a photon at the threshold energy, $E_\gamma \sim E_\gamma^{\mathrm{t}} = 2m_ec^2$, then

$$\begin{aligned} \mathrm{KE}_A &\simeq \frac{E_\gamma^2}{2m_Ac^2} = 2\left(\frac{m_e}{m_A}\right)m_ec^2 \\ &\sim 1.022\ (\mathrm{MeV})\frac{1/1836}{100} \sim 5.6\ \mathrm{eV} \end{aligned}$$

a tiny contribution to the overall energy balance.

Alternatively, pair production may proceed in the Coulomb field of an electron, for which our balance equations become

$$\begin{aligned} E_\gamma &= 2m_ec^2 + \mathrm{KE}_{e^-} + \mathrm{KE}_{e^+} + \mathrm{KE}_e \\ \mathbf{p}_\gamma &= \mathbf{p}_{e^-} + \mathbf{p}_{e^+} + \mathbf{p}_e \end{aligned} \tag{6.36}$$

where the unsuperscripted notation refers to the target atomic electron. It can be readily shown that E_γ is minimized for this case when $\mathbf{p}_{e^-} = \mathbf{p}_{e^+} = \mathbf{p}_e$ so that

$$p_\gamma^{\mathrm{t}} = 3p_e$$

and

$$E_\gamma^{\mathrm{t}} = 2m_ec^2 + 3\mathrm{KE}_e$$

Squaring the momentum equation and treating the electrons relativistically, we find

$$\mathrm{KE}_e^2 + 2\mathrm{KE}_e m_ec^2 = \frac{1}{9}(E_\gamma^{\mathrm{t}})^2$$

which, combined with the energy balance equation, yields directly

$$E_\gamma^{\mathrm{t}} = 4m_ec^2 \sim 2.044\ \mathrm{MeV}$$

exactly twice that found when $\mathrm{X} = A$. The difference is that the recoil energy is no longer negligible when $\mathrm{X} = e$, and all three products participate in momentum conservation. At threshold

$$\begin{aligned} \mathrm{KE}_e = \mathrm{KE}_{e^-} = \mathrm{KE}_{e^+} &= \frac{1}{3}(E_\gamma^{\mathrm{t}} - 2m_ec^2) \\ &= \frac{2}{3}m_ec^2 = \frac{1}{6}E_\gamma^{\mathrm{t}} \end{aligned}$$

which cannot be ignored.

6.2.3.1 Pair Production Cross Section

As with the Compton cross section, pair production cross section determination is greatly complicated by quantum and relativistic effects. In fact, so much so for pair formation that a simple analytic form for the differential pair production cross section is not available. Instead, let's devote our attention to the total cross section only.

In the presence of an atom, the pair production interaction has total cross section[Marmier69] for $E_\gamma \gg m_e c^2$

$$\sigma_{pp_A} = \alpha r_e^2 Z^2 \left[\frac{28}{9} \ln \frac{2E_\gamma}{m_e c^2} - \frac{218}{27} \right] \tag{6.37}$$

having the dimensions of cm^2/atom, where, again, α is the fine structure constant and r_e^2 is the classical electron radius. Obviously, when $E_\gamma < 2m_e c^2$ then $\sigma_{pp_A} = 0$. For pair formation interactions on electrons

$$\sigma_{pp_e} = \alpha r_e^2 \left[\frac{28}{9} \ln \frac{2E_\gamma}{m_e c^2} - 11.3 \right] \tag{6.38}$$

having dimensions of cm^2/electron. The expression is valid only for $E_\gamma > 4m_e c^2$ and is zero below this threshold. The composite pair production cross section is then

$$\sigma_{pp} = \sigma_{pp_A} + Z\sigma_{pp_e}$$

Since $\sigma_{pp_A} > Z^2 \sigma_{pp_e}$ and $\frac{\sigma_{pp_A}}{Z\sigma_{pp_e}} \sim Z$, we remain without significant error in approximating $\sigma_{pp} \sim \sigma_{pp_A}$ in our energy range of interest for decay and reactor relevant problems, $0.01 \lesssim E_\gamma \lesssim 10$ MeV.

6.2.3.2 Annihilation Radiation

As mentioned previously, the ultimate fate of all positrons in matter is annihilation, the complete destruction of the anti-matter positron with its matter cousin, the negatron, to produce electromagnetic quanta and liberate $2m_e c^2$ energy. In this way, annihilation can be viewed as inverse pair production.

There are several possible modes of annihilation. The two most likely are one and two quantum annihilation. The lower probability single quantum annihilation event, sometimes called "in flight annihilation," annihilates an atomic electron bound to nucleus X that recoils to conserve momentum

$$\mathrm{e}^+ + \mathrm{e}^- + X \longrightarrow X + \gamma$$

where $E_\gamma|_{\min} = 2m_e c^2 = 1.022$ MeV. With much greater likelihood, however, is the two quantum event. Energetic positrons more readily slow down in matter to low energy; thereafter, they are destined to form the metastable, pseudo-atom positronium by forming a mutually rotating, two body, hydrogen-like object with an electron. This entity has an extremely short mean life ($\sim 10^{-10}$ s) and decays by two quantum annihilation

$$\mathrm{e}^+ + \mathrm{e}^- \longrightarrow \gamma_1 + \gamma_2$$

Since the reactants possess little kinetic energy at annihilation, the two "annihilation photons" are of energy $E_\gamma \simeq 0.511$ MeV and are oppositely directed to conserve momentum. Because even an energetic positron's range in most materials is very short (< few mm), its kinetic energy is spent locally, whereas the annihilation photon energy is deposited elsewhere.

6.3 Photon Attenuation

Like neutrons, photons interact with individual particles (atoms, nuclei, or electrons) in matter. The interaction probability per unit length for uncollided photons is constant, implying exponential removal from the uncollided beam

$$\frac{I(x)}{I_o} = e^{-\mu x}$$

where $I(x)$ is the uncollided photon beam intensity ($= E_\gamma \varphi_\gamma(x)$, MeV/cm^2 s) at one-dimensional position x, I_o is the incident photon intensity at $x = 0$, and μ is the macroscopic photon cross section having dimension cm^{-1} (cf: neutron slab attenuation, sec. 3.7). The latter is most often referred to as the "linear attenuation coefficient" for photons and is, by definition, the sum over linear attenuation coefficients for all available mechanisms, *i.e.*

$$\begin{aligned} \mu &= \mu_{pe} + \mu_C + \mu_{pp} \\ &= N(\sigma_{pe} + Z\sigma_C + \sigma_{pp}) \end{aligned}$$

where N is the atom density in the target material and pe, C, and pp refer to photoelectric, Compton, and pair production interactions, respectively. Implicit in writing this is the assumption of independence, *i.e.* the probability of interaction of type $j(= pe, C, pp)$ is independent of both photon flux and other interactions. Consistent with our treatment of neutron interactions, independence requires that photons do not interact among themselves (only with target particles), two or more photons do not interact simultaneously with the same target, and photon interactions do not significantly alter the medium in which they interact. Though the latter may be correctly treated by time dependent analysis, these conditions are all essentially satisfied when the photon density, $n_\gamma \ll N$, almost always an excellent assumption.

Under these conditions, the probability that a photon will survive to position x without experiencing a photoelectric interaction is $e^{-\mu_{pe}x}$, without experiencing a Compton interaction is $e^{-\mu_C x}$, and without experiencing a pair formation interaction is $e^{-\mu_{pp}x}$. Then, the probability of survival to x without any interaction is $e^{-x(\mu_{pe}+\mu_C+\mu_{pp})} = e^{-\mu x}$.

A related quantity is the "mass attenuation coefficient"

$$\frac{\mu}{\rho} = \frac{\mu_{pe}}{\rho} + \frac{\mu_C}{\rho} + \frac{\mu_{pp}}{\rho} \tag{6.39}$$

having dimension, cm^2/g. The rationale for presenting photon attenuation data this way is twofold. Firstly, the quantity μ/ρ is independent of physical density (but, of course, is still a function of Z, E_γ) which may vary greatly, hence, μ/ρ is more convenient for tabulation. Secondly, in the energy range where Compton interactions are dominant, this quantity

$$\frac{\mu}{\rho} \sim \frac{\mu_C}{\rho} \sim N\frac{Z\sigma_C}{\rho} \sim N_A\sigma_C\frac{Z}{A}$$

is at best only a weak function of Z since $Z/A \sim$ const. Again, where Compton is dominant, and taking into account variations in Z/A, the mass attenuation coefficient in material b relative to that in material a is

$$\left(\frac{\mu_C}{\rho}\right)_b = \left(\frac{\mu_C}{\rho}\right)_a \frac{(Z/A)_b}{(Z/A)_a} \tag{6.40}$$

Expressing attenuation data as mass attenuation coefficient requires that the argument in the exponential attenuation expression be written

$$\left(\frac{\mu}{\rho}\right)\rho x$$

where the quantity ρx (dimension, g/cm^2) is called the "mass thickness" of the material and is a convenient parameter for gauging photon attenuation ability for materials of differing physical density.

Mass attenuation coefficient data are shown in Figs. 6.15 and 6.16 for Pb and Al, respectively, on the energy range $0.01 \leq E_\gamma \leq 10$ MeV. (The calculations are from the ITS code[Halbleib92].) The envelope is the total mass attenuation coefficient, μ/ρ. The Compton (incoherent) coefficient includes contribution from electron binding as discussed in sec. 6.2.2.2. Extensive data sets for mass attenuation coefficients covering a wide range of elements and mixtures can be found in [Faw93], [Hubbell82], and [Storm70].

For mixtures where the composition is most conveniently expressed in weight fraction f_w^i, the mass attenuation coefficient for interactions of type

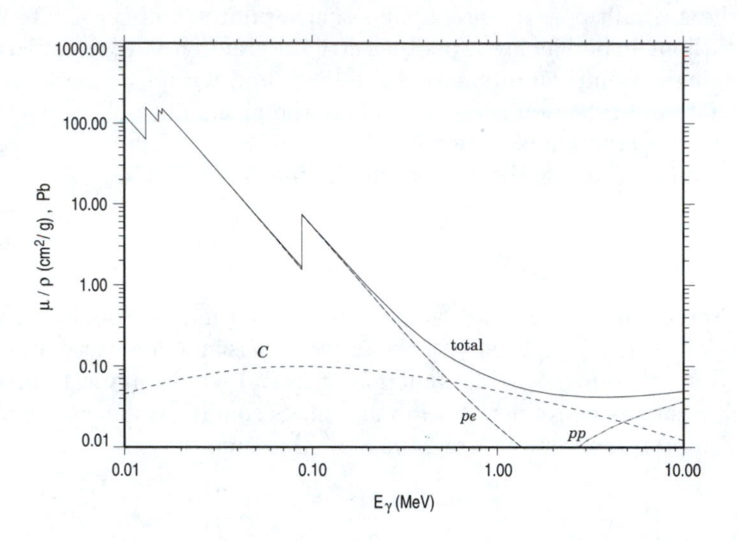

Figure 6.15: Mass attenuation coefficients for Pb vs. $E_\gamma = [0.01, 10]$ MeV. The envelope is the total mass attenuation coefficient, μ/ρ.

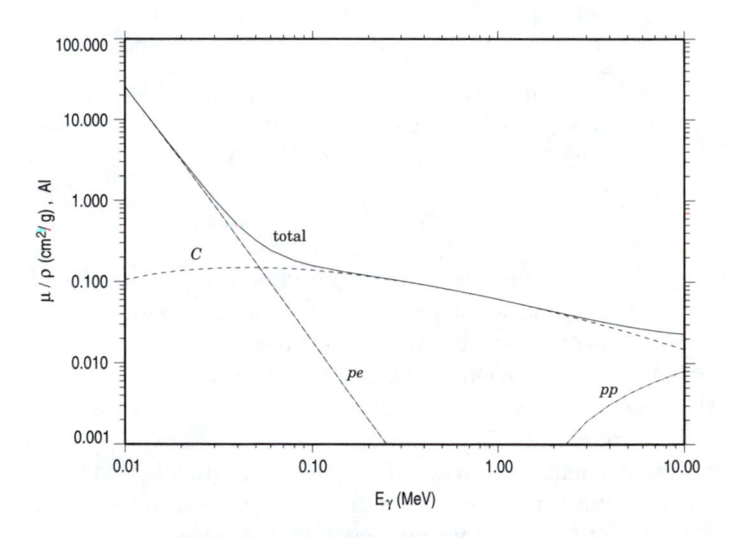

Figure 6.16: Mass attenuation coefficients for Al vs. $E_\gamma = [0.01, 10]$ MeV. The envelope is the total mass attenuation coefficient, μ/ρ.

j in the mixture can be found by applying Eq.(3.29)

$$\frac{\mu_j^m}{\rho} = \sum_i f_w^i \frac{\mu_j^i}{\rho_i} \tag{6.41}$$

where ρ_i is the mass density of the ith constituent in the mixture. Then the mixture total mass attenuation coefficient follows by summation over all interaction types

$$\frac{\mu^m}{\rho} = \sum_i f_w^i \frac{1}{\rho_i}\left(\mu_{pe} + \mu_C + \mu_{pp}\right) = \sum_i f_w^i \frac{\mu^i}{\rho_i}$$

The mean free path for photons in matter is defined the same as that for neutrons (sec. 3.7.2). In an infinite medium, then, we can identify the photon mean free path, $\lambda_\gamma = 1/\mu = \frac{1}{\rho}(\frac{\mu}{\rho})^{-1}$.

6.4 Photon Energy Absorption

Damage to materials including biological organisms is a function of the energy deposited by incident radiation. In the photon attenuation problem, we can identify the quantity

$$E_\gamma \mathcal{R}(\mathbf{r}, t) = \mu E_\gamma \varphi_\gamma(\mathbf{r}, t) = \mu I(\mathbf{r}, t) \quad = \quad \text{rate at which photon energy is removed from the uncollided flux (MeV/cm}^3\text{ s)}$$

a straightforward calculation. A more important quantity, and more difficult to compute, is the rate at which photon energy is absorbed locally in the medium. This task is made difficult by non-local energy deposition. Photon interactions transfer the bulk of E_γ to electrons and scattered photons (or characteristic, or annihilation photons). Because electron range is short ($\lesssim$ few mm, sec. 6.1.3), it is only that fraction of E_γ converted to electron kinetic energy that is deposited locally. The remainder, in the form of characteristic X-rays following photoelectric interactions, $E_{\gamma'}$ from Compton scattering, or annihilation photons following pair interactions, may be transported some distance (because of long λ_γ) from the point of origin. Contemporary computer algorithms have been developed to follow scattered and secondary radiation. This subject, referred to as "computational radiation transport," although quite powerful and enlightening, is well beyond the scope of our discussion.

Let's instead pursue analytic treatment as we have done throughout. Although this provides only very approximate models, it does, however, introduce some important concepts. It is true by construction that at spatial

Example 39: Photon Mean Free Path
Find the mean free path of 0.1 and 1.0 MeV photons in Al and Pb.

Solution:
From the data of Fig. 6.16, we find for Al, $(\frac{\mu}{\rho}) \sim 0.16$ and $0.06 \text{ cm}^2/\text{g}$, at 0.1 and 1.0 MeV, respectively. Since $\rho_{\text{Al}} = 2.7 \text{ g/cm}^3$, then

$$\lambda_{\gamma_{\text{Al}}}(0.1 \text{ MeV}) \sim 2.3 \text{ cm} \quad \text{and} \quad \lambda_{\gamma_{\text{Al}}}(1 \text{ MeV}) \sim 6.2 \text{ cm}$$

For Pb, we find from Fig. 6.15, $(\frac{\mu}{\rho}) \sim 5.3$ and 0.068 for 0.1 and 1.0 MeV photons, respectively, so that with $\rho_{\text{Pb}} = 11.3 \text{ g/cm}^3$

$$\lambda_{\gamma_{\text{Pb}}}(0.1 \text{ MeV}) \sim 0.017 \text{ cm} \quad \text{and} \quad \lambda_{\gamma_{\text{Pb}}}(1 \text{ MeV}) \sim 1.3 \text{ cm}$$

Discussion:
Since photon attenuation in Al, air, and H_2O is dominated by Compton collisions in the energy range [0.1, 1.0] MeV, we may employ our connection expression (6.40) to estimate

$$\left(\frac{\mu}{\rho}\right)_{\text{Air}} \sim \left(\frac{\mu}{\rho}\right)_{\text{Al}} \frac{(Z/A)_{\text{Air}}}{(Z/A)_{\text{Al}}} \sim 0.16\frac{(7/14)}{(13/27)} \sim 0.166 \text{ cm}^2/\text{g}$$

Notice, it has been assumed that nitrogen is representative of air, but O, C, and Ar all have about the same Z/A. We may then estimate

$$\lambda_{\gamma_{\text{Air}}}(0.1 \text{ MeV}) \sim 46.6 \text{ m} \quad \text{and} \quad \lambda_{\gamma_{\text{Air}}}(1 \text{ MeV}) \sim 124 \text{ m}$$

Applying the same approach to H_2O with $A = 18$ and $Z = 10$, we find

$$\lambda_{\gamma_{\text{H}_2\text{O}}}(0.1 \text{ MeV}) \sim 5.4 \text{ cm} \quad \text{and} \quad \lambda_{\gamma_{\text{H}_2\text{O}}}(1 \text{ MeV}) \sim 14.4 \text{ cm}$$

position $\mathbf{r}$ at instant of time t for the uncollided photon intensity can be decomposed

$$\mu I(\mathbf{r},t) = \mu_{pe} I(\mathbf{r},t) + \mu_C I(\mathbf{r},t) + \mu_{pp} I(\mathbf{r},t)$$

Separating contributions to E_γ, the above expression can be rewritten

$$\begin{aligned} \mu E_\gamma \varphi_\gamma &= \left[\mathrm{KE}_e + (1 - y_f^{\mathrm{K}})\mathrm{KE}_{e_A} + y_f^{\mathrm{K}} \mathrm{BE}_e^{\mathrm{K}}\right] \mu_{pe} \varphi_\gamma \\ &+ \left[\langle \mathrm{KE}_e \rangle + \langle E_{\gamma'} \rangle\right] \mu_C \varphi_\gamma \\ &+ \left[\mathrm{KE}_{e^-} + \mathrm{KE}_{e^+} + 2 f_l m_e c^2 + 2 f_r m_e c^2\right] \mu_{pp} \varphi_\gamma \end{aligned} \quad (6.42)$$

where we are dropping the notation for spatial and temporal dependence for brevity and recognize that all quantities in square brackets sum to E_γ. The terms $\langle \mathrm{KE}_e \rangle$ and $\langle E_{\gamma'} \rangle$ are defined below so that obviously, $\mathrm{KE}_e|_{\min} \leq \langle \mathrm{KE}_e \rangle \leq \mathrm{KE}_e|_{\max}$ and $E_{\gamma'}|_{\min} \leq \langle E_{\gamma'} \rangle \leq E_{\gamma'}|_{\max}$. In the first term, KE_{e_A} represents the Auger electron kinetic energy (sec. 6.2.1.2). The terms $(1 - y_f^{\mathrm{K}})\mathrm{KE}_{e_A}$ and $y_f^{\mathrm{K}} \mathrm{BE}_e^{\mathrm{K}}$ contribute a total of $h\nu_K$ per photoelectric interaction. It is almost always acceptable to consider this energy locally deposited, even for the fluorescence X-rays, since their mean free path is short. Hence, in the standard model, all *pe* energy is considered locally deposited. In *pp* interactions, the pair kinetic energy is locally deposited, but only a fraction, f_l, of the annihilation energy is as well. The remainder, $f_r = 1 - f_l$, is remotely deposited by the 0.511 MeV annihilation photons since they have long λ_γ. However, making the approximation $f_l = 1$ (so that $f_r = 0$) usually allows computation without serious error. The reason being that at low energy, below ~ 3 MeV, Compton interactions dominate so that the error in neglecting annihilation photon energy is insignificant. At higher energy, the ratio $2m_e c^2/E_\gamma$ becomes small so that the fractional error is again small. We may, therefore, assume that all of E_γ is deposited locally in *pp* without incurring substantial error.

Treatment of Compton scattering is a bit more involved. Ignoring the scattered photon energy would result in significant error. The usual approximation in the standard model is to consider that only the fraction in the electron component, $\langle \mathrm{KE}_e \rangle$, to be locally deposited. Finally, an expression for local energy deposition rate in this approximation is

$$E_\gamma \varphi_\gamma \left[\mu - \frac{\langle E_{\gamma'} \rangle}{E_\gamma} \mu_C\right] = E_\gamma \varphi_\gamma \left[\mu_{pe} + \frac{\langle \mathrm{KE}_e \rangle}{E_\gamma} \mu_C + \mu_{pp}\right] \quad (6.43)$$

These modified Compton attenuation coefficients appear so regularly in such computations of local energy deposition that they are regarded as having their own identity

$$\mu_{Cs} = \mu_C \frac{\langle E_{\gamma'} \rangle}{E_\gamma}$$

as the linear Compton energy scattering coefficient, and

$$\mu_{Ca} = \mu_C \frac{\langle \mathrm{KE}_e \rangle}{E_\gamma}$$

as the linear Compton energy absorption coefficient. The Compton energy averages are defined

$$\langle E_{\gamma'} \rangle = \frac{\int E_{\gamma'} \sigma_C(\theta)\, d\Omega}{\int \sigma_C(\theta)\, d\Omega}$$

and

$$\langle \mathrm{KE}_e \rangle = \frac{\int \mathrm{KE}_e \sigma_C(\theta)\, d\Omega}{\int \sigma_C(\theta)\, d\Omega}$$

for which the exact solutions are rather involved, but can be found in closed form[Evans55]. For high energy photons, the quantity $\sigma_C(\theta)\, d\Omega = \sigma_C(\mathrm{KE}_e)\, d\mathrm{KE}_e$ is a weak function of KE_e at intermediate energies (*i.e.* away from $\theta \sim 0$ and $\theta \sim \pi$), so that the averages may be approximated

$$\begin{aligned}
\langle E_{\gamma'} \rangle &\simeq \frac{\int E_{\gamma'}\, d\Omega}{\int d\Omega} = \frac{1}{4\pi} \int E_{\gamma'}\, 2\pi \sin\theta\, d\theta \\
&= \frac{m_e c^2}{2E_\gamma} \ln\left(1 + 2E_\gamma / m_e c^2\right)
\end{aligned}$$

Regardless, some general properties can be extracted. Since $\mathrm{KE}_e = E_\gamma - E_{\gamma'}$, it is immediately clear that

$$\langle \mathrm{KE}_e \rangle + \langle E_{\gamma'} \rangle = E_\gamma$$

as we had written above in Eq.(6.42), so that the contributions to μ_C sum in the expected fashion

$$\mu_{Cs} + \mu_{Ca} = \mu_C$$

From Eq.(6.43), we can identify

$$\mu - \mu_{Cs} = \mu_{pe} + \mu_{Ca} + \mu_{pp} = \mu_a$$

as the linear energy absorption coefficient. The literature is replete with this terminology. Data for μ_a and μ_a/ρ are often represented with attenuation coefficients.

The following rates are then identified:

$\mu I(\mathbf{r}, t)$ = rate at which photon energy is removed by all mechanisms (MeV/cm^3 s)

$\mu_a I(\mathbf{r}, t)$ = rate at which photon energy is deposited in the medium by all mechanisms (MeV/cm^3 s)

$\mu_{Ca} I(\mathbf{r}, t)$ = rate at which photon energy is deposited in the medium by Compton electrons (MeV/cm^3 s)

and obviously

$$E_\gamma \mu_{Ca} \varphi_\gamma < E_\gamma \mu_a \varphi_\gamma < E_\gamma \mu \varphi_\gamma$$

Among the many deficiencies of this simple model, we should recognize the most serious, its inability to resolve fine spatial details in the absorption profile (on scales short compared to electron range and X-ray mean free path) or its inability to easily incorporate the potentially important contribution from scattered photons.

6.5 Radiation Exposure and Dose

The effects of ionizing radiation on materials depends sensitively on the type and energy of the radiation particles and on the composition of the material. Damage is a function of the radiation-deposited energy density. In order to quantify damage and identify the relative importance of radiation particles in affecting damage, we need a carefully defined set of radiation units. Units of radiation energy deposition play an important role in assessing and managing potential hazard.

Since all such radiation particles produce ionization as their principle energy deposition mechanism, our radiation units are referenced to their ionizing ability. The unit of radiation "exposure," the roentgen (R), quantifies the ionizing ability of X and γ radiation in air. In the words of the International Commission on Radiological Units[ICRU54], "**The roentgen** shall be the quantity of X- or γ-radiation such that the associated corpuscular emission per 0.001293 gram of air produces, in air, ions carrying 1 electrostatic unit of quantity of electricity of either sign." This mass of air is equivalent to 1 cm^3 volume of dry air at 1 atm (= 760 mm Hg) and 0°C. The electrostatic unit (esu) is the CGS (sec. 1.3.1) unit of electric charge equal to $\frac{1}{3\times10^9}$ C so that

$$\begin{aligned} 1\ \mathrm{R} &= 2.083 \times 10^9 \frac{\mathrm{e-i\ pairs}}{\mathrm{cm}^3} \\ &= 1.61 \times 10^{12} \frac{\mathrm{e-i\ pairs}}{\mathrm{g}} \end{aligned}$$

in air under those conditions. In our preferred SI system of units, the roentgen definition becomes

$$1 \text{ R} = 2.58 \times 10^{-4} \text{ C/kg}$$

Though simple, this definition is very specific and brimming with meaning. Firstly, this unit identifies a quantity of ionization only, and only for X and γ radiation in air. In this way, it allows the quantification of the damaging effects of radiation. It is not, however, a measure of radiation flux or radiation energy, though these parameters are certainly influential in determining the ionizing ability of a quantity of radiation.

Recall that when energetic photons interact with matter, the result is always energetic primary electrons ($e^{\pm}$). The slowing of these primary electrons eventually dissipates E_γ through secondary ionizations. In air, $\sim$ 34 eV is dissipated, on average, in liberating a single electron-ion (e-i) pair. An incident 1 MeV photon, for example, will deposit its energy with a single or few primary electrons which, in turn, produce nearly 30,000 secondary electrons. The overwhelming numbers of secondary electrons makes them the principle energy dissipators. We can consider in this context the R unit of exposure as an effective quantity of energy dissipation by X or γ radiation. The dissipated energy is absorbed by the medium (air). Hence, the roentgen as a unit of "absorbed dose" for X or γ radiation is

$$\begin{aligned} 1 \text{ R} &= 1.61 \times 10^{12} \frac{\text{e} - \text{i}}{\text{g}} \times \frac{34 \text{ eV}}{\text{e} - \text{i}} \times \frac{1.6022 \times 10^{-19} \text{ J}}{\text{eV}} \times \frac{10^7 \text{ erg}}{\text{J}} \\ &= 87.7 \frac{\text{erg}}{\text{g}} = 8.77 \times 10^{-3} \frac{\text{J}}{\text{kg}} \end{aligned}$$

The definition of the R unit is independent of the time required for charge to be accumulated or energy to be dissipated. The rate at which an organism or material receives X or γ dose (the "dose rate") is commonly measured in R/h or mR/h.

Since it is generally realized that radiation damage is dependent on the specific energy deposition (deposition per unit mass), the concept of dose rate is extended to radiation particle other than X or γ and to materials other than air. The quantity "radiation absorbed dose" (rad) is thus introduced

$$1 \text{ rad} = 100 \frac{\text{erg}}{\text{g}} = 0.01 \text{ J/kg}$$

This quantity of absorbed dose is numerically very close to the absorbed dose received in 1 g of water (or tissue) exposed to 1 R of X or γ radiation (*i.e.* $\sim$ 93 erg). When radiation intensity is known, the dose rate may be estimated at that locale

$$D = I \left(\frac{\mu_a}{\rho} \right) \tag{6.44}$$

The accepted SI unit of absorbed dose is the gray (Gy)

$$1 \text{ Gy} = 1 \text{ J/kg} = 100 \text{ rad}$$

Example 40: Radiation Exposure and Dose

What is the exposure rate and dose rate received by an individual standing in the path of a beam of collimated 0.1 MeV photons with $\varphi_\gamma = 10^7 \frac{\gamma}{\text{cm}^2\ \text{s}}$?

Solution:

Energy absorption coefficients can be found in [Hubbell82]. For air at 0.1 MeV, $\mu_a/\rho \sim 0.0232$ cm^2/g and for tissue at the same energy, $\mu_a/\rho \sim 0.025$ cm^2/g. Therefore,

$$\begin{aligned} D_{\text{air}} &= 10^7 \text{ cm}^{-2}\text{ s}^{-1} \times 0.1 \text{ MeV} \times \frac{1.6022 \times 10^{-19} \text{ J}}{\text{MeV}} \\ &\quad \times \frac{10^7 \text{ erg}}{\text{J}} \times 0.0232 \frac{\text{cm}^2}{\text{g}} \\ &= 0.037 \frac{\text{erg}}{\text{gs}} \times \frac{3600 \text{ s}}{\text{h}} \times \frac{1\text{R}}{87.7 \text{ erg/g}} = 1.52 \text{ R/h} \end{aligned}$$

and

$$D_{\text{tissue}} = 1.44 \text{ rad/h}$$

In biological organisms, the deleterious effects of ionizing radiation may vary among sources of radiation producing the same absorbed dose. This is because some radiation particles (especially heavy charged particles and neutrons, which produce energetic charged particles) have a high LET (sec. 6.1.2). The particles deposit their energy in short paths with extremely high charge concentration, thereby concentrating damage. To quantify this difference, the dimensionless quantity "relative biological effectiveness" (RBE) is established employing a 0.2 MeV photon standard

RBE = (absorbed dose of 0.2 MeV X radiation producing a given biological effect)/
(absorbed dose from any other form of radiation required to produce the same biological effect)

so that RBE ≥ 1. The RBE is a function of the organism and biological effect under investigation. A conservative approach for assessing the relative hazard to humans is the adoption of the upper limit to RBE for biological effects of importance to man. This upper bound to human RBE is called the "quality factor" (QF). Some representative values of QF for radiation particles of interest are provided in table 6.2. To quantify the increased hazard from higher RBE radiation, we introduce a new quantity "roentgen equivalent man" (rem) for the "dose equivalent" (D_e) such that

$$D_e(\text{rem}) = QF \times D(\text{rad})$$

The SI unit of dose equivalent is the sievert (Sv)

$$1 \text{ Sv} = 100 \text{ rem}$$

Table 6.2: Quality Factors (QF) for Radiation Particles of Interest[Cember69]

Radiation Type	QF
X, γ	1
$\beta^{\pm}$ ($Q_\beta \lesssim 0.03$ MeV)	1
$\beta^{\pm}$ ($Q_\beta \gtrsim 0.03$ MeV)	1.7
n (thermal)	3
n (fast)	10
p	10
α	10
A^{+} (heavy ions)	20

To place these quantities in perspective, the equivalent dose of 0.2 to 0.4 rem/y is typical for the dose received due to background radiation (natural radiation and cosmic rays) by individuals living in the United States depending on location and altitude. Medical X-ray imaging contributes another 1–2 rem/y for many people in the western hemisphere, by far the largest radiation dose to the general population. By comparison, tens of rem received at once are required to observe noticeable influence of radiation in humans. Instantaneous doses of about 400 rem will induce death in 50% of a human population within 30 days, while thousands to millions of rem may be required to terminate lower forms like insects, viruses, and bacteria.

Knowledge of radiation effects allows the setting of tolerance limits below which the risk from radiation induced effects is minimal. A limit of

5 rem/y (or about 100 mrem/wk) is set for occupational radiation workers (radiation laboratory personnel, nuclear power industry personnel, ...), while the actual doses received are usually only a tiny fraction of that allowable through programs employing the concept of ALARA (as low as reasonably achievable). Radiation dose to the general public from nuclear activities is limited to 10% of those for occupational workers.

References

[Cember69] Cember, H., *Introduction to Health Physics*, Pergamon Press, Oxford, 1969.

[Čerenkov37] Čerenkov, P.A., *Phys. Rev.*, **52** (1937) 378.

[Evans55] Evans, R. D., *The Atomic Nucleus*, McGraw Hill, Inc., NY, 1955.

[Faw93] Faw, R. E., and Shultis, J. K., *Radiological Assessment Sources and Exposures*, Prentice-Hall, Inc., Englewood Cliffs, NJ, 1993.

[Halbleib92] Halbleib, J. A., *et al.*, *ITS Version 3.0: The Integrated TIGER Series of Coupled Electron/Photon Monte Carlo Transport Codes*, Sandia National Laboratories Report SAND91-1634, Albuquerque, NM, 1992.

[Heitler54] Heitler, W., *The Quantum Theory of Radiation*, 3^{ed}, Oxford Univ. Press, NY, 1954.

[Hubbell82] Hubbell, J. H., *Int. J. Appl. Radiat. Isot.*, **33** (1982) 1269.

[Hubert90] Hubert, F., Bimbot, R., and Gauvin, H., *Atomic Data and Nuclear Data Tables*, **46** (1990) 11.

[ICRU54] International Commission on Radiation Units and Measurements, *Nucleonics*, **12** (1954) 11.

[ICRU71] *Radiation Quantities and Units*, International Commission on Radiation Units and Measurements, Report 17, Washington, DC, 1971.

[ICRU84] *Stopping Powers for Electrons and Positrons*, Report 37, International Commission on Radiation Units and Measurements, Washington, DC, 1984.

[ICRU93] *Stopping Powers and Ranges for Protons and Alpha Particles*, Report 49, International Commission on Radiation Units and Measurements, Washington, DC, 1993.

[Jackson75] Jackson, J. D., *Classical Electrodynamics*, John Wiley & Sons, NY, 1975.

[Janni82] Janni, J. F., *Atomic Data and Nuclear Data Tables*, **27** (1982) 147.

[Katz52] Katz, L., and Penfold, A.S., *Rev. Mod. Phys.*, **24** (1952) 28.

[Klein29] Klein, O., and Nishina, Y., *Z. Physik*, **52** (1929) 853.

[Knoll89] Knoll, G. F., *Radiation Detection and Measurement*, 2^{ed}, John Wiley & Sons, NY, 1989.

[Krane88] Krane, K. S., *Introductory Nuclear Physics*, John Wiley & Sons, NY, 1988.

[Marmier69] Marmier, P., and Sheldon, E., *Physics of Nuclei and Particles*, Academic Press, Inc., NY, 1969.

[Moller32] Moller, C., *Ann. Physik*, **14** (1932) 531.

[RHH70] *Radiological Health Handbook*, U.S. Department of Health, Education, and Welfare, U.S. Government Printing Office, Washington, DC, 1970.

[Shultis96] Shultis, J. K., and Faw, R. E., *Radiation Shielding*, Prentice-Hall, Inc., Upper Saddle River, NJ, 1996.

[Storm70] Storm, E., and Israel, H. I., *Nucl. Data Tables*, **A7** (1970) 565.

Problems and Questions

1. Derive an expression for the target scattering angle, θ, as a function of ϕ, for the Coulomb scattering problem. Use this expression to plot θ vs. ϕ for p-e and α-e interactions, with $0 < \phi < \pi/2$.

2. Show that ions cannot be scattered backward on interactions with electrons.

3. Find the maximum final speed of a baseball, initially at rest, struck by a heavy bat at speed v_b. Assume the bat to be infinitely massive and the collision elastic. Show all your steps.

4. Determine the range of the α particle emitted from ^{210}Po in the following materials: gold ($\rho = 19.33$ g/cm^3), aluminum ($\rho = 2.7$ g/cm^3), beryllium ($\rho = 1.85$ g/cm^3), water, and air at STP. Find the ranges of equally energetic tritons in these five materials.

5. Derive the energy transfer expression (6.5).

6. Find the range of ^{90}Sr β^-'s in: lead ($\rho = 11.34$ g/cm^3), iron ($\rho = 7.87$ g/cm^3), Plexiglas ($\rho = 1.18$ g/cm^3), water, and air at STP.

7. Recalling that Tritium is a pure β^- emitter, determine the thickness of air ($\rho = 0.001293$ g/cm^3), Plexiglas ($\rho = 1.18$ g/cm^3), or lead ($\rho = 11.34$ g/cm^3) required to stop all the β^-'s.

8. Consider an experiment to measure the range of β^- particles from ^{10}Be. Your detector window has a density thickness of 0.0017 g/cm^2. What is the minimum amount of Plexiglas ($\rho = 1.18$ g/cm^3) that would be needed between the source and detector to eliminate all β^- particles from this source entering the detector?

9. Provide an estimate for the number of head-on collisions required to slow a 1 MeV proton to rest on interactions with electrons. Describe four major differences between electron and heavy charged particle slowing down.

10. Obtain expression (6.16) as the non-relativistic limit of Eq.(6.15).

11. Sketch and briefly describe three photon interaction mechanisms and show the conservation of energy-mass and momentum equations for each interaction.

12. Show that the photoelectric effect cannot occur in free space, but only in the presence of an atom.

13. Explain why Compton scattering can occur with a free electron while both the photoelectric effect and pair production require the presence of a non-zero rest mass particle.

14. What is the maximum Compton scattered electron kinetic energy for a 2 MeV incident γ? What is the minimum $E_{\gamma'}$?

15. From the differential Compton cross section, Eq.(6.34), find the total Compton cross section, σ_C.

16. Explain why the pair production threshold is $2m_ec^2$ in the presence of an atom but $4m_ec^2$ in the presence of an electron.

17. Show that the threshold energy, E_γ^{t}, for the photoelectric effect is equal to the binding energy of the electron, BE_e, as long as $2\mathrm{BE}_e \ll m_ec^2$.

18. Retrace the steps in the Compton scattering formalism to show that the energy of the scattered photon is correctly expressed by Eq.(6.28).

19. Describe the processes you would expect to occur when photons from a ^{60}Co source interact with matter. Determine the range of charged particle energies to be expected from these interactions. Recall that two γ-rays (1.17 and 1.33 MeV) are emitted in the decay of ^{60}Co.

20. Determine the quantity of locally deposited energy for:

    (a) a 300 keV γ interacting via the photoelectric effect,

    (b) a 1.5 MeV γ interacting via Compton scattering, and

    (c) a 2.0 MeV γ interacting via pair production.

21. Show that in the 2-quantum annihilation interaction with an at-rest electron/position pair:

    (a) the two emitted gammas are of equal and opposite momenta, and

    (b) the energy of each gamma is 0.511 MeV.

22. For a Compton scattering interaction, determine:

    (a) the Compton edge energy for 0.1, 1.0, and 10.0 MeV incident photons,

    (b) the Compton gap energy (energy separation between the Compton edge and the incident photon energy) for the same three incident photon energies above,

(c) the Compton gap energy/E_γ for all three cases, and

(d) discuss the behavior of these results as the incident photon energy is increased.

23. For photoelectric interactions, plot the threshold energy normalized to the atom rest energy as a function of the atomic binding energy relative to the atom rest energy when $x = \mathrm{BE}_e/m_A c^2$ is not small, from $0 < x < 1.0$ on a log scale. (Do not use the binomial approximation.)

24. An experimental researcher wishes to transport a large ^{77}Ge source to a new laboratory. This source emits 2.2 MeV β^- particles and is encapsulated with a thin plastic covering (1 mill) ($\rho_{\text{plastic}} = 1.7$ g/cm^3). How much additional Plexiglas ($\rho_{\text{plex}} = 1.18$ g/cm^3) must be provided to stop these β^- particles?

    The attenuation of these β^- particles in the Plexiglas shield produces bremsstrahlung γs. What thickness Pb shield is required to limit the uncollided photon flux to 5% of its original value?

    Assume: a) There is no γ interaction in the Plexiglas.
    b) All the γ's are emitted perpendicular to the interface.

25. The figure below depicts an experimental setup used to measure the thickness of an aluminum slab. The Al thickness can be determined by detecting the uncollided flux of 0.662 MeV γ-rays from a ^{137}Cs source before and after attenuation in the Al slab. The number of counts measured at position 1 (in front of the slab) is 50400, and at position 2 (in back of the slab) is 30540 in two equal duration measurements with a radiation detector. Find the thickness of the Al slab, x in cm ($\rho_{\text{Al}} = 2.7$ g/cm^3). You may assume there is no photon attenuation in air.

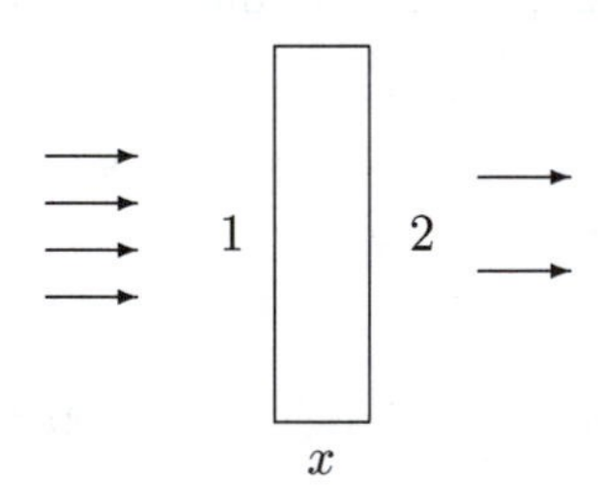

26. Plot the photoelectric, Compton, and pair production cross sections as a function of E_γ from 0.01 to 10 MeV in aluminum using the expressions provided in the text. Compare these results with those of Fig. 6.16.

27. Plot the photoelectric, Compton, pair production, and total mass attenuation coefficients against E_γ energy from 0.01 to 10 MeV in Fe ($\rho = 7.87$ g/cm^3). Compare these against the data in [Hubbell82].

28. A mysterious truck is parked in front of the federal building in your town. The mayor has evacuated the area and asks you to deploy a remotely manipulated radiation detection system that can detect β^- and γ radiation. Expecting a fertilizer bomb, you decide to look for radiation from ^{32}P (pure β^- emitter) and ^{40}K (β^- and γ emitter).

    (a) If the truck walls are 1/4" thick aluminum ($\rho = 2.7$ g/cm^3) and your remote system cannot enter the vehicle, quantify whether or not you will be able to measure any β^- radiation from either source.

    (b) Describe the degree of attenuation expected for uncollided photons.

29. Estimate the uncollided γ exposure and tissue dose rates you would receive 1 m from a 1 Ci ^{60}Co source with 10 cm Pb between you and the source. (Assume a point source and ignore attenuation in air.)

Appendix A

Unit Conversion Factors

Length
1 mill = 10^{-3} in. = 0.0254 mm
1 in. = 2.54 cm = 0.0254 m
1 ft = 12 in. = 30.48 cm = 0.3048 m
1 yd = 3 ft = 91.44 cm = 0.9144 m
1 mi = 5280 ft = 1609 m = 1.609 km
1 ly = 9.46×10^{12} km = 5.88×10^{12} mi ("light year")
1 pc = 3.09×10^{13} km = 1.92×10^{13} mi ("par-sec")

Area
1 in.^{2} = 6.4516 cm^2 = 6.4516×10^{-4} m^2
1 ft^2 = 144 in.^{2} = 929 cm^2 = 0.0929 m^2
1 yd^2 = 9 ft^2 = 0.836 m^2
1 mi^2 = 640 acre = 2.59 km^2
1 acre = $\frac{1}{640}$ mi^2 = 43,560 ft^2

Volume
1 in.^{3} = 16.39 cm^3 = 0.0164 l
1 ft^3 = 1728 in.^{3} = 28,322 cm^3 = 28.32 l
1 l = 1000 cm^3 = 10^{-3} m^3 ("liter")
1 oz = 1.805 in.^{3} = 29.58 cm^3 ("ounce")
1 pt = 16 oz = 0.473 l ("pint")
1 qt = 32 oz = 0.947 l ("quart")
1 gal = 128 oz = 3.79 l ("gallon")

Time
1 y = 365.25 d = 8766 h = 525,960 m = 3.156×10^7 s

Speed
1 ft/s = 30.48 cm/s = 0.3048 m/s
1 mi/h = 1.467 ft/s = 1.609 km/h = 44.7 cm/s

Acceleration
1 ft/s^2 = 30.48 cm/s^2
1 m/s^2 = 3.28 ft/s^2
9.807 m/s^2 = 32.17 ft/s^2 ("1 g")

Mass
1 oz = 28.35 g = 0.02835 kg
1 lb = 16 oz = 453.6 g = 0.4536 kg
1 slug = 514.8 oz = 14.59 kg
1 ton = 2000 lb = 907.2 kg
1 metric ton = 1000 kg = 2204.6 lb
1 amu = 1.66054×10^{-27} kg

Force
1 lb$_f$ = 32.17 pdl = 4.448 N ("pound force")
1 pdl = 0.031 lb$_f$ = 0.138 N ("poundal")

Pressure
1 atm = 760 mm Hg = 760 Torr = 1.013×10^5 Pa = 14.7 lb/in.^{2}
1 bar = 10^5 Pa

Energy
1 Btu = 252 cal = 2.93×10^{-4} kW-h = 1055 J ("British thermal unit")
1 cal = 1.163×10^{-6} kW-h = 4.186 J ("calorie")
1 erg = 2.39×10^{-8} cal = 10^{-7} J
1 fp = 0.324 cal = 1.356 J ("foot pound")
1 kW-h = 3412.3 Btu = 3.6×10^6 J
1 eV = 1.6022×10^{-19} J

Power
1 Btu/h = 0.07 cal/s = 2.93×10^{-4} kW
1 hp = 2545 Btu/h = 171.1 cal/s = 0.7457 kW ("horse power")

Magnetic Induction
1 G = 10^{-4} V-s/m^2 =10^{-4} T ("Gauss")
1 T = 10 kG = 1 Wb/m^2 ("Tesla")

Appendix B

Table of Atomic Masses

The table below comprises the most up-to-date listing of atomic masses for all the known isotopes from *The 1995 Update to the Atomic Mass Evaluation*, by G. Audi and A. H. Wapstra, Nuclear Physics A595, **4** (1995) 409.

N	Z	A	El	mass (amu)	N	Z	A	El	mass (amu)
1	0	1	n	1.008664923	5	8	13	O	13.024810400
0	1	1	H	1.007825032	10	4	14	Be	14.042815522
1	1	2	H	2.014101778	9	5	14	B	14.025404064
2	1	3	H	3.016049268	8	6	14	C	14.003241988
1	2	3	He	3.016029310	7	7	14	N	14.003074005
3	1	4	H	4.027834627	6	8	14	O	14.008595285
2	2	4	He	4.002603250	5	9	14	F	14.036080
1	3	4	Li	4.027182329	10	5	15	B	15.031097291
4	1	5	H	5.039542911	9	6	15	C	15.010599258
3	2	5	He	5.012223628	8	7	15	N	15.000108898
2	3	5	Li	5.012537796	7	8	15	O	15.003065386
1	4	5	Be	5.040790	6	9	15	F	15.018010856
5	1	6	H	6.044942608	11	5	16	B	16.039808836
4	2	6	He	6.018888072	10	6	16	C	16.014701243
3	3	6	Li	6.015122281	9	7	16	N	16.006101417
2	4	6	Be	6.019725804	8	8	16	O	15.994914622
5	2	7	He	7.028030527	7	9	16	F	16.011465730
4	3	7	Li	7.016004049	6	10	16	Ne	16.025756907
3	4	7	Be	7.016929246	12	5	17	B	17.046931399
2	5	7	B	7.029917389	11	6	17	C	17.022583712
6	2	8	He	8.033921838	10	7	17	N	17.008449673
5	3	8	Li	8.022486670	9	8	17	O	16.999131501
4	4	8	Be	8.005305094	8	9	17	F	17.002095238
3	5	8	B	8.024606713	7	10	17	Ne	17.017697565
2	6	8	C	8.037675026	13	5	18	B	18.056170
7	2	9	He	9.043820323	12	6	18	C	18.026757058
6	3	9	Li	9.026789122	11	7	18	N	18.014081827
5	4	9	Be	9.012182135	10	8	18	O	17.999160419
4	5	9	B	9.013328806	9	9	18	F	18.000937667
3	6	9	C	9.031040087	8	10	18	Ne	18.005697066
8	2	10	He	10.052399713	7	11	18	Na	18.027180
7	3	10	Li	10.035480884	14	5	19	B	19.063730
6	4	10	Be	10.013533720	13	6	19	C	19.035248094
5	5	10	B	10.012937027	12	7	19	N	19.017026896
4	6	10	C	10.016853110	11	8	19	O	19.003578730
3	7	10	N	10.042618	10	9	19	F	18.998403205
8	3	11	Li	11.043796166	9	10	19	Ne	19.001879839
7	4	11	Be	11.021657653	8	11	19	Na	19.013879450
6	5	11	B	11.009305466	14	6	20	C	20.040322395
5	6	11	C	11.011433818	13	7	20	N	20.023367295
4	7	11	N	11.026796226	12	8	20	O	20.004076150
9	3	12	Li	12.053780	11	9	20	F	19.999981324
8	4	12	Be	12.026920631	10	10	20	Ne	19.992440176
7	5	12	B	12.014352109	9	11	20	Na	20.007348260
6	6	12	C	12.0000000	8	12	20	Mg	20.018862744
5	7	12	N	12.018613202	15	6	21	C	21.049340
4	8	12	O	12.034404776	14	7	21	N	21.027087574
9	4	13	Be	13.036133834	13	8	21	O	21.008654631
8	5	13	B	13.017780267	12	9	21	F	20.999948921
7	6	13	C	13.003354838	11	10	21	Ne	20.993846744
6	7	13	N	13.005738584	10	11	21	Na	20.997655099

N	Z	A	El	mass (amu)	N	Z	A	El	mass (amu)
9	12	21	Mg	21.011714174	12	15	27	P	26.999191645
8	13	21	Al	21.028040	11	16	27	S	27.018795
16	6	22	C	22.056450	19	9	28	F	28.035670
15	7	22	N	22.034440259	18	10	28	Ne	28.012108072
14	8	22	O	22.009967157	17	11	28	Na	27.998890410
13	9	22	F	22.002999250	16	12	28	Mg	27.983876703
12	10	22	Ne	21.991385510	15	13	28	Al	27.981910184
11	11	22	Na	21.994436782	14	14	28	Si	27.976926533
10	12	22	Mg	21.999574055	13	15	28	P	27.992312330
9	13	22	Al	22.019520	12	16	28	S	28.004372661
8	14	22	Si	22.034530	11	17	28	Cl	28.028510
16	7	23	N	23.040510	20	9	29	F	29.043260
15	8	23	O	23.015691325	19	10	29	Ne	29.019345902
14	9	23	F	23.003574385	18	11	29	Na	29.002811301
13	10	23	Ne	22.994467337	17	12	29	Mg	28.988554743
12	11	23	Na	22.989769675	16	13	29	Al	28.980444848
11	12	23	Mg	22.994124850	15	14	29	Si	28.976494719
10	13	23	Al	23.007264900	14	15	29	P	28.981801376
9	14	23	Si	23.025520	13	16	29	S	28.996608805
17	7	24	N	24.050500	12	17	29	Cl	29.014110
16	8	24	O	24.020369922	20	10	30	Ne	30.023872000
15	9	24	F	24.008099371	19	11	30	Na	30.009226487
14	10	24	Ne	23.993615074	18	12	30	Mg	29.990464529
13	11	24	Na	23.990963332	17	13	30	Al	29.982960304
12	12	24	Mg	23.985041898	16	14	30	Si	29.973770218
11	13	24	Al	23.999940911	15	15	30	P	29.978313807
10	14	24	Si	24.011545711	14	16	30	S	29.984902954
9	15	24	P	24.034350	13	17	30	Cl	30.004770
17	8	25	O	25.029140	12	18	30	Ar	30.021560
16	9	25	F	25.012094963	21	10	31	Ne	31.033110
15	10	25	Ne	24.997789899	20	11	31	Na	31.013595108
14	11	25	Na	24.989954352	19	12	31	Mg	30.996548459
13	12	25	Mg	24.985837023	18	13	31	Al	30.983946023
12	13	25	Al	24.990428555	17	14	31	Si	30.975363275
11	14	25	Si	25.004106640	16	15	31	P	30.973761512
10	15	25	P	25.020260	15	16	31	S	30.979554421
18	8	26	O	26.037750	14	17	31	Cl	30.992416014
17	9	26	F	26.019633157	13	18	31	Ar	31.012126
16	10	26	Ne	26.000461498	22	10	32	Ne	32.039910
15	11	26	Na	25.992589898	21	11	32	Na	32.019649792
14	12	26	Mg	25.982593040	20	12	32	Mg	31.999145889
13	13	26	Al	25.986891659	19	13	32	Al	31.988124379
12	14	26	Si	25.992329935	18	14	32	Si	31.974148129
11	15	26	P	26.011780	17	15	32	P	31.973907163
10	16	26	S	26.027880	16	16	32	S	31.972070690
18	9	27	F	27.026892316	15	17	32	Cl	31.985688908
17	10	27	Ne	27.007615200	14	18	32	Ar	31.997660660
16	11	27	Na	26.994008702	13	19	32	K	32.021920
15	12	27	Mg	26.984340742	22	11	33	Na	33.027386000
14	13	27	Al	26.981538441	21	12	33	Mg	33.005586975
13	14	27	Si	26.986704764	20	13	33	Al	32.990869587

N	Z	A	El	mass (amu)	N	Z	A	El	mass (amu)
19	14	33	Si	32.978000520	20	18	38	Ar	37.962732161
18	15	33	P	32.971725281	19	19	38	K	37.969080107
17	16	33	S	32.971458497	18	20	38	Ca	37.976318637
16	17	33	Cl	32.977451798	17	21	38	Sc	37.994700
15	18	33	Ar	32.989928719	16	22	38	Ti	38.009770
14	19	33	K	33.007260	26	13	39	Al	39.021900
23	11	34	Na	34.034900	25	14	39	Si	39.002300
22	12	34	Mg	34.009072440	24	15	39	P	38.986420000
21	13	34	Al	33.996927255	23	16	39	S	38.975135275
20	14	34	Si	33.978575745	22	17	39	Cl	38.968007677
19	15	34	P	33.973636381	21	18	39	Ar	38.964313413
18	16	34	S	33.967866831	20	19	39	K	38.963706861
17	17	34	Cl	33.973761967	19	20	39	Ca	38.970717729
16	18	34	Ar	33.980270118	18	21	39	Sc	38.984790009
15	19	34	K	33.998410	17	22	39	Ti	39.001323
14	20	34	Ca	34.014120	26	14	40	Si	40.005800
24	11	35	Na	35.044180	25	15	40	P	39.991050000
23	12	35	Mg	35.017490	24	16	40	S	39.975470000
22	13	35	Al	34.999937650	23	17	40	Cl	39.970415555
21	14	35	Si	34.984584158	22	18	40	Ar	39.962383123
20	15	35	P	34.973314249	21	19	40	K	39.963998672
19	16	35	S	34.969032140	20	20	40	Ca	39.962591155
18	17	35	Cl	34.968852707	19	21	40	Sc	39.977964014
17	18	35	Ar	34.975256726	18	22	40	Ti	39.990498907
16	19	35	K	34.988011615	17	23	40	V	40.011090
15	20	35	Ca	35.004765	27	14	41	Si	41.012700
24	12	36	Mg	36.022450	26	15	41	P	40.994800000
23	13	36	Al	36.006351501	25	16	41	S	40.980030000
22	14	36	Si	35.986687363	24	17	41	Cl	40.970650212
21	15	36	P	35.978259824	23	18	41	Ar	40.964500828
20	16	36	S	35.967080880	22	19	41	K	40.961825972
19	17	36	Cl	35.968306945	21	20	41	Ca	40.962278349
18	18	36	Ar	35.967546282	20	21	41	Sc	40.969251316
17	19	36	K	35.981293405	19	22	41	Ti	40.983131
16	20	36	Ca	35.993087234	18	23	41	V	40.999740
15	21	36	Sc	36.014920	28	14	42	Si	42.016100
25	12	37	Mg	37.031240	27	15	42	P	42.000090
24	13	37	Al	37.010310000	26	16	42	S	41.981490000
23	14	37	Si	36.992995990	25	17	42	Cl	41.973174994
22	15	37	P	36.979608338	24	18	42	Ar	41.963046386
21	16	37	S	36.971125716	23	19	42	K	41.962403059
20	17	37	Cl	36.965902600	22	20	42	Ca	41.958618337
19	18	37	Ar	36.966775912	21	21	42	Sc	41.965516761
18	19	37	K	36.973376915	20	22	42	Ti	41.973031622
17	20	37	Ca	36.985871505	19	23	42	V	41.991230
16	21	37	Sc	37.003050	18	24	42	Cr	42.006430
25	13	38	Al	38.016900	28	15	43	P	43.003310
24	14	38	Si	37.995980000	27	16	43	S	42.986600000
23	15	38	P	37.984470000	26	17	43	Cl	42.974203385
22	16	38	S	37.971163443	25	18	43	Ar	42.965670701
21	17	38	Cl	37.968010550	24	19	43	K	42.960715746

N	Z	A	El	mass (amu)	N	Z	A	El	mass (amu)
23	20	43	Ca	42.958766833	32	16	48	S	48.012990
22	21	43	Sc	42.961150980	31	17	48	Cl	47.994850
21	22	43	Ti	42.968523342	30	18	48	Ar	47.975070
20	23	43	V	42.980650	29	19	48	K	47.965512946
19	24	43	Cr	42.997707	28	20	48	Ca	47.952533512
29	15	44	P	44.009880	27	21	48	Sc	47.952234991
28	16	44	S	43.988320	26	22	48	Ti	47.947947053
27	17	44	Cl	43.978538712	25	23	48	V	47.952254480
26	18	44	Ar	43.965365269	24	24	48	Cr	47.954035861
25	19	44	K	43.961556146	23	25	48	Mn	47.968870
24	20	44	Ca	43.955481094	22	26	48	Fe	47.980560
23	21	44	Sc	43.959403048	21	27	48	Co	48.001760
22	22	44	Ti	43.959690235	33	16	49	S	49.022010
21	23	44	V	43.974400	32	17	49	Cl	48.999890
20	24	44	Cr	43.985470	31	18	49	Ar	48.982180
19	25	44	Mn	44.006870	30	19	49	K	48.967450084
30	15	45	P	45.015140	29	20	49	Ca	48.955673302
29	16	45	S	44.994820	28	21	49	Sc	48.950024065
28	17	45	Cl	44.979700000	27	22	49	Ti	48.947870789
27	18	45	Ar	44.968094979	26	23	49	V	48.948516914
26	19	45	K	44.960699658	25	24	49	Cr	48.951341135
25	20	45	Ca	44.956185938	24	25	49	Mn	48.959623415
24	21	45	Sc	44.955910243	23	26	49	Fe	48.973610
23	22	45	Ti	44.958124349	22	27	49	Co	48.989720
22	23	45	V	44.965782286	33	17	50	Cl	50.007730
21	24	45	Cr	44.979160	32	18	50	Ar	49.985940
20	25	45	Mn	44.994510	31	19	50	K	49.972782832
19	26	45	Fe	45.014560	30	20	50	Ca	49.957518286
31	15	46	P	46.023830	29	21	50	Sc	49.952187008
30	16	46	S	45.999570	28	22	50	Ti	49.944792069
29	17	46	Cl	45.984120	27	23	50	V	49.947162792
28	18	46	Ar	45.968093467	26	24	50	Cr	49.946049607
27	19	46	K	45.961976203	25	25	50	Mn	49.954243960
26	20	46	Ca	45.953692759	24	26	50	Fe	49.962993316
25	21	46	Sc	45.955170250	23	27	50	Co	49.981540
24	22	46	Ti	45.952629491	22	28	50	Ni	49.995930
23	23	46	V	45.960199491	34	17	51	Cl	51.013530
22	24	46	Cr	45.968361649	33	18	51	Ar	50.993240
21	25	46	Mn	45.986720	32	19	51	K	50.976380
20	26	46	Fe	46.000810	31	20	51	Ca	50.961474238
31	16	47	S	47.007620	30	21	51	Sc	50.953602700
30	17	47	Cl	46.987950	29	22	51	Ti	50.946616017
29	18	47	Ar	46.972186238	28	23	51	V	50.943963675
28	19	47	K	46.961677807	27	24	51	Cr	50.944771767
27	20	47	Ca	46.954546459	26	25	51	Mn	50.948215487
26	21	47	Sc	46.952408027	25	26	51	Fe	50.956824936
25	22	47	Ti	46.951763792	24	27	51	Co	50.970720
24	23	47	V	46.954906918	23	28	51	Ni	50.987720
23	24	47	Cr	46.962906512	34	18	52	Ar	51.998170
22	25	47	Mn	46.976100	33	19	52	K	51.982610
21	26	47	Fe	46.992890	32	20	52	Ca	51.965100000

N	Z	A	El	mass (amu)	N	Z	A	El	mass (amu)
31	21	52	Sc	51.956650000	30	26	56	Fe	55.934942133
30	22	52	Ti	51.946898175	29	27	56	Co	55.939843937
29	23	52	V	51.944779658	28	28	56	Ni	55.942136339
28	24	52	Cr	51.940511904	27	29	56	Cu	55.958560
27	25	52	Mn	51.945570079	26	30	56	Zn	55.972380
26	26	52	Fe	51.948116526	25	31	56	Ga	55.994910
25	27	52	Co	51.963590	37	20	57	Ca	56.992356
24	28	52	Ni	51.975680	36	21	57	Sc	56.977040
23	29	52	Cu	51.997180	35	22	57	Ti	56.962900
35	18	53	Ar	53.006227	34	23	57	V	56.952360000
34	19	53	K	52.987120	33	24	57	Cr	56.943753800
33	20	53	Ca	52.970050	32	25	57	Mn	56.938287458
32	21	53	Sc	52.959240	31	26	57	Fe	56.935398707
31	22	53	Ti	52.949731709	30	27	57	Co	56.936296235
30	23	53	V	52.944342517	29	28	57	Ni	56.939800489
29	24	53	Cr	52.940653781	28	29	57	Cu	56.949215695
28	25	53	Mn	52.941294702	27	30	57	Zn	56.964910
27	26	53	Fe	52.945312282	26	31	57	Ga	56.982930
26	27	53	Co	52.954224985	37	21	58	Sc	57.983070
25	28	53	Ni	52.968460	36	22	58	Ti	57.966110
24	29	53	Cu	52.985550	35	23	58	V	57.956650000
35	19	54	K	53.993990	34	24	58	Cr	57.944250000
34	20	54	Ca	53.974680	33	25	58	Mn	57.939986451
33	21	54	Sc	53.963000000	32	26	58	Fe	57.933280458
32	22	54	Ti	53.950870000	31	27	58	Co	57.935757571
31	23	54	V	53.946444381	30	28	58	Ni	57.935347922
30	24	54	Cr	53.938884921	29	29	58	Cu	57.944540734
29	25	54	Mn	53.940363247	28	30	58	Zn	57.954596465
28	26	54	Fe	53.939614836	27	31	58	Ga	57.974250
27	27	54	Co	53.948464147	26	32	58	Ge	57.991010
26	28	54	Ni	53.957910508	38	21	59	Sc	58.988041
25	29	54	Cu	53.976710	37	22	59	Ti	58.971960
24	30	54	Zn	53.992950	36	23	59	V	58.959300000
36	19	55	K	54.999388	35	24	59	Cr	58.948630000
35	20	55	Ca	54.980550	34	25	59	Mn	58.940447166
34	21	55	Sc	54.967430	33	26	59	Fe	58.934880493
33	22	55	Ti	54.955120000	32	27	59	Co	58.933200194
32	23	55	V	54.947238194	31	28	59	Ni	58.934351553
31	24	55	Cr	54.940844164	30	29	59	Cu	58.939504114
30	25	55	Mn	54.938049636	29	30	59	Zn	58.949267074
29	26	55	Fe	54.938298029	28	31	59	Ga	58.963370
28	27	55	Co	54.942003149	27	32	59	Ge	58.981750
27	28	55	Ni	54.951336329	38	22	60	Ti	59.975640
26	29	55	Cu	54.966050	37	23	60	V	59.964500000
25	30	55	Zn	54.983980	36	24	60	Cr	59.949730000
36	20	56	Ca	55.985790	35	25	60	Mn	59.943193998
35	21	56	Sc	55.972660	34	26	60	Fe	59.934076943
34	22	56	Ti	55.957990000	33	27	60	Co	59.933822196
33	23	56	V	55.950360000	32	28	60	Ni	59.930790633
32	24	56	Cr	55.940645238	31	29	60	Cu	59.937368123
31	25	56	Mn	55.938909366	30	30	60	Zn	59.941832031

N	Z	A	El	mass (amu)	N	Z	A	El	mass (amu)
29	31	60	Ga	59.957060	37	28	65	Ni	64.930088013
28	32	60	Ge	59.970190	36	29	65	Cu	64.927793707
27	33	60	As	59.993130	35	30	65	Zn	64.929245079
39	22	61	Ti	60.982018	34	31	65	Ga	64.932739322
38	23	61	V	60.967410	33	32	65	Ge	64.939440762
37	24	61	Cr	60.954090000	32	33	65	As	64.949484
36	25	61	Mn	60.944460000	31	34	65	Se	64.964660
35	26	61	Fe	60.936749461	41	25	66	Mn	65.960820
34	27	61	Co	60.932479381	40	26	66	Fe	65.945980000
33	28	61	Ni	60.931060442	39	27	66	Co	65.939825412
32	29	61	Cu	60.933462181	38	28	66	Ni	65.929115232
31	30	61	Zn	60.939513907	37	29	66	Cu	65.928873041
30	31	61	Ga	60.949170	36	30	66	Zn	65.926036763
29	32	61	Ge	60.963790	35	31	66	Ga	65.931592355
28	33	61	As	60.980620	34	32	66	Ge	65.933846798
39	23	62	V	61.973140	33	33	66	As	65.944368
38	24	62	Cr	61.955800000	32	34	66	Se	65.955210
37	25	62	Mn	61.947970000	42	25	67	Mn	66.963820
36	26	62	Fe	61.936770495	41	26	67	Fe	66.950000000
35	27	62	Co	61.934054212	40	27	67	Co	66.940610000
34	28	62	Ni	61.928348763	39	28	67	Ni	66.931569638
33	29	62	Cu	61.932587299	38	29	67	Cu	66.927750294
32	30	62	Zn	61.934334132	37	30	67	Zn	66.927130859
31	31	62	Ga	61.944179608	36	31	67	Ga	66.928204915
30	32	62	Ge	61.954650	35	32	67	Ge	66.932738415
29	33	62	As	61.973200	34	33	67	As	66.939190417
40	23	63	V	62.976750	33	34	67	Se	66.950090
39	24	63	Cr	62.961860	32	35	67	Br	66.964790
38	25	63	Mn	62.949810000	42	26	68	Fe	67.952510
37	26	63	Fe	62.940118442	41	27	68	Co	67.944360000
36	27	63	Co	62.933615218	40	28	68	Ni	67.931844932
35	28	63	Ni	62.929672948	39	29	68	Cu	67.929637875
34	29	63	Cu	62.929601079	38	30	68	Zn	67.924847566
33	30	63	Zn	62.933215563	37	31	68	Ga	67.927983497
32	31	63	Ga	62.939141527	36	32	68	Ge	67.928097266
31	32	63	Ge	62.949640	35	33	68	As	67.936792976
30	33	63	As	62.963690	34	34	68	Se	67.941870
40	24	64	Cr	63.964200	33	35	68	Br	67.958248
39	25	64	Mn	63.953730000	43	26	69	Fe	68.957700
38	26	64	Fe	63.940870000	42	27	69	Co	68.945200000
37	27	64	Co	63.935813523	41	28	69	Ni	68.935181837
36	28	64	Ni	63.927969574	40	29	69	Cu	68.929425281
35	29	64	Cu	63.929767865	39	30	69	Zn	68.926553538
34	30	64	Zn	63.929146578	38	31	69	Ga	68.925580912
33	31	64	Ga	63.936838307	37	32	69	Ge	68.927972002
32	32	64	Ge	63.941572638	36	33	69	As	68.932280154
31	33	64	As	63.957572	35	34	69	Se	68.939562155
41	24	65	Cr	64.970370	34	35	69	Br	68.950178
40	25	65	Mn	64.956100000	33	36	69	Kr	68.965320
39	26	65	Fe	64.944940000	43	27	70	Co	69.949810
38	27	65	Co	64.936484581	42	28	70	Ni	69.936140000

N	Z	A	El	mass (amu)	N	Z	A	El	mass (amu)
41	29	70	Cu	69.932409287	36	38	74	Sr	73.956310
40	30	70	Zn	69.925324870	47	28	75	Ni	74.952970
39	31	70	Ga	69.926027741	46	29	75	Cu	74.941700
38	32	70	Ge	69.924250365	45	30	75	Zn	74.932937379
37	33	70	As	69.930927811	44	31	75	Ga	74.926500645
36	34	70	Se	69.933504	43	32	75	Ge	74.922859494
35	35	70	Br	69.944616	42	33	75	As	74.921596417
34	36	70	Kr	69.956010	41	34	75	Se	74.922523571
44	27	71	Co	70.951730	40	35	75	Br	74.925776410
43	28	71	Ni	70.940000000	39	36	75	Kr	74.931033794
42	29	71	Cu	70.932619818	38	37	75	Rb	74.938569199
41	30	71	Zn	70.927727195	37	38	75	Sr	74.949920
40	31	71	Ga	70.924705010	48	28	76	Ni	75.955330
39	32	71	Ge	70.924953991	47	29	76	Cu	75.945990
38	33	71	As	70.927114724	46	30	76	Zn	75.933394207
37	34	71	Se	70.932268	45	31	76	Ga	75.928928262
36	35	71	Br	70.939246	44	32	76	Ge	75.921402716
35	36	71	Kr	70.950510	43	33	76	As	75.922393933
34	37	71	Rb	70.965320	42	34	76	Se	75.919214107
45	27	72	Co	71.956410	41	35	76	Br	75.924541974
44	28	72	Ni	71.941300000	40	36	76	Kr	75.925948304
43	29	72	Cu	71.935520	39	37	76	Rb	75.935071448
42	30	72	Zn	71.926861122	38	38	76	Sr	75.941610
41	31	72	Ga	71.926369350	49	28	77	Ni	76.960830
40	32	72	Ge	71.922076184	48	29	77	Cu	76.947950
39	33	72	As	71.926752647	47	30	77	Zn	76.937085857
38	34	72	Se	71.927112313	46	31	77	Ga	76.929281189
37	35	72	Br	71.936496876	45	32	77	Ge	76.923548462
36	36	72	Kr	71.941907540	44	33	77	As	76.920647703
35	37	72	Rb	71.959080	43	34	77	Se	76.919914610
45	28	73	Ni	72.946080	42	35	77	Br	76.921380123
44	29	73	Cu	72.936490	41	36	77	Kr	76.924667880
43	30	73	Zn	72.929779469	40	37	77	Rb	76.930406599
42	31	73	Ga	72.925169832	39	38	77	Sr	76.937761511
41	32	73	Ge	72.923459361	38	39	77	Y	76.949620
40	33	73	As	72.923825288	50	28	78	Ni	77.963800
39	34	73	Se	72.926766800	49	29	78	Cu	77.952810
38	35	73	Br	72.931794889	48	30	78	Zn	77.938569576
37	36	73	Kr	72.938931115	47	31	78	Ga	77.931655950
36	37	73	Rb	72.950366	46	32	78	Ge	77.922852886
35	38	73	Sr	72.965970	45	33	78	As	77.921828577
46	28	74	Ni	73.947910	44	34	78	Se	77.917309522
45	29	74	Cu	73.940200	43	35	78	Br	77.921146130
44	30	74	Zn	73.929458261	42	36	78	Kr	77.920386271
43	31	74	Ga	73.926940999	41	37	78	Rb	77.928141485
42	32	74	Ge	73.921178213	40	38	78	Sr	77.932179362
41	33	74	As	73.923929076	39	39	78	Y	77.943500
40	34	74	Se	73.922476561	50	29	79	Cu	78.955280
39	35	74	Br	73.929891152	49	30	79	Zn	78.942675
38	36	74	Kr	73.933258225	48	31	79	Ga	78.932916371
37	37	74	Rb	73.944470376	47	32	79	Ge	78.925401560

N	Z	A	El	mass (amu)	N	Z	A	El	mass (amu)
46	33	79	As	78.920948498	45	38	83	Sr	82.917555029
45	34	79	Se	78.918499802	44	39	83	Y	82.922352572
44	35	79	Br	78.918337647	43	40	83	Zr	82.928652130
43	36	79	Kr	78.920082992	42	41	83	Nb	82.936703713
42	37	79	Rb	78.923996719	41	42	83	Mo	82.948740
41	38	79	Sr	78.929707076	53	31	84	Ga	83.952340
40	39	79	Y	78.937350712	52	32	84	Ge	83.937310
39	40	79	Zr	78.949160	51	33	84	As	83.929060
51	29	80	Cu	79.961890	50	34	84	Se	83.918464523
50	30	80	Zn	79.944414722	49	35	84	Br	83.916503685
49	31	80	Ga	79.936588154	48	36	84	Kr	83.911506627
48	32	80	Ge	79.925444764	47	37	84	Rb	83.914384676
47	33	80	As	79.922578162	46	38	84	Sr	83.913424778
46	34	80	Se	79.916521828	45	39	84	Y	83.920387768
45	35	80	Br	79.918529952	44	40	84	Zr	83.923250
44	36	80	Kr	79.916378040	43	41	84	Nb	83.933570
43	37	80	Rb	79.922519322	42	42	84	Mo	83.940090
42	38	80	Sr	79.924524588	53	32	85	Ge	84.942690
41	39	80	Y	79.934337	52	33	85	As	84.931810
40	40	80	Zr	79.940550	51	34	85	Se	84.922244678
51	30	81	Zn	80.950480	50	35	85	Br	84.915608027
50	31	81	Ga	80.937752955	49	36	85	Kr	84.912526954
49	32	81	Ge	80.928821065	48	37	85	Rb	84.911789341
48	33	81	As	80.922132884	47	38	85	Sr	84.912932689
47	34	81	Se	80.917992931	46	39	85	Y	84.916427076
46	35	81	Br	80.916291060	45	40	85	Zr	84.921465220
45	36	81	Kr	80.916592419	44	41	85	Nb	84.927906486
44	37	81	Rb	80.918994165	43	42	85	Mo	84.936590
43	38	81	Sr	80.923213095	42	43	85	Tc	84.948940
42	39	81	Y	80.929128719	54	32	86	Ge	85.946270
41	40	81	Zr	80.936815296	53	33	86	As	85.936230
40	41	81	Nb	80.949050	52	34	86	Se	85.924271165
52	30	82	Zn	81.954840	51	35	86	Br	85.918797162
51	31	82	Ga	81.943160	50	36	86	Kr	85.910610313
50	32	82	Ge	81.929550326	49	37	86	Rb	85.911167080
49	33	82	As	81.924504668	48	38	86	Sr	85.909262351
48	34	82	Se	81.916700000	47	39	86	Y	85.914887724
47	35	82	Br	81.916804666	46	40	86	Zr	85.916472851
46	36	82	Kr	81.913484601	45	41	86	Nb	85.925037588
45	37	82	Rb	81.918207691	44	42	86	Mo	85.930695167
44	38	82	Sr	81.918401258	43	43	86	Tc	85.942880
43	39	82	Y	81.926792071	54	33	87	As	86.939580
42	40	82	Zr	81.931086249	53	34	87	Se	86.928520749
41	41	82	Nb	81.943130	52	35	87	Br	86.920710713
52	31	83	Ga	82.946870	51	36	87	Kr	86.913354251
51	32	83	Ge	82.934510	50	37	87	Rb	86.909183465
50	33	83	As	82.924980625	49	38	87	Sr	86.908879316
49	34	83	Se	82.919119072	48	39	87	Y	86.910877833
48	35	83	Br	82.915180219	47	40	87	Zr	86.914816578
47	36	83	Kr	82.914135952	46	41	87	Nb	86.920361435
46	37	83	Rb	82.915111951	45	42	87	Mo	86.927326830

N	Z	A	El	mass (amu)	N	Z	A	El	mass (amu)
44	43	87	Tc	86.936530	45	46	91	Pd	90.949480
43	44	87	Ru	86.949180	58	34	92	Se	91.949330
55	33	88	As	87.944560	57	35	92	Br	91.939255258
54	34	88	Se	87.931423982	56	36	92	Kr	91.926152752
53	35	88	Br	87.924065908	55	37	92	Rb	91.919725442
52	36	88	Kr	87.914446951	54	38	92	Sr	91.911029895
51	37	88	Rb	87.911318556	53	39	92	Y	91.908946832
50	38	88	Sr	87.905614339	52	40	92	Zr	91.905040106
49	39	88	Y	87.909503361	51	41	92	Nb	91.907193214
48	40	88	Zr	87.910226179	50	42	92	Mo	91.906810480
47	41	88	Nb	87.917956	49	43	92	Tc	91.915259655
46	42	88	Mo	87.921952728	48	44	92	Ru	91.920120
45	43	88	Tc	87.932830	47	45	92	Rh	91.931980
44	44	88	Ru	87.940420	46	46	92	Pd	91.940420
56	33	89	As	88.949230	58	35	93	Br	92.943100
55	34	89	Se	88.936020	57	36	93	Kr	92.931265246
54	35	89	Br	88.926387260	56	37	93	Rb	92.922032765
53	36	89	Kr	88.917632505	55	38	93	Sr	92.914022410
52	37	89	Rb	88.912279939	54	39	93	Y	92.909581582
51	38	89	Sr	88.907452906	53	40	93	Zr	92.906475627
50	39	89	Y	88.905847902	52	41	93	Nb	92.906377543
49	40	89	Zr	88.908888916	51	42	93	Mo	92.906812213
48	41	89	Nb	88.913495503	50	43	93	Tc	92.910248473
47	42	89	Mo	88.919480562	49	44	93	Ru	92.917051523
46	43	89	Tc	88.927542880	48	45	93	Rh	92.925740
45	44	89	Ru	88.936110	47	46	93	Pd	92.935910
44	45	89	Rh	88.949380	59	35	94	Br	93.948680
56	34	90	Se	89.939420	58	36	94	Kr	93.934362
55	35	90	Br	89.930634988	57	37	94	Rb	93.926407326
54	36	90	Kr	89.919523803	56	38	94	Sr	93.915359856
53	37	90	Rb	89.914808941	55	39	94	Y	93.911594008
52	38	90	Sr	89.907737596	54	40	94	Zr	93.906315765
51	39	90	Y	89.907151443	53	41	94	Nb	93.907283457
50	40	90	Zr	89.904703679	52	42	94	Mo	93.905087578
49	41	90	Nb	89.911264109	51	43	94	Tc	93.909656309
48	42	90	Mo	89.913936161	50	44	94	Ru	93.911359569
47	43	90	Tc	89.923555830	49	45	94	Rh	93.921698
46	44	90	Ru	89.929780	48	46	94	Pd	93.928770
45	45	90	Rh	89.942870	47	47	94	Ag	93.942780
57	34	91	Se	90.945370	59	36	95	Kr	94.939840
56	35	91	Br	90.933965300	58	37	95	Rb	94.929319260
55	36	91	Kr	90.923442418	57	38	95	Sr	94.919358213
54	37	91	Rb	90.916534160	56	39	95	Y	94.912823709
53	38	91	Sr	90.910209845	55	40	95	Zr	94.908042739
52	39	91	Y	90.907303415	54	41	95	Nb	94.906835178
51	40	91	Zr	90.905644968	53	42	95	Mo	94.905841487
50	41	91	Nb	90.906990538	52	43	95	Tc	94.907656454
49	42	91	Mo	90.911750754	51	44	95	Ru	94.910412729
48	43	91	Tc	90.918428200	50	45	95	Rh	94.915898541
47	44	91	Ru	90.926377434	49	46	95	Pd	94.924690
46	45	91	Rh	90.936550	48	47	95	Ag	94.935480

N	Z	A	El	mass (amu)	N	Z	A	El	mass (amu)
60	36	96	Kr	95.943070	50	49	99	In	98.934610
59	37	96	Rb	95.934283962	63	37	100	Rb	99.949870
58	38	96	Sr	95.921680473	62	38	100	Sr	99.935351729
57	39	96	Y	95.915897787	61	39	100	Y	99.927756402
56	40	96	Zr	95.908275675	60	40	100	Zr	99.917761704
55	41	96	Nb	95.908100076	59	41	100	Nb	99.914181434
54	42	96	Mo	95.904678904	58	42	100	Mo	99.907477149
53	43	96	Tc	95.907870803	57	43	100	Tc	99.907657594
52	44	96	Ru	95.907597681	56	44	100	Ru	99.904219664
51	45	96	Rh	95.914518212	55	45	100	Rh	99.908116630
50	46	96	Pd	95.918221940	54	46	100	Pd	99.908504596
49	47	96	Ag	95.930680	53	47	100	Ag	99.916069387
48	48	96	Cd	95.939770	52	48	100	Cd	99.920230232
61	36	97	Kr	96.948560	51	49	100	In	99.931149033
60	37	97	Rb	96.937342863	50	50	100	Sn	99.938954
59	38	97	Sr	96.926148757	64	37	101	Rb	100.953195994
58	39	97	Y	96.918131017	63	38	101	Sr	100.940517434
57	40	97	Zr	96.910950716	62	39	101	Y	100.930313395
56	41	97	Nb	96.908097144	61	40	101	Zr	100.921139958
55	42	97	Mo	96.906021033	60	41	101	Nb	100.915251567
54	43	97	Tc	96.906364843	59	42	101	Mo	100.910346543
53	44	97	Ru	96.907554546	58	43	101	Tc	100.907314380
52	45	97	Rh	96.911336643	57	44	101	Ru	100.905582219
51	46	97	Pd	96.916478921	56	45	101	Rh	100.906163526
50	47	97	Ag	96.924000	55	46	101	Pd	100.908289144
49	48	97	Cd	96.934940	54	47	101	Ag	100.912802135
61	37	98	Rb	97.941703557	53	48	101	Cd	100.918681442
60	38	98	Sr	97.928471177	52	49	101	In	100.926560
59	39	98	Y	97.922219525	51	50	101	Sn	100.936060
58	40	98	Zr	97.912746366	65	37	102	Rb	101.959210
57	41	98	Nb	97.910330690	64	38	102	Sr	101.943018795
56	42	98	Mo	97.905407846	63	39	102	Y	101.933555501
55	43	98	Tc	97.907215692	62	40	102	Zr	101.922981089
54	44	98	Ru	97.905287111	61	41	102	Nb	101.918037417
53	45	98	Rh	97.910716431	60	42	102	Mo	101.910297162
52	46	98	Pd	97.912720751	59	43	102	Tc	101.909212938
51	47	98	Ag	97.921759995	58	44	102	Ru	101.904349503
50	48	98	Cd	97.927579	57	45	102	Rh	101.906842845
49	49	98	In	97.942240	56	46	102	Pd	101.905607716
62	37	99	Rb	98.945420616	55	47	102	Ag	101.911999996
61	38	99	Sr	98.933315038	54	48	102	Cd	101.914777255
60	39	99	Y	98.924634736	53	49	102	In	101.924707541
59	40	99	Zr	98.916511084	52	50	102	Sn	101.930490
58	41	99	Nb	98.911617864	65	38	103	Sr	102.948950
57	42	99	Mo	98.907711598	64	39	103	Y	102.936940
56	43	99	Tc	98.906254554	63	40	103	Zr	102.926597062
55	44	99	Ru	98.905939307	62	41	103	Nb	102.919141297
54	45	99	Rh	98.908132101	61	42	103	Mo	102.913204596
53	46	99	Pd	98.911767757	60	43	103	Tc	102.909178805
52	47	99	Ag	98.917597103	59	44	103	Ru	102.906323677
51	48	99	Cd	98.925010	58	45	103	Rh	102.905504182

N	Z	A	El	mass (amu)	N	Z	A	El	mass (amu)
57	46	103	Pd	102.906087204	63	44	107	Ru	106.909907207
56	47	103	Ag	102.908972453	62	45	107	Rh	106.906750540
55	48	103	Cd	102.913418952	61	46	107	Pd	106.905128453
54	49	103	In	102.919913896	60	47	107	Ag	106.905093020
53	50	103	Sn	102.928130	59	48	107	Cd	106.906614232
52	51	103	Sb	102.940120	58	49	107	In	106.910292195
66	38	104	Sr	103.952330	57	50	107	Sn	106.915666702
65	39	104	Y	103.941450	56	51	107	Sb	106.924150
64	40	104	Zr	103.928780	55	52	107	Te	106.935036
63	41	104	Nb	103.922459464	68	40	108	Zr	107.944280
62	42	104	Mo	103.913758387	67	41	108	Nb	107.935010
61	43	104	Tc	103.911444898	66	42	108	Mo	107.923579
60	44	104	Ru	103.905430145	65	43	108	Tc	107.918479973
59	45	104	Rh	103.906655315	64	44	108	Ru	107.910192211
58	46	104	Pd	103.904034912	63	45	108	Rh	107.908730768
57	47	104	Ag	103.908628228	62	46	108	Pd	107.903894451
56	48	104	Cd	103.909848091	61	47	108	Ag	107.905953705
55	49	104	In	103.918338416	60	48	108	Cd	107.904183403
54	50	104	Sn	103.923185469	59	49	108	In	107.909719683
53	51	104	Sb	103.936287	58	50	108	Sn	107.911965339
66	39	105	Y	104.945090	57	51	108	Sb	107.922160
65	40	105	Zr	104.933050	56	52	108	Te	107.929486838
64	41	105	Nb	104.923934023	55	53	108	I	107.943291
63	42	105	Mo	104.916972087	68	41	109	Nb	108.937630
62	43	105	Tc	104.911658043	67	42	109	Mo	108.927810
61	44	105	Ru	104.907750341	66	43	109	Tc	108.919627
60	45	105	Rh	104.905692444	65	44	109	Ru	108.913201565
59	46	105	Pd	104.905084046	64	45	109	Rh	108.908735621
58	47	105	Ag	104.906528234	63	46	109	Pd	108.905953535
57	48	105	Cd	104.909467818	62	47	109	Ag	108.904755514
56	49	105	In	104.914673434	61	48	109	Cd	108.904985569
55	50	105	Sn	104.921390409	60	49	109	In	108.907154078
54	51	105	Sb	104.931528593	59	50	109	Sn	108.911286879
67	39	106	Y	105.950220	58	51	109	Sb	108.918136092
66	40	106	Zr	105.935910	57	52	109	Te	108.927456483
65	41	106	Nb	105.928190	56	53	109	I	108.938191658
64	42	106	Mo	105.918134284	69	41	110	Nb	109.942680
63	43	106	Tc	105.914355408	68	42	110	Mo	109.929730
62	44	106	Ru	105.907326913	67	43	110	Tc	109.923390
61	45	106	Rh	105.907284615	66	44	110	Ru	109.913966185
60	46	106	Pd	105.903483087	65	45	110	Rh	109.910949525
59	47	106	Ag	105.906666431	64	46	110	Pd	109.905152385
58	48	106	Cd	105.906458007	63	47	110	Ag	109.906110460
57	49	106	In	105.913461134	62	48	110	Cd	109.903005578
56	50	106	Sn	105.916880472	61	49	110	In	109.907168783
55	51	106	Sb	105.928763	60	50	110	Sn	109.907852688
54	52	106	Te	105.937702	59	51	110	Sb	109.916763
67	40	107	Zr	106.940860	58	52	110	Te	109.922407164
66	41	107	Nb	106.930310	57	53	110	I	109.935214
65	42	107	Mo	106.921694724	56	54	110	Xe	109.944476
64	43	107	Tc	106.915081691	69	42	111	Mo	110.934510

N	Z	A	El	mass (amu)	N	Z	A	El	mass (amu)
68	43	111	Tc	110.925050	60	54	114	Xe	113.928145
67	44	111	Ru	110.917560	59	55	114	Cs	113.941421
66	45	111	Rh	110.911660	58	56	114	Ba	113.950941
65	46	111	Pd	110.907643952	72	43	115	Tc	114.938280
64	47	111	Ag	110.905294679	71	44	115	Ru	114.928310
63	48	111	Cd	110.904181628	70	45	115	Rh	114.920124676
62	49	111	In	110.905110677	69	46	115	Pd	114.913683410
61	50	111	Sn	110.907735404	68	47	115	Ag	114.908762282
60	51	111	Sb	110.913210	67	48	115	Cd	114.905430553
59	52	111	Te	110.921120589	66	49	115	In	114.903878328
58	53	111	I	110.930276	65	50	115	Sn	114.903345973
57	54	111	Xe	110.941632	64	51	115	Sb	114.906598812
70	42	112	Mo	111.936840	63	52	115	Te	114.911578627
69	43	112	Tc	111.929240	62	53	115	I	114.917918
68	44	112	Ru	111.918553	61	54	115	Xe	114.926538
67	45	112	Rh	111.914613	60	55	115	Cs	114.935939
66	46	112	Pd	111.907313277	59	56	115	Ba	114.947710
65	47	112	Ag	111.907004132	72	44	116	Ru	115.930160
64	48	112	Cd	111.902757226	71	45	116	Rh	115.923713
63	49	112	In	111.905533338	70	46	116	Pd	115.914158288
62	50	112	Sn	111.904820810	69	47	116	Ag	115.911359558
61	51	112	Sb	111.912394640	68	48	116	Cd	115.904755434
60	52	112	Te	111.917061617	67	49	116	In	115.905259995
59	53	112	I	111.927970	66	50	116	Sn	115.901744149
58	54	112	Xe	111.935665350	65	51	116	Sb	115.906797235
57	55	112	Cs	111.950331	64	52	116	Te	115.908420253
71	42	113	Mo	112.942030	63	53	116	I	115.916735014
70	43	113	Tc	112.931330	62	54	116	Xe	115.921738
69	44	113	Ru	112.922540	61	55	116	Cs	115.932914152
68	45	113	Rh	112.915420	60	56	116	Ba	115.941680
67	46	113	Pd	112.910151346	73	44	117	Ru	116.934790
66	47	113	Ag	112.906565708	72	45	117	Rh	116.925350
65	48	113	Cd	112.904400947	71	46	117	Pd	116.917840
64	49	113	In	112.904061223	70	47	117	Ag	116.911684187
63	50	113	Sn	112.905173373	69	48	117	Cd	116.907218242
62	51	113	Sb	112.909377941	68	49	117	In	116.904515731
61	52	113	Te	112.915927	67	50	117	Sn	116.902953765
60	53	113	I	112.923644245	66	51	117	Sb	116.904839590
59	54	113	Xe	112.933382836	65	52	117	Te	116.908634180
58	55	113	Cs	112.944535512	64	53	117	I	116.913647692
71	43	114	Tc	113.935880	63	54	117	Xe	116.920564355
70	44	114	Ru	113.923999	62	55	117	Cs	116.928639484
69	45	114	Rh	113.918846	61	56	117	Ba	116.938860
68	46	114	Pd	113.910365322	60	57	117	La	116.950010
67	47	114	Ag	113.908807907	74	44	118	Ru	117.937030
66	48	114	Cd	113.903358121	73	45	118	Rh	117.929430
65	49	114	In	113.904916758	72	46	118	Pd	117.918983915
64	50	114	Sn	113.902781816	71	47	118	Ag	117.914582383
63	51	114	Sb	113.909095876	70	48	118	Cd	117.906914144
62	52	114	Te	113.912057	69	49	118	In	117.906354623
61	53	114	I	113.921850	68	50	118	Sn	117.901606328

N	Z	A	El	mass (amu)	N	Z	A	El	mass (amu)
67	51	118	Sb	117.905531885	75	47	122	Ag	121.923320
66	52	118	Te	117.905825187	74	48	122	Cd	121.913500
65	53	118	I	117.913375230	73	49	122	In	121.910277103
64	54	118	Xe	117.916570920	72	50	122	Sn	121.903440138
63	55	118	Cs	117.926554883	71	51	122	Sb	121.905175415
62	56	118	Ba	117.933440	70	52	122	Te	121.903047064
61	57	118	La	117.946570	69	53	122	I	121.907592451
74	45	119	Rh	118.931360	68	54	122	Xe	121.908548396
73	46	119	Pd	118.922680	67	55	122	Cs	121.916121946
72	47	119	Ag	118.915666045	66	56	122	Ba	121.920260
71	48	119	Cd	118.909922582	65	57	122	La	121.930710
70	49	119	In	118.905846334	64	58	122	Ce	121.938010
69	50	119	Sn	118.903308880	63	59	122	Pr	121.951650
68	51	119	Sb	118.903946460	77	46	123	Pd	122.934260
67	52	119	Te	118.906408110	76	47	123	Ag	122.924900
66	53	119	I	118.910180837	75	48	123	Cd	122.917003675
65	54	119	Xe	118.915554295	74	49	123	In	122.910438951
64	55	119	Cs	118.922370879	73	50	123	Sn	122.905721901
63	56	119	Ba	118.931051927	72	51	123	Sb	122.904215696
62	57	119	La	118.940990	71	52	123	Te	122.904272951
61	58	119	Ce	118.952760	70	53	123	I	122.905597944
75	45	120	Rh	119.935780	69	54	123	Xe	122.908470748
74	46	120	Pd	119.924030	68	55	123	Cs	122.912990168
73	47	120	Ag	119.918788609	67	56	123	Ba	122.918850
72	48	120	Cd	119.909851352	66	57	123	La	122.926240
71	49	120	In	119.907961505	65	58	123	Ce	122.935510
70	50	120	Sn	119.902196571	64	59	123	Pr	122.945960
69	51	120	Sb	119.905074315	77	47	124	Ag	123.928530
68	52	120	Te	119.904019891	76	48	124	Cd	123.917648302
67	53	120	I	119.910047843	75	49	124	In	123.913175916
66	54	120	Xe	119.912151990	74	50	124	Sn	123.905274630
65	55	120	Cs	119.920678219	73	51	124	Sb	123.905937525
64	56	120	Ba	119.926045941	72	52	124	Te	123.902819466
63	57	120	La	119.938070	71	53	124	I	123.906211423
62	58	120	Ce	119.946640	70	54	124	Xe	123.905895774
76	45	121	Rh	120.938080	69	55	124	Cs	123.912245731
75	46	121	Pd	120.928180	68	56	124	Ba	123.915088437
74	47	121	Ag	120.919851074	67	57	124	La	123.924530
73	48	121	Cd	120.912980390	66	58	124	Ce	123.930520
72	49	121	In	120.907848847	65	59	124	Pr	123.942960
71	50	121	Sn	120.904236867	78	47	125	Ag	124.930540
70	51	121	Sb	120.903818044	77	48	125	Cd	124.921247170
69	52	121	Te	120.904929815	76	49	125	In	124.913601387
68	53	121	I	120.907366063	75	50	125	Sn	124.907784924
67	54	121	Xe	120.911386497	74	51	125	Sb	124.905247804
66	55	121	Cs	120.917183637	73	52	125	Te	124.904424718
65	56	121	Ba	120.924485908	72	53	125	I	124.904624150
64	57	121	La	120.933010	71	54	125	Xe	124.906398236
63	58	121	Ce	120.943670	70	55	125	Cs	124.909724871
62	59	121	Pr	120.955364	69	56	125	Ba	124.914620234
76	46	122	Pd	121.929800	68	57	125	La	124.920670

N	Z	A	El	mass (amu)	N	Z	A	El	mass (amu)
67	58	125	Ce	124.928540	74	55	129	Cs	128.906063369
66	59	125	Pr	124.937830	73	56	129	Ba	128.908673749
79	47	126	Ag	125.934500	72	57	129	La	128.912667334
78	48	126	Cd	125.922353996	71	58	129	Ce	128.918089
77	49	126	In	125.916464532	70	59	129	Pr	128.924860
76	50	126	Sn	125.907653953	69	60	129	Nd	128.933254
75	51	126	Sb	125.907248153	68	61	129	Pm	128.943160
74	52	126	Te	125.903305543	82	48	130	Cd	129.933980
73	53	126	I	125.905619387	81	49	130	In	129.924854941
72	54	126	Xe	125.904268868	80	50	130	Sn	129.913852185
71	55	126	Cs	125.909447953	79	51	130	Sb	129.911546459
70	56	126	Ba	125.911244146	78	52	130	Te	129.906222753
69	57	126	La	125.919370	77	53	130	I	129.906674018
68	58	126	Ce	125.924100	76	54	130	Xe	129.903507903
67	59	126	Pr	125.935310	75	55	130	Cs	129.906706163
66	60	126	Nd	125.943070	74	56	130	Ba	129.906310478
80	47	127	Ag	126.936880	73	57	130	La	129.912320
79	48	127	Cd	126.926434822	72	58	130	Ce	129.914694
78	49	127	In	126.917344048	71	59	130	Pr	129.923380
77	50	127	Sn	126.910350980	70	60	130	Nd	129.928780
76	51	127	Sb	126.906914564	69	61	130	Pm	129.940450
75	52	127	Te	126.905217290	68	62	130	Sm	129.948630
74	53	127	I	126.904468420	82	49	131	In	130.926767408
73	54	127	Xe	126.905179581	81	50	131	Sn	130.916919144
72	55	127	Cs	126.907417600	80	51	131	Sb	130.911946487
71	56	127	Ba	126.911121328	79	52	131	Te	130.908521880
70	57	127	La	126.916160	78	53	131	I	130.906124168
69	58	127	Ce	126.922750	77	54	131	Xe	130.905081920
68	59	127	Pr	126.930830	76	55	131	Cs	130.905460232
67	60	127	Nd	126.940500	75	56	131	Ba	130.906930798
80	48	128	Cd	127.927760617	74	57	131	La	130.910108489
79	49	128	In	127.920170658	73	58	131	Ce	130.914424137
78	50	128	Sn	127.910534953	72	59	131	Pr	130.920060245
77	51	128	Sb	127.909167330	71	60	131	Nd	130.927102697
76	52	128	Te	127.904461383	70	61	131	Pm	130.935800
75	53	128	I	127.905805254	69	62	131	Sm	130.945890
74	54	128	Xe	127.903530436	83	49	132	In	131.932919005
73	55	128	Cs	127.907747919	82	50	132	Sn	131.917744455
72	56	128	Ba	127.908308870	81	51	132	Sb	131.914413247
71	57	128	La	127.915447940	80	52	132	Te	131.908523782
70	58	128	Ce	127.918870	79	53	132	I	131.907994525
69	59	128	Pr	127.928800	78	54	132	Xe	131.904154457
68	60	128	Nd	127.935390	77	55	132	Cs	131.906429799
67	61	128	Pm	127.948260	76	56	132	Ba	131.905056152
81	48	129	Cd	128.932260	75	57	132	La	131.910110399
80	49	129	In	128.921657958	74	58	132	Ce	131.911490
79	50	129	Sn	128.913439976	73	59	132	Pr	131.919120
78	51	129	Sb	128.909150092	72	60	132	Nd	131.923120
77	52	129	Te	128.906595593	71	61	132	Pm	131.933750
76	53	129	I	128.904987487	70	62	132	Sm	131.940820
75	54	129	Xe	128.904779458	69	63	132	Eu	131.954160

N	Z	A	El	mass (amu)	N	Z	A	El	mass (amu)
84	49	133	In	132.938340	79	57	136	La	135.907651181
83	50	133	Sn	132.923814085	78	58	136	Ce	135.907143574
82	51	133	Sb	132.915236466	77	59	136	Pr	135.912646935
81	52	133	Te	132.910939068	76	60	136	Nd	135.915020542
80	53	133	I	132.907806465	75	61	136	Pm	135.923447865
79	54	133	Xe	132.905905660	74	62	136	Sm	135.928300
78	55	133	Cs	132.905446870	73	63	136	Eu	135.939500
77	56	133	Ba	132.906002368	72	64	136	Gd	135.947070
76	57	133	La	132.908396372	87	50	137	Sn	136.945790
75	58	133	Ce	132.911550	86	51	137	Sb	136.935310
74	59	133	Pr	132.916200	85	52	137	Te	136.925324769
73	60	133	Nd	132.922210	84	53	137	I	136.917872653
72	61	133	Pm	132.929720	83	54	137	Xe	136.911562939
71	62	133	Sm	132.938730	82	55	137	Cs	136.907083505
70	63	133	Eu	132.948900	81	56	137	Ba	136.905821414
85	49	134	In	133.944660	80	57	137	La	136.906465656
84	50	134	Sn	133.928463576	79	58	137	Ce	136.907777634
83	51	134	Sb	133.920551554	78	59	137	Pr	136.910678351
82	52	134	Te	133.911540546	77	60	137	Nd	136.914639730
81	53	134	I	133.909876552	76	61	137	Pm	136.920713
80	54	134	Xe	133.905394504	75	62	137	Sm	136.927046709
79	55	134	Cs	133.906713419	74	63	137	Eu	136.935210
78	56	134	Ba	133.904503347	73	64	137	Gd	136.944650
77	57	134	La	133.908489607	87	51	138	Sb	137.940960
76	58	134	Ce	133.909026379	86	52	138	Te	137.929220
75	59	134	Pr	133.915672	85	53	138	I	137.922383666
74	60	134	Nd	133.918645	84	54	138	Xe	137.913988549
73	61	134	Pm	133.928490	83	55	138	Cs	137.911010537
72	62	134	Sm	133.934020	82	56	138	Ba	137.905241273
71	63	134	Eu	133.946320	81	57	138	La	137.907106826
85	50	135	Sn	134.934730	80	58	138	Ce	137.905985574
84	51	135	Sb	134.925167962	79	59	138	Pr	137.910748891
83	52	135	Te	134.916450782	78	60	138	Nd	137.911930
82	53	135	I	134.910050310	77	61	138	Pm	137.919445
81	54	135	Xe	134.907207499	76	62	138	Sm	137.923540
80	55	135	Cs	134.905971903	75	63	138	Eu	137.933450
79	56	135	Ba	134.905682749	74	64	138	Gd	137.939970
78	57	135	La	134.906971003	73	65	138	Tb	137.952870
77	58	135	Ce	134.909145555	88	51	139	Sb	138.945710
76	59	135	Pr	134.913139140	87	52	139	Te	138.934730
75	60	135	Nd	134.918240	86	53	139	I	138.926093402
74	61	135	Pm	134.924617	85	54	139	Xe	138.918786859
73	62	135	Sm	134.932350	84	55	139	Cs	138.913357921
72	63	135	Eu	134.941720	83	56	139	Ba	138.908835384
86	50	136	Sn	135.939340	82	57	139	La	138.906348160
85	51	136	Sb	135.930660	81	58	139	Ce	138.906646605
84	52	136	Te	135.920103155	80	59	139	Pr	138.908932181
83	53	136	I	135.914655105	79	60	139	Nd	138.911924150
82	54	136	Xe	135.907219526	78	61	139	Pm	138.916759814
81	55	136	Cs	135.907305741	77	62	139	Sm	138.922302000
80	56	136	Ba	135.904570109	76	63	139	Eu	138.929838

N	Z	A	El	mass (amu)	N	Z	A	El	mass (amu)
75	64	139	Gd	138.938080	87	56	143	Ba	142.920617184
74	65	139	Tb	138.948030	86	57	143	La	142.916058646
88	52	140	Te	139.938700	85	58	143	Ce	142.912381158
87	53	140	I	139.931210	84	59	143	Pr	142.910812233
86	54	140	Xe	139.921635665	83	60	143	Nd	142.909809626
85	55	140	Cs	139.917277075	82	61	143	Pm	142.910927571
84	56	140	Ba	139.910599485	81	62	143	Sm	142.914623555
83	57	140	La	139.909472552	80	63	143	Eu	142.920286634
82	58	140	Ce	139.905434035	79	64	143	Gd	142.926738636
81	59	140	Pr	139.909071204	78	65	143	Tb	142.934750
80	60	140	Nd	139.909309824	77	66	143	Dy	142.943830
79	61	140	Pm	139.915801649	76	67	143	Ho	142.954690
78	62	140	Sm	139.918991000	91	53	144	I	143.949610
77	63	140	Eu	139.928083921	90	54	144	Xe	143.938230
76	64	140	Gd	139.933945	89	55	144	Cs	143.932027373
75	65	140	Tb	139.945540	88	56	144	Ba	143.922940468
74	66	140	Dy	139.953790	87	57	144	La	143.919591666
89	52	141	Te	140.944390	86	58	144	Ce	143.913642686
88	53	141	I	140.934830	85	59	144	Pr	143.913300595
87	54	141	Xe	140.926646282	84	60	144	Nd	143.910082629
86	55	141	Cs	140.920043984	83	61	144	Pm	143.912585768
85	56	141	Ba	140.914406439	82	62	144	Sm	143.911994730
84	57	141	La	140.910957016	81	63	144	Eu	143.918774116
83	58	141	Ce	140.908271103	80	64	144	Gd	143.922789
82	59	141	Pr	140.907647726	79	65	144	Tb	143.932530
81	60	141	Nd	140.909604800	78	66	144	Dy	143.939070
80	61	141	Pm	140.913606636	77	67	144	Ho	143.951640
79	62	141	Sm	140.918468512	76	68	144	Er	143.960590
78	63	141	Eu	140.924885867	91	54	145	Xe	144.943670
77	64	141	Gd	140.932210	90	55	145	Cs	144.935388226
76	65	141	Tb	140.941160	89	56	145	Ba	144.926923807
75	66	141	Dy	140.951190	88	57	145	La	144.921638370
90	52	142	Te	141.948500	87	58	145	Ce	144.917227871
89	53	142	I	141.940180	86	59	145	Pr	144.914506897
88	54	142	Xe	141.929702981	85	60	145	Nd	144.912568847
87	55	142	Cs	141.924292317	84	61	145	Pm	144.912743879
86	56	142	Ba	141.916448175	83	62	145	Sm	144.913405611
85	57	142	La	141.914074489	82	63	145	Eu	144.916261285
84	58	142	Ce	141.909239733	81	64	145	Gd	144.921687498
83	59	142	Pr	141.910039865	80	65	145	Tb	144.928880
82	60	142	Nd	141.907718643	79	66	145	Dy	144.936953
81	61	142	Pm	141.912950738	78	67	145	Ho	144.946880
80	62	142	Sm	141.915193274	77	68	145	Er	144.957460
79	63	142	Eu	141.923400033	92	54	146	Xe	145.947300
78	64	142	Gd	141.928231	91	55	146	Cs	145.940162028
77	65	142	Tb	141.938859	90	56	146	Ba	145.930106645
76	66	142	Dy	141.946267	89	57	146	La	145.925700146
75	67	142	Ho	141.959860	88	58	146	Ce	145.918689722
90	53	143	I	142.944070	87	59	146	Pr	145.917588016
89	54	143	Xe	142.934890	86	60	146	Nd	145.913112139
88	55	143	Cs	142.927330292	85	61	146	Pm	145.914692165

N	Z	A	El	mass (amu)	N	Z	A	El	mass (amu)
84	62	146	Sm	145.913036760	83	66	149	Dy	148.927333981
83	63	146	Eu	145.917199714	82	67	149	Ho	148.933789944
82	64	146	Gd	145.918305344	81	68	149	Er	148.942174
81	65	146	Tb	145.927180629	80	69	149	Tm	148.952650
80	66	146	Dy	145.932720118	79	70	149	Yb	148.963480
79	67	146	Ho	145.944100	95	55	150	Cs	149.957970
78	68	146	Er	145.952120	94	56	150	Ba	149.945620
77	69	146	Tm	145.966495	93	57	150	La	149.938570
93	54	147	Xe	146.953010	92	58	150	Ce	149.930226399
92	55	147	Cs	146.943864435	91	59	150	Pr	149.926995031
91	56	147	Ba	146.933992519	90	60	150	Nd	149.920886563
90	57	147	La	146.927819639	89	61	150	Pm	149.920979477
89	58	147	Ce	146.922510962	88	62	150	Sm	149.917271454
88	59	147	Pr	146.918979001	87	63	150	Eu	149.919698294
87	60	147	Nd	146.916095794	86	64	150	Gd	149.918655455
86	61	147	Pm	146.915133898	85	65	150	Tb	149.923654158
85	62	147	Sm	146.914893275	84	66	150	Dy	149.925579728
84	63	147	Eu	146.916741206	83	67	150	Ho	149.933352
83	64	147	Gd	146.919089446	82	68	150	Er	149.937762
82	65	147	Tb	146.924037176	81	69	150	Tm	149.949670
81	66	147	Dy	146.930878496	80	70	150	Yb	149.957990
80	67	147	Ho	146.939840	79	71	150	Lu	149.972668
79	68	147	Er	146.949310	96	55	151	Cs	150.962000
78	69	147	Tm	146.961081	95	56	151	Ba	150.950700
93	55	148	Cs	147.948899539	94	57	151	La	150.941560
92	56	148	Ba	147.937682377	93	58	151	Ce	150.934040
91	57	148	La	147.932191197	92	59	151	Pr	150.928227869
90	58	148	Ce	147.924394738	91	60	151	Nd	150.923824739
89	59	148	Pr	147.922183237	90	61	151	Pm	150.921202693
88	60	148	Nd	147.916888516	89	62	151	Sm	150.919928351
87	61	148	Pm	147.917467786	88	63	151	Eu	150.919846022
86	62	148	Sm	147.914817914	87	64	151	Gd	150.920344273
85	63	148	Eu	147.918153775	86	65	151	Tb	150.923098169
84	64	148	Gd	147.918109771	85	66	151	Dy	150.926179630
83	65	148	Tb	147.924298636	84	67	151	Ho	150.931680791
82	66	148	Dy	147.927177882	83	68	151	Er	150.937460
81	67	148	Ho	147.937269	82	69	151	Tm	150.945434
80	68	148	Er	147.944440	81	70	151	Yb	150.955249
79	69	148	Tm	147.957550	80	71	151	Lu	150.967147
78	70	148	Yb	147.966760	96	56	152	Ba	151.954160
94	55	149	Cs	148.952720	95	57	152	La	151.946110
93	56	149	Ba	148.942460	94	58	152	Ce	151.936380
92	57	149	La	148.934370	93	59	152	Pr	151.931600
91	58	149	Ce	148.928289207	92	60	152	Nd	151.924682428
90	59	149	Pr	148.923791056	91	61	152	Pm	151.923490557
89	60	149	Nd	148.920144190	90	62	152	Sm	151.919728244
88	61	149	Pm	148.918329195	89	63	152	Eu	151.921740399
87	62	149	Sm	148.917179521	88	64	152	Gd	151.919787882
86	63	149	Eu	148.917925922	87	65	152	Tb	151.924071324
85	64	149	Gd	148.919336427	86	66	152	Dy	151.924713874
84	65	149	Tb	148.923241630	85	67	152	Ho	151.931740598

N	Z	A	El	mass (amu)	N	Z	A	El	mass (amu)
84	68	152	Er	151.935078452	83	72	155	Hf	154.962760
83	69	152	Tm	151.944300	98	58	156	Ce	155.951260
82	70	152	Yb	151.950167	97	59	156	Pr	155.944120
81	71	152	Lu	151.963610	96	60	156	Nd	155.935200
97	56	153	Ba	152.959610	95	61	156	Pm	155.931060357
96	57	153	La	152.949450	94	62	156	Sm	155.925526236
95	58	153	Ce	152.940580	93	63	156	Eu	155.924750855
94	59	153	Pr	152.933650	92	64	156	Gd	155.922119552
93	60	153	Nd	152.927694534	91	65	156	Tb	155.924743749
92	61	153	Pm	152.924113189	90	66	156	Dy	155.924278273
91	62	153	Sm	152.922093907	89	67	156	Ho	155.929710
90	63	153	Eu	152.921226219	88	68	156	Er	155.931015001
89	64	153	Gd	152.921746283	87	69	156	Tm	155.939006895
88	65	153	Tb	152.923430858	86	70	156	Yb	155.942847109
87	66	153	Dy	152.925760865	85	71	156	Lu	155.952907
86	67	153	Ho	152.930194506	84	72	156	Hf	155.959247
85	68	153	Er	152.935093125	83	73	156	Ta	155.971689
84	69	153	Tm	152.942027631	99	58	157	Ce	156.956340
83	70	153	Yb	152.949210	98	59	157	Pr	156.947170
82	71	153	Lu	152.958690	97	60	157	Nd	156.939270
97	57	154	La	153.954400	96	61	157	Pm	156.933200
96	58	154	Ce	153.943320	95	62	157	Sm	156.928354506
95	59	154	Pr	153.937390	94	63	157	Eu	156.925419435
94	60	154	Nd	153.929483295	93	64	157	Gd	156.923956686
93	61	154	Pm	153.926547019	92	65	157	Tb	156.924021155
92	62	154	Sm	153.922205303	91	66	157	Dy	156.925461256
91	63	154	Eu	153.922975386	90	67	157	Ho	156.928188059
90	64	154	Gd	153.920862271	89	68	157	Er	156.931945517
89	65	154	Tb	153.924686236	88	69	157	Tm	156.936756069
88	66	154	Dy	153.924422046	87	70	157	Yb	156.942658650
87	67	154	Ho	153.930596268	86	71	157	Lu	156.950101536
86	68	154	Er	153.932777294	85	72	157	Hf	156.958127
85	69	154	Tm	153.941424	84	73	157	Ta	156.968145
84	70	154	Yb	153.946242	99	59	158	Pr	157.951780
83	71	154	Lu	153.957100	98	60	158	Nd	157.941870
82	72	154	Hf	153.964250	97	61	158	Pm	157.936690
98	57	155	La	154.958130	96	62	158	Sm	157.929987938
97	58	155	Ce	154.948040	95	63	158	Eu	157.927841923
96	59	155	Pr	154.939990	94	64	158	Gd	157.924100533
95	60	155	Nd	154.932629551	93	65	158	Tb	157.925410260
94	61	155	Pm	154.928097047	92	66	158	Dy	157.924404637
93	62	155	Sm	154.924635940	91	67	158	Ho	157.928945730
92	63	155	Eu	154.922889429	90	68	158	Er	157.929912
91	64	155	Gd	154.922618801	89	69	158	Tm	157.936996
90	65	155	Tb	154.923500411	88	70	158	Yb	157.939857897
89	66	155	Dy	154.925748950	87	71	158	Lu	157.949169
88	67	155	Ho	154.929079084	86	72	158	Hf	157.954647
87	68	155	Er	154.933204273	85	73	158	Ta	157.966368
86	69	155	Tm	154.939191562	84	74	158	W	157.973939
85	70	155	Yb	154.945792	100	59	159	Pr	158.955230
84	71	155	Lu	154.954233	99	60	159	Nd	158.946390

N	Z	A	El	mass (amu)	N	Z	A	El	mass (amu)
98	61	159	Pm	158.939130	96	66	162	Dy	161.926794731
97	62	159	Sm	158.933200	95	67	162	Ho	161.929092420
96	63	159	Eu	158.929084500	94	68	162	Er	161.928774923
95	64	159	Gd	158.926385075	93	69	162	Tm	161.933970147
94	65	159	Tb	158.925343135	92	70	162	Yb	161.935750
93	66	159	Dy	158.925735660	91	71	162	Lu	161.943222
92	67	159	Ho	158.927708537	90	72	162	Hf	161.947202977
91	68	159	Er	158.930680718	89	73	162	Ta	161.957148
90	69	159	Tm	158.934808966	88	74	162	W	161.963342
89	70	159	Yb	158.940153735	87	75	162	Re	161.975707
88	71	159	Lu	158.946615113	86	76	162	Os	161.983819
87	72	159	Hf	158.954003	102	61	163	Pm	162.953520
86	73	159	Ta	158.962914	101	62	163	Sm	162.945360
85	74	159	W	158.972280	100	63	163	Eu	162.939210
100	60	160	Nd	159.949390	99	64	163	Gd	162.933990
99	61	160	Pm	159.942990	98	65	163	Tb	162.930643942
98	62	160	Sm	159.935140	97	66	163	Dy	162.928727532
97	63	160	Eu	159.931967	96	67	163	Ho	162.928730286
96	64	160	Gd	159.927050616	95	68	163	Er	162.930029273
95	65	160	Tb	159.927164021	94	69	163	Tm	162.932647648
94	66	160	Dy	159.925193718	93	70	163	Yb	162.936265492
93	67	160	Ho	159.928725679	92	71	163	Lu	162.941203796
92	68	160	Er	159.929078924	91	72	163	Hf	162.947057
91	69	160	Tm	159.935090772	90	73	163	Ta	162.954316650
90	70	160	Yb	159.937560	89	74	163	W	162.962532
89	71	160	Lu	159.946019	88	75	163	Re	162.971967
88	72	160	Hf	159.950713588	87	76	163	Os	162.982048
87	73	160	Ta	159.961358	102	62	164	Sm	163.948280
86	74	160	W	159.968369	101	63	164	Eu	163.942990
85	75	160	Re	159.981485	100	64	164	Gd	163.935860
101	60	161	Nd	160.954330	99	65	164	Tb	163.933347253
100	61	161	Pm	160.945860	98	66	164	Dy	163.929171165
99	62	161	Sm	160.938830	97	67	164	Ho	163.930230577
98	63	161	Eu	160.933680	96	68	164	Er	163.929196996
97	64	161	Gd	160.929665688	95	69	164	Tm	163.933450972
96	65	161	Tb	160.927566289	94	70	164	Yb	163.934520
95	66	161	Dy	160.926929595	93	71	164	Lu	163.941215
94	67	161	Ho	160.927851662	92	72	164	Hf	163.944422
93	68	161	Er	160.930001348	91	73	164	Ta	163.953570
92	69	161	Tm	160.933398042	90	74	164	W	163.958983810
91	70	161	Yb	160.937853	89	75	164	Re	163.970319
90	71	161	Lu	160.943543	88	76	164	Os	163.977927
89	72	161	Hf	160.950330852	103	62	165	Sm	164.952980
88	73	161	Ta	160.958372992	102	63	165	Eu	164.945720
87	74	161	W	160.967089	101	64	165	Gd	164.939380
86	75	161	Re	160.977661	100	65	165	Tb	164.934880
101	61	162	Pm	161.950290	99	66	165	Dy	164.931699828
100	62	162	Sm	161.941220	98	67	165	Ho	164.930319169
99	63	162	Eu	161.937040	97	68	165	Er	164.930722800
98	64	162	Gd	161.930981211	96	69	165	Tm	164.932432463
97	65	162	Tb	161.929484803	95	70	165	Yb	164.935397592

N	Z	A	El	mass (amu)	N	Z	A	El	mass (amu)
94	71	165	Lu	164.939605886	90	78	168	Pt	167.988035
93	72	165	Hf	164.944540	105	64	169	Gd	168.952870
92	73	165	Ta	164.950817	104	65	169	Tb	168.946220
91	74	165	W	164.958335962	103	66	169	Dy	168.940303648
90	75	165	Re	164.967050268	102	67	169	Ho	168.936868306
89	76	165	Os	164.976475	101	68	169	Er	168.934588082
88	77	165	Ir	164.987580	100	69	169	Tm	168.934211117
103	63	166	Eu	165.949970	99	70	169	Yb	168.935187120
102	64	166	Gd	165.941600	98	71	169	Lu	168.937648757
101	65	166	Tb	165.938050	97	72	169	Hf	168.941158567
100	66	166	Dy	165.932803241	96	73	169	Ta	168.945920
99	67	166	Ho	165.932281267	95	74	169	W	168.951759
98	68	166	Er	165.930289970	94	75	169	Re	168.958830
97	69	166	Tm	165.933553133	93	76	169	Os	168.967076205
96	70	166	Yb	165.933879623	92	77	169	Ir	168.976390868
95	71	166	Lu	165.939762646	91	78	169	Pt	168.986421
94	72	166	Hf	165.942250	105	65	170	Tb	169.950250
93	73	166	Ta	165.950470	104	66	170	Dy	169.942670
92	74	166	W	165.955019896	103	67	170	Ho	169.939614951
91	75	166	Re	165.965803	102	68	170	Er	169.935460334
90	76	166	Os	165.972526	101	69	170	Tm	169.935797877
89	77	166	Ir	165.985506	100	70	170	Yb	169.934758652
104	63	167	Eu	166.953050	99	71	170	Lu	169.938472190
103	64	167	Gd	166.945570	98	72	170	Hf	169.939650
102	65	167	Tb	166.940050	97	73	170	Ta	169.946090
101	66	167	Dy	166.935649025	96	74	170	W	169.949290
100	67	167	Ho	166.933126195	95	75	170	Re	169.958163
99	68	167	Er	166.932045448	94	76	170	Os	169.963569716
98	69	167	Tm	166.932848844	93	77	170	Ir	169.975033
97	70	167	Yb	166.934946862	92	78	170	Pt	169.982326
96	71	167	Lu	166.938307056	106	65	171	Tb	170.953300
95	72	167	Hf	166.942600	105	66	171	Dy	170.946480
94	73	167	Ta	166.947973	104	67	171	Ho	170.941461227
93	74	167	W	166.954672	103	68	171	Er	170.938025885
92	75	167	Re	166.962564	102	69	171	Tm	170.936425817
91	76	167	Os	166.971554	101	70	171	Yb	170.936322297
90	77	167	Ir	166.981543	100	71	171	Lu	170.937909903
104	64	168	Gd	167.948360	99	72	171	Hf	170.940490
103	65	168	Tb	167.943640	98	73	171	Ta	170.944460
102	66	168	Dy	167.937230	97	74	171	W	170.949460
101	67	168	Ho	167.935496424	96	75	171	Re	170.955547
100	68	168	Er	167.932367781	95	76	171	Os	170.963040
99	69	168	Tm	167.934170375	94	77	171	Ir	170.971779
98	70	168	Yb	167.933894465	93	78	171	Pt	170.981251
97	71	168	Lu	167.938698576	92	79	171	Au	170.991774
96	72	168	Hf	167.940630	106	66	172	Dy	171.949110
95	73	168	Ta	167.947787	105	67	172	Ho	171.944820
94	74	168	W	167.951863	104	68	172	Er	171.939352149
93	75	168	Re	167.961609	103	69	172	Tm	171.938396118
92	76	168	Os	167.967832911	102	70	172	Yb	171.936377696
91	77	168	Ir	167.979966	101	71	172	Lu	171.939082239

N	Z	A	El	mass (amu)	N	Z	A	El	mass (amu)
100	72	172	Hf	171.939457980	106	70	176	Yb	175.942568409
99	73	172	Ta	171.944739818	105	71	176	Lu	175.942682399
98	74	172	W	171.947424	104	72	176	Hf	175.941401828
97	75	172	Re	171.955285	103	73	176	Ta	175.944740551
96	76	172	Os	171.960078	102	74	176	W	175.945590
95	77	172	Ir	171.970643	101	75	176	Re	175.951570
94	78	172	Pt	171.977376138	100	76	176	Os	175.954950
93	79	172	Au	171.990109	99	77	176	Ir	175.963511
107	66	173	Dy	172.953440	98	78	176	Pt	175.969000
106	67	173	Ho	172.947290	97	79	176	Au	175.980269
105	68	173	Er	172.942400	96	80	176	Hg	175.987413248
104	69	173	Tm	172.939600336	109	68	177	Er	176.954370
103	70	173	Yb	172.938206756	108	69	177	Tm	176.949040
102	71	173	Lu	172.938926901	107	70	177	Yb	176.945257126
101	72	173	Hf	172.940650	106	71	177	Lu	176.943754987
100	73	173	Ta	172.943538	105	72	177	Hf	176.943220013
99	74	173	W	172.947832	104	73	177	Ta	176.944471766
98	75	173	Re	172.953062	103	74	177	W	176.946620
97	76	173	Os	172.959791	102	75	177	Re	176.950270
96	77	173	Ir	172.967707	101	76	177	Os	176.955045
95	78	173	Pt	172.976499642	100	77	177	Ir	176.961170
94	79	173	Au	172.986398138	99	78	177	Pt	176.968453
107	67	174	Ho	173.951150	98	79	177	Au	176.977215
106	68	174	Er	173.944340	97	80	177	Hg	176.986336874
105	69	174	Tm	173.942164618	96	81	177	Tl	176.996881
104	70	174	Yb	173.938858101	109	69	178	Tm	177.952640
103	71	174	Lu	173.940333522	108	70	178	Yb	177.946643396
102	72	174	Hf	173.940040159	107	71	178	Lu	177.945951366
101	73	174	Ta	173.944167937	106	72	178	Hf	177.943697732
100	74	174	W	173.946160	105	73	178	Ta	177.945750349
99	75	174	Re	173.953112	104	74	178	W	177.945848364
98	76	174	Os	173.957124	103	75	178	Re	177.950851081
97	77	174	Ir	173.966804	102	76	178	Os	177.953348225
96	78	174	Pt	173.972811276	101	77	178	Ir	177.961083
95	79	174	Au	173.984917	100	78	178	Pt	177.965710
108	67	175	Ho	174.954050	99	79	178	Au	177.975975
107	68	175	Er	174.947930	98	80	178	Hg	177.982476325
106	69	175	Tm	174.943832897	97	81	178	Tl	177.995228
105	70	175	Yb	174.941272494	110	69	179	Tm	178.955340
104	71	175	Lu	174.940767904	109	70	179	Yb	178.950170
103	72	175	Hf	174.941502991	108	71	179	Lu	178.947324216
102	73	175	Ta	174.943650	107	72	179	Hf	178.945815073
101	74	175	W	174.946770	106	73	179	Ta	178.945934113
100	75	175	Re	174.951393	105	74	179	W	178.947071733
99	76	175	Os	174.957080	104	75	179	Re	178.949981038
98	77	175	Ir	174.964279	103	76	179	Os	178.953951
97	78	175	Pt	174.972276	102	77	179	Ir	178.959150
96	79	175	Au	174.981552	101	78	179	Pt	178.965475
95	80	175	Hg	174.991411	100	79	179	Au	178.973412
108	68	176	Er	175.950290	99	80	179	Hg	178.981783
107	69	176	Tm	175.946991412	98	81	179	Tl	178.991466

N	Z	A	El	mass (amu)	N	Z	A	El	mass (amu)
110	70	180	Yb	179.952330	111	73	184	Ta	183.954009331
109	71	180	Lu	179.949879968	110	74	184	W	183.950932553
108	72	180	Hf	179.946548760	109	75	184	Re	183.952524289
107	73	180	Ta	179.947465655	108	76	184	Os	183.952490808
106	74	180	W	179.946705734	107	77	184	Ir	183.957388318
105	75	180	Re	179.950787680	106	78	184	Pt	183.959895
104	76	180	Os	179.952351	105	79	184	Au	183.967474
103	77	180	Ir	179.959253	104	80	184	Hg	183.971897
102	78	180	Pt	179.963215	103	81	184	Tl	183.981760
101	79	180	Au	179.972396	102	82	184	Pb	183.988198
100	80	180	Hg	179.978322	113	72	185	Hf	184.958780
99	81	180	Tl	179.990194	112	73	185	Ta	184.955559086
111	70	181	Yb	180.956150	111	74	185	W	184.953420586
110	71	181	Lu	180.951970	110	75	185	Re	184.952955747
109	72	181	Hf	180.949099124	109	76	185	Os	184.954043023
108	73	181	Ta	180.947996346	108	77	185	Ir	184.956590
107	74	181	W	180.948198054	107	78	185	Pt	184.960753782
106	75	181	Re	180.950064596	106	79	185	Au	184.965806956
105	76	181	Os	180.953274494	105	80	185	Hg	184.971983
104	77	181	Ir	180.957642156	104	81	185	Tl	184.979100
103	78	181	Pt	180.963177	103	82	185	Pb	184.987580
102	79	181	Au	180.969948	102	83	185	Bi	184.997708
101	80	181	Hg	180.977806	114	72	186	Hf	185.960920
100	81	181	Tl	180.986904	113	73	186	Ta	185.958550100
99	82	181	Pb	180.996714	112	74	186	W	185.954362204
111	71	182	Lu	181.955210	111	75	186	Re	185.954986529
110	72	182	Hf	181.950552893	110	76	186	Os	185.953838355
109	73	182	Ta	181.950152414	109	77	186	Ir	185.957951104
108	74	182	W	181.948205519	108	78	186	Pt	185.959432346
107	75	182	Re	181.951211444	107	79	186	Au	185.965997671
106	76	182	Os	181.952186222	106	80	186	Hg	185.969460021
105	77	182	Ir	181.958127689	105	81	186	Tl	185.978548
104	78	182	Pt	181.961267637	104	82	186	Pb	185.984301
103	79	182	Au	181.969620	103	83	186	Bi	185.996480
102	80	182	Hg	181.974751	114	73	187	Ta	186.960410
101	81	182	Tl	181.985610	113	74	187	W	186.957158365
100	82	182	Pb	181.992676101	112	75	187	Re	186.955750787
112	71	183	Lu	182.957570	111	76	187	Os	186.955747928
111	72	183	Hf	182.953531012	110	77	187	Ir	186.957360830
110	73	183	Ta	182.951373188	109	78	187	Pt	186.960558
109	74	183	W	182.950224458	108	79	187	Au	186.964562
108	75	183	Re	182.950821349	107	80	187	Hg	186.969785
107	76	183	Os	182.953110	106	81	187	Tl	186.976170
106	77	183	Ir	182.956814	105	82	187	Pb	186.984030
105	78	183	Pt	182.961729	104	83	187	Bi	186.993458
104	79	183	Au	182.967620	115	73	188	Ta	187.963710
103	80	183	Hg	182.974561	114	74	188	W	187.958486954
102	81	183	Tl	182.982697	113	75	188	Re	187.958112287
101	82	183	Pb	182.991930	112	76	188	Os	187.955835993
113	71	184	Lu	183.961170	111	77	188	Ir	187.958851962
112	72	184	Hf	183.955447880	110	78	188	Pt	187.959395697

N	Z	A	El	mass (amu)	N	Z	A	El	mass (amu)
109	79	188	Au	187.965085	112	81	193	Tl	192.970548
108	80	188	Hg	187.967555	111	82	193	Pb	192.976080
107	81	188	Tl	187.975920	110	83	193	Bi	192.983061
106	82	188	Pb	187.981061	109	84	193	Po	192.991102
105	83	188	Bi	187.992173	108	85	193	At	193.000188
115	74	189	W	188.961912220	118	76	194	Os	193.965179314
114	75	189	Re	188.959228359	117	77	194	Ir	193.965075610
113	76	189	Os	188.958144866	116	78	194	Pt	193.962663581
112	77	189	Ir	188.958716473	115	79	194	Au	193.965338890
111	78	189	Pt	188.960831900	114	80	194	Hg	193.965381832
110	79	189	Au	188.963892	113	81	194	Tl	193.971053
109	80	189	Hg	188.968132	112	82	194	Pb	193.973966
108	81	189	Tl	188.973689	111	83	194	Bi	193.982748
107	82	189	Pb	188.980880	110	84	194	Po	193.988284107
106	83	189	Bi	188.989505	109	85	194	At	193.998971
116	74	190	W	189.963179541	119	76	195	Os	194.968123889
115	75	190	Re	189.961816139	118	77	195	Ir	194.965976800
114	76	190	Os	189.958445210	117	78	195	Pt	194.964774449
113	77	190	Ir	189.960592299	116	79	195	Au	194.965017928
112	78	190	Pt	189.959930073	115	80	195	Hg	194.966638981
111	79	190	Au	189.964698757	114	81	195	Tl	194.969650
110	80	190	Hg	189.966277	113	82	195	Pb	194.974471
109	81	190	Tl	189.973792	112	83	195	Bi	194.980751
108	82	190	Pb	189.978180008	111	84	195	Po	194.988045
107	83	190	Bi	189.988518	110	85	195	At	194.996554
106	84	190	Po	189.995110	120	76	196	Os	195.969622550
116	75	191	Re	190.963123592	119	77	196	Ir	195.968379906
115	76	191	Os	190.960927951	118	78	196	Pt	195.964934884
114	77	191	Ir	190.960591191	117	79	196	Au	195.966551315
113	78	191	Pt	190.961684653	116	80	196	Hg	195.965814846
112	79	191	Au	190.963649239	115	81	196	Tl	195.970515
111	80	191	Hg	190.967063110	114	82	196	Pb	195.972710
110	81	191	Tl	190.971886	113	83	196	Bi	195.980608
109	82	191	Pb	190.978200	112	84	196	Po	195.985512
108	83	191	Bi	190.986053	111	85	196	At	195.995702
107	84	191	Po	190.994653	110	86	196	Rn	196.002309
117	75	192	Re	191.965960	120	77	197	Ir	196.969636496
116	76	192	Os	191.961479047	119	78	197	Pt	196.967323401
115	77	192	Ir	191.962602198	118	79	197	Au	196.966551609
114	78	192	Pt	191.961035158	117	80	197	Hg	196.967195333
113	79	192	Au	191.964810107	116	81	197	Tl	196.969536200
112	80	192	Hg	191.965572	115	82	197	Pb	196.973380
111	81	192	Tl	191.972142	114	83	197	Bi	196.978934287
110	82	192	Pb	191.975763	113	84	197	Po	196.985567
109	83	192	Bi	191.985368	112	85	197	At	196.993290
108	84	192	Po	191.991522	111	86	197	Rn	197.001661
117	76	193	Os	192.964148083	121	77	198	Ir	197.972280
116	77	193	Ir	192.962923700	120	78	198	Pt	197.967876009
115	78	193	Pt	192.962984504	119	79	198	Au	197.968225244
114	79	193	Au	192.964131745	118	80	198	Hg	197.966751830
113	80	193	Hg	192.966644169	117	81	198	Tl	197.970466294

N	Z	A	El	mass (amu)	N	Z	A	El	mass (amu)
116	82	198	Pb	197.971980	118	85	203	At	202.986847216
115	83	198	Bi	197.979024396	117	86	203	Rn	202.993316
114	84	198	Po	197.983343	116	87	203	Fr	203.001048
113	85	198	At	197.992752	115	88	203	Ra	203.009210
112	86	198	Rn	197.998779978	125	79	204	Au	203.977705
122	77	199	Ir	198.973787159	124	80	204	Hg	203.973475640
121	78	199	Pt	198.970576213	123	81	204	Tl	203.973848646
120	79	199	Au	198.968748016	122	82	204	Pb	203.973028761
119	80	199	Hg	198.968262489	121	83	204	Bi	203.977805161
118	81	199	Tl	198.969813837	120	84	204	Po	203.980307113
117	82	199	Pb	198.972909384	119	85	204	At	203.987261559
116	83	199	Bi	198.977576953	118	86	204	Rn	203.991365
115	84	199	Po	198.983595	117	87	204	Fr	204.000593
114	85	199	At	198.990633	116	88	204	Ra	204.006477
113	86	199	Rn	198.998309	126	79	205	Au	204.979610
122	78	200	Pt	199.971423885	125	80	205	Hg	204.976056104
121	79	200	Au	199.970717886	124	81	205	Tl	204.974412270
120	80	200	Hg	199.968308726	123	82	205	Pb	204.974467112
119	81	200	Tl	199.970945394	122	83	205	Bi	204.977374688
118	82	200	Pb	199.971815560	121	84	205	Po	204.981165396
117	83	200	Bi	199.978141983	120	85	205	At	204.986036352
116	84	200	Po	199.981735	119	86	205	Rn	204.991668
115	85	200	At	199.990293	118	87	205	Fr	204.998663961
114	86	200	Rn	199.995677	117	88	205	Ra	205.006187
113	87	200	Fr	200.006499	126	80	206	Hg	205.977498672
123	78	201	Pt	200.974496467	125	81	206	Tl	205.976095321
122	79	201	Au	200.971640839	124	82	206	Pb	205.974449002
121	80	201	Hg	200.970285275	123	83	206	Bi	205.978482854
120	81	201	Tl	200.970803770	122	84	206	Po	205.980465241
119	82	201	Pb	200.972846589	121	85	206	At	205.986599242
118	83	201	Bi	200.976970721	120	86	206	Rn	205.990160
117	84	201	Po	200.982209	119	87	206	Fr	205.998486886
116	85	201	At	200.988486908	118	88	206	Ra	206.003782
115	86	201	Rn	200.995535	127	80	207	Hg	206.982577025
114	87	201	Fr	201.003986	126	81	207	Tl	206.977407908
124	78	202	Pt	201.975740	125	82	207	Pb	206.975880605
123	79	202	Au	201.973788431	124	83	207	Bi	206.978455217
122	80	202	Hg	201.970625604	123	84	207	Po	206.981578228
121	81	202	Tl	201.972090569	122	85	207	At	206.985775861
120	82	202	Pb	201.972143786	121	86	207	Rn	206.990726826
119	83	202	Bi	201.977674504	120	87	207	Fr	206.996859385
118	84	202	Po	201.980704	119	88	207	Ra	207.003727
117	85	202	At	201.988448629	118	89	207	Ac	207.012094
116	86	202	Rn	201.993218	128	80	208	Hg	207.985940
115	87	202	Fr	202.003287	127	81	208	Tl	207.982004653
124	79	203	Au	202.975137256	126	82	208	Pb	207.976635850
123	80	203	Hg	202.972857096	125	83	208	Bi	207.979726699
122	81	203	Tl	202.972329088	124	84	208	Po	207.981231059
121	82	203	Pb	202.973375491	123	85	208	At	207.986582508
120	83	203	Bi	202.976868118	122	86	208	Rn	207.989631237
119	84	203	Po	202.981412863	121	87	208	Fr	207.997133849

N	Z	A	El	mass (amu)	N	Z	A	El	mass (amu)
120	88	208	Ra	208.001776	130	84	214	Po	213.995185949
119	89	208	Ac	208.011485	129	85	214	At	213.996356412
128	81	209	Tl	208.985349125	128	86	214	Rn	213.995346275
127	82	209	Pb	208.981074801	127	87	214	Fr	213.998954740
126	83	209	Bi	208.980383241	126	88	214	Ra	214.000091141
125	84	209	Po	208.982415788	125	89	214	Ac	214.006893072
124	85	209	At	208.986158678	124	90	214	Th	214.011451
123	86	209	Rn	208.990376634	123	91	214	Pa	214.020739230
122	87	209	Fr	208.995915421	132	83	215	Bi	215.001832349
121	88	209	Ra	209.001944	131	84	215	Po	214.999414609
120	89	209	Ac	209.009568736	130	85	215	At	214.998641245
129	81	210	Tl	209.990065574	129	86	215	Rn	214.998729195
128	82	210	Pb	209.984173129	128	87	215	Fr	215.000326029
127	83	210	Bi	209.984104944	127	88	215	Ra	215.002704195
126	84	210	Po	209.982857396	126	89	215	Ac	215.006450832
125	85	210	At	209.987131308	125	90	215	Th	215.011726597
124	86	210	Rn	209.989679862	124	91	215	Pa	215.019097612
123	87	210	Fr	209.996398327	133	83	216	Bi	216.006199
122	88	210	Ra	210.000446	132	84	216	Po	216.001905198
121	89	210	Ac	210.009256802	131	85	216	At	216.002408839
120	90	210	Th	210.015030	130	86	216	Rn	216.000258153
129	82	211	Pb	210.988731474	129	87	216	Fr	216.003187873
128	83	211	Bi	210.987258139	128	88	216	Ra	216.003518402
127	84	211	Po	210.986636869	127	89	216	Ac	216.008721268
126	85	211	At	210.987480806	126	90	216	Th	216.011050963
125	86	211	Rn	210.990585410	125	91	216	Pa	216.019109649
124	87	211	Fr	210.995529332	133	84	217	Po	217.006253
123	88	211	Ra	211.000893996	132	85	217	At	217.004709619
122	89	211	Ac	211.007648196	131	86	217	Rn	217.003914555
121	90	211	Th	211.014858	130	87	217	Fr	217.004616452
130	82	212	Pb	211.991887495	129	88	217	Ra	217.006306010
129	83	212	Bi	211.991271542	128	89	217	Ac	217.009332676
128	84	212	Po	211.988851755	127	90	217	Th	217.013066169
127	85	212	At	211.990734657	126	91	217	Pa	217.018288571
126	86	212	Rn	211.990688899	134	84	218	Po	218.008965773
125	87	212	Fr	211.996194988	133	85	218	At	218.008681458
124	88	212	Ra	211.999783492	132	86	218	Rn	218.005586315
123	89	212	Ac	212.007811441	131	87	218	Fr	218.007563326
122	90	212	Th	212.012916	130	88	218	Ra	218.007123948
131	82	213	Pb	212.996500	129	89	218	Ac	218.011625045
130	83	213	Bi	212.994374836	128	90	218	Th	218.013267744
129	84	213	Po	212.992842522	127	91	218	Pa	218.020007906
128	85	213	At	212.992921150	126	92	218	U	218.023487
127	86	213	Rn	212.993868354	134	85	219	At	219.011296478
126	87	213	Fr	212.996174845	133	86	219	Rn	219.009474831
125	88	213	Ra	213.000345847	132	87	219	Fr	219.009240843
124	89	213	Ac	213.006573689	131	88	219	Ra	219.010068787
123	90	213	Th	213.012962	130	89	219	Ac	219.012404918
122	91	213	Pa	213.021183209	129	90	219	Th	219.015521253
132	82	214	Pb	213.999798147	128	91	219	Pa	219.019880348
131	83	214	Bi	213.998698664	127	92	219	U	219.024915423

N	Z	A	El	mass (amu)	N	Z	A	El	mass (amu)
135	85	220	At	220.015301	136	90	226	Th	226.024890681
134	86	220	Rn	220.011384149	135	91	226	Pa	226.027932750
133	87	220	Fr	220.012312978	134	92	226	U	226.029339750
132	88	220	Ra	220.011014669	133	93	226	Np	226.035129
131	89	220	Ac	220.014752105	141	86	227	Rn	227.035407
130	90	220	Th	220.015733126	140	87	227	Fr	227.031833167
129	91	220	Pa	220.021876493	139	88	227	Ra	227.029170677
128	92	220	U	220.024712	138	89	227	Ac	227.027746979
136	85	221	At	221.018140	137	90	227	Th	227.027698859
135	86	221	Rn	221.015455	136	91	227	Pa	227.028793151
134	87	221	Fr	221.014245654	135	92	227	U	227.031140069
133	88	221	Ra	221.013907762	134	93	227	Np	227.034958261
132	89	221	Ac	221.015575746	142	86	228	Rn	228.038084
131	90	221	Th	221.018171499	141	87	228	Fr	228.035723
130	91	221	Pa	221.021863742	140	88	228	Ra	228.031064101
129	92	221	U	221.026351	139	89	228	Ac	228.031014825
137	85	222	At	222.022330	138	90	228	Th	228.028731348
136	86	222	Rn	222.017570472	137	91	228	Pa	228.031036942
135	87	222	Fr	222.017543957	136	92	228	U	228.031366357
134	88	222	Ra	222.015361820	135	93	228	Np	228.036180
133	89	222	Ac	222.017828852	134	94	228	Pu	228.038727686
132	90	222	Th	222.018454131	142	87	229	Fr	229.038426
131	91	222	Pa	222.023726	141	88	229	Ra	229.034820309
130	92	222	U	222.026070	140	89	229	Ac	229.032930871
138	85	223	At	223.025340	139	90	229	Th	229.031755340
137	86	223	Rn	223.021790	138	91	229	Pa	229.032088601
136	87	223	Fr	223.019730712	137	92	229	U	229.033496137
135	88	223	Ra	223.018497140	136	93	229	Np	229.036246866
134	89	223	Ac	223.019126030	135	94	229	Pu	229.040138934
133	90	223	Th	223.020795153	143	87	230	Fr	230.042510
132	91	223	Pa	223.023963748	142	88	230	Ra	230.037084774
131	92	223	U	223.027722956	141	89	230	Ac	230.036025144
138	86	224	Rn	224.024090	140	90	230	Th	230.033126574
137	87	224	Fr	224.023235513	139	91	230	Pa	230.034532562
136	88	224	Ra	224.020202004	138	92	230	U	230.033927392
135	89	224	Ac	224.021708435	137	93	230	Np	230.037812591
134	90	224	Th	224.021459250	136	94	230	Pu	230.039645603
133	91	224	Pa	224.025614854	144	87	231	Fr	231.045407
132	92	224	U	224.027590139	143	88	231	Ra	231.041220
139	86	225	Rn	225.028440	142	89	231	Ac	231.038551503
138	87	225	Fr	225.025606914	141	90	231	Th	231.036297060
137	88	225	Ra	225.023604463	140	91	231	Pa	231.035878898
136	89	225	Ac	225.023220576	139	92	231	U	231.036289158
135	90	225	Th	225.023941441	138	93	231	Np	231.038233161
134	91	225	Pa	225.026115172	137	94	231	Pu	231.041258
133	92	225	U	225.029384369	136	95	231	Am	231.045560
132	93	225	Np	225.033899689	145	87	232	Fr	232.049654
140	86	226	Rn	226.030890	144	88	232	Ra	232.043693
139	87	226	Fr	226.029343423	143	89	232	Ac	232.042022474
138	88	226	Ra	226.025402555	142	90	232	Th	232.038050360
137	89	226	Ac	226.026089848	141	91	232	Pa	232.038581720

N	Z	A	El	mass (amu)	N	Z	A	El	mass (amu)
140	92	232	U	232.037146280	146	92	238	U	238.050782583
139	93	232	Np	232.040099	145	93	238	Np	238.050940464
138	94	232	Pu	232.041179445	144	94	238	Pu	238.049553400
137	95	232	Am	232.046590	143	95	238	Am	238.051977839
145	88	233	Ra	233.047995	142	96	238	Cm	238.053016298
144	89	233	Ac	233.044550	141	97	238	Bk	238.058266
143	90	233	Th	233.041576923	140	98	238	Cf	238.061410
142	91	233	Pa	233.040240235	148	91	239	Pa	239.057130
141	92	233	U	233.039628196	147	92	239	U	239.054287777
140	93	233	Np	233.040732350	146	93	239	Np	239.052931399
139	94	233	Pu	233.042987570	145	94	239	Pu	239.052156519
138	95	233	Am	233.046472	144	95	239	Am	239.053018481
137	96	233	Cm	233.050800	143	96	239	Cm	239.054951
146	88	234	Ra	234.050547	142	97	239	Bk	239.058362
145	89	234	Ac	234.048420	141	98	239	Cf	239.062579
144	90	234	Th	234.043595497	149	91	240	Pa	240.060980
143	91	234	Pa	234.043302325	148	92	240	U	240.056585734
142	92	234	U	234.040945606	147	93	240	Np	240.056168828
141	93	234	Np	234.042888556	146	94	240	Pu	240.053807460
140	94	234	Pu	234.043304681	145	95	240	Am	240.055287826
139	95	234	Am	234.047794	144	96	240	Cm	240.055519046
138	96	234	Cm	234.050240	143	97	240	Bk	240.059749
146	89	235	Ac	235.051102	142	98	240	Cf	240.062295
145	90	235	Th	235.047504420	141	99	240	Es	240.068920
144	91	235	Pa	235.045436759	149	92	241	U	241.060330
143	92	235	U	235.043923062	148	93	241	Np	241.058246266
142	93	235	Np	235.044055876	147	94	241	Pu	241.056845291
141	94	235	Pu	235.045281500	146	95	241	Am	241.056822944
140	95	235	Am	235.048029	145	96	241	Cm	241.057646736
139	96	235	Cm	235.051591	144	97	241	Bk	241.060223
138	97	235	Bk	235.056580	143	98	241	Cf	241.063716
147	89	236	Ac	236.055178	142	99	241	Es	241.068662
146	90	236	Th	236.049710	150	92	242	U	242.062925
145	91	236	Pa	236.048675176	149	93	242	Np	242.061635
144	92	236	U	236.045561897	148	94	242	Pu	242.058736847
143	93	236	Np	236.046559724	147	95	242	Am	242.059543039
142	94	236	Pu	236.046048088	146	96	242	Cm	242.058829326
141	95	236	Am	236.049569	145	97	242	Bk	242.062050
140	96	236	Cm	236.051405	144	98	242	Cf	242.063688713
139	97	236	Bk	236.057330	143	99	242	Es	242.069699
147	90	237	Th	237.053894	142	100	242	Fm	242.073430
146	91	237	Pa	237.051139430	150	93	243	Np	243.064273
145	92	237	U	237.048723955	149	94	243	Pu	243.061997013
144	93	237	Np	237.048167253	148	95	243	Am	243.061372686
143	94	237	Pu	237.048403774	147	96	243	Cm	243.061382249
142	95	237	Am	237.049970748	146	97	243	Bk	243.063001570
141	96	237	Cm	237.052891	145	98	243	Cf	243.065421
140	97	237	Bk	237.057127	144	99	243	Es	243.069631
139	98	237	Cf	237.062070	143	100	243	Fm	243.074510
148	90	238	Th	238.056243	151	93	244	Np	244.067850
147	91	238	Pa	238.054497046	150	94	244	Pu	244.064197650

N	Z	A	El	mass (amu)	N	Z	A	El	mass (amu)
149	95	244	Am	244.064279429	148	102	250	No	250.087493
148	96	244	Cm	244.062746349	155	96	251	Cm	251.082277873
147	97	244	Bk	244.065167882	154	97	251	Bk	251.080753440
146	98	244	Cf	244.065990390	153	98	251	Cf	251.079580056
145	99	244	Es	244.070969	152	99	251	Es	251.079983592
144	100	244	Fm	244.074077	151	100	251	Fm	251.081566467
151	94	245	Pu	245.067738657	150	101	251	Md	251.084919
150	95	245	Am	245.066445398	149	102	251	No	251.088960
149	96	245	Cm	245.065485586	148	103	251	Lr	251.094360
148	97	245	Bk	245.066355386	156	96	252	Cm	252.084870
147	98	245	Cf	245.068039	155	97	252	Bk	252.084303
146	99	245	Es	245.071317	154	98	252	Cf	252.081619582
145	100	245	Fm	245.075375	153	99	252	Es	252.082972247
144	101	245	Md	245.081017	152	100	252	Fm	252.082460071
152	94	246	Pu	246.070198429	151	101	252	Md	252.086630
151	95	246	Am	246.069768438	150	102	252	No	252.088965909
150	96	246	Cm	246.067217551	149	103	252	Lr	252.095330
149	97	246	Bk	246.068666836	156	97	253	Bk	253.086880
148	98	246	Cf	246.068798807	155	98	253	Cf	253.085126791
147	99	246	Es	246.072965	154	99	253	Es	253.084817974
146	100	246	Fm	246.075281634	153	100	253	Fm	253.085176259
145	101	246	Md	246.081933	152	101	253	Md	253.087280
153	94	247	Pu	247.074070	151	102	253	No	253.090649
152	95	247	Am	247.072086	150	103	253	Lr	253.095258
151	96	247	Cm	247.070346811	149	104	253	Db	253.100679
150	97	247	Bk	247.070298533	157	97	254	Bk	254.090600
149	98	247	Cf	247.070992043	156	98	254	Cf	254.087316198
148	99	247	Es	247.073650	155	99	254	Es	254.088016026
147	100	247	Fm	247.076819	154	100	254	Fm	254.086847795
146	101	247	Md	247.081804	153	101	254	Md	254.089725
153	95	248	Am	248.075745	152	102	254	No	254.090948746
152	96	248	Cm	248.072342247	151	103	254	Lr	254.096587
151	97	248	Bk	248.073080	150	104	254	Db	254.100166
150	98	248	Cf	248.072178080	157	98	255	Cf	255.091039
149	99	248	Es	248.075458	156	99	255	Es	255.090266386
148	100	248	Fm	248.077184411	155	100	255	Fm	255.089955466
147	101	248	Md	248.082909	154	101	255	Md	255.091075196
154	95	249	Am	249.078480	153	102	255	No	255.093232449
153	96	249	Cm	249.075947062	152	103	255	Lr	255.096769
152	97	249	Bk	249.074979937	151	104	255	Db	255.101492
151	98	249	Cf	249.074846818	150	105	255	Jl	255.107398
150	99	249	Es	249.076405	158	98	256	Cf	256.093440
149	100	249	Fm	249.079024	157	99	256	Es	256.093592
148	101	249	Md	249.083002	156	100	256	Fm	256.091766522
147	102	249	No	249.087823	155	101	256	Md	256.094052757
154	96	250	Cm	250.078350687	154	102	256	No	256.094275879
153	97	250	Bk	250.078310529	153	103	256	Lr	256.098763
152	98	250	Cf	250.076399951	152	104	256	Db	256.101179573
151	99	250	Es	250.078654	151	105	256	Jl	256.108110
150	100	250	Fm	250.079514759	158	99	257	Es	257.095979
149	101	250	Md	250.084488	157	100	257	Fm	257.095098635

N	Z	A	El	mass (amu)	N	Z	A	El	mass (amu)
156	101	257	Md	257.095534643	158	107	265	Bh	265.125198
155	102	257	No	257.096852778	157	108	265	Hn	265.130001
154	103	257	Lr	257.099606	156	109	265	Mt	265.136567
153	104	257	Db	257.103072	160	106	266	Rf	266.121928
152	105	257	Jl	257.107858	159	107	266	Bh	266.127009
158	100	258	Fm	258.097069	158	108	266	Hn	266.130042
157	101	258	Md	258.098425321	157	109	266	Mt	266.137940
156	102	258	No	258.098200	160	107	267	Bh	267.127740
155	103	258	Lr	258.101883	159	108	267	Hn	267.131774
154	104	258	Db	258.103568	158	109	267	Mt	267.137526
153	105	258	Jl	258.109438	157	110	267	Xa	267.143956
152	106	258	Rf	258.113151	160	108	268	Hn	268.132156
159	100	259	Fm	259.100588	159	109	268	Mt	268.138816
158	101	259	Md	259.100503	158	110	268	Xa	268.143529
157	102	259	No	259.101024	161	108	269	Hn	269.134114
156	103	259	Lr	259.102990	160	109	269	Mt	269.139106
155	104	259	Db	259.105628	159	110	269	Xa	269.145144
154	105	259	Jl	259.109721	161	109	270	Mt	270.140723
153	106	259	Rf	259.114652	160	110	270	Xa	270.144626
159	101	260	Md	260.103645	162	109	271	Mt	271.141229
158	102	260	No	260.102636	161	110	271	Xa	271.146078
157	103	260	Lr	260.105572	162	110	272	Xa	272.146310
156	104	260	Db	260.106434	161	111	272	Xb	272.153477
155	105	260	Jl	260.111427	163	110	273	Xa	273.149245
154	106	260	Rf	260.114435447					
153	107	260	Bh	260.121803					
159	102	261	No	261.105743					
158	103	261	Lr	261.106941					
157	104	261	Db	261.108752					
156	105	261	Jl	261.112106					
155	106	261	Rf	261.116199					
154	107	261	Bh	261.121800					
160	102	262	No	262.107520					
159	103	262	Lr	262.109692					
158	104	262	Db	262.109918					
157	105	262	Jl	262.114153					
156	106	262	Rf	262.116477					
155	107	262	Bh	262.123009					
160	103	263	Lr	263.111394					
159	104	263	Db	263.112540					
158	105	263	Jl	263.115078					
157	106	263	Rf	263.118313					
156	107	263	Bh	263.123146					
155	108	263	Hn	263.128710					
160	104	264	Db	264.113978					
159	105	264	Jl	264.117473					
158	106	264	Rf	264.118924					
157	107	264	Bh	264.124730					
156	108	264	Hn	264.128408258					
160	105	265	Jl	265.118659					
159	106	265	Rf	265.121066					

Appendix C

Neutron Scattering Cross Sections

On the following pages are 2200 m/s neutron scattering cross section data for natural elements, and some compounds. Molecular weight and number densities are for natural abundances. Number densities for compounds have the units molecules/cm^3. Missing data are indicated by a dash. [Data from *Reactor Physics Constants*, ANL-5800, 1963.] Absorption and fission cross section data can be found in [General Electric Co., *Nuclides and Isotopes*, 15ed., 1996.]

Z	Element or Compound	Atomic or Molecular Weight (g/mol)	ρ (g/cm^3)	$N \times 10^{24}$ (cm^{-3})	σ_s (b)
1	H	1.008	0.000089	0.000053	38
	H_2O	18.016	1.0	0.0335	103
	D_2O	20.03	1.1	0.0331	13.6
2	He	4.003	0.000178	0.000026	0.8
3	Li	6.94	0.534	0.0463	1.4
4	Be	9.012	1.85	0.1236	7.0
	BeO	25.01	3.025	0.0728	6.8
5	B	10.81	2.5	0.1393	4
6	C (Graphite)	12.011	2.21	0.1108	4.8
7	N	14.007	0.0013	0.000053	10
8	O	15.9994	0.0014	0.000053	4.2
9	F	18.998	0.0017	0.000053	3.9
10	Ne	20.18	0.0009	0.000026	2.4
11	Na	22.99	0.971	0.0254	4
12	Mg	24.305	1.74	0.0431	3.6
13	Al	26.98	2.7	0.0602	1.4
14	Si	28.086	2.33	0.04996	1.7
15	P	30.974	1.82	0.0354	5
16	S	32.066	2.07	0.0389	1.1
17	Cl	35.453	0.0032	0.000053	16
18	A	39.948	0.0018	0.000026	1.5
19	K	39.099	0.87	0.0134	1.5
20	Ca	40.078	1.55	0.0233	3.0
21	Sc	44.956	2.5	0.0335	24
22	Ti	47.867	4.5	0.0566	4
23	V	50.942	6.1	0.0721	5
24	Cr	51.996	7.16	0.0829	3
25	Mn	54.94	7.2	0.0789	2.3
26	Fe	55.845	7.87	0.0849	11
27	Co	58.933	8.9	0.091	7
28	Ni	58.693	8.9	0.0913	17.5
29	Cu	63.546	8.93	0.0846	7.2
30	Zn	65.39	7.14	0.0658	3.6
31	Ga	69.72	5.91	0.0511	4
32	Ge	72.61	5.36	0.0445	3
33	As	74.92	5.73	0.0461	6
34	Se	78.96	4.8	0.0366	11
35	Br	79.904	3.12	0.0235	6
36	Kr	83.8	0.0037	0.000026	7.2

Z	Element or Compound	Atomic or Molecular Weight (g/mol)	ρ (g/cm^3)	$N \times 10^{24}$ (cm^{-3})	σ_s (b)
37	Rb	85.47	1.53	0.0108	12
38	Sr	87.62	2.54	0.0175	10
39	Y	88.906	5.51	0.0373	3
40	Zr	91.224	6.57	0.0434	8
41	Nb	92.906	8.57	0.0555	5
42	Mo	95.94	10.24	0.0643	7
44	Ru	101.07	12.2	0.0727	6
45	Rh	102.906	12.45	0.0729	5
46	Pd	106.42	12.02	0.068	3.6
47	Ag	107.87	10.5	0.0586	6
48	Cd	112.41	8.65	0.0464	7
49	In	114.82	7.28	0.0382	2.2
50	Sn	118.71	7.3	0.037	4
51	Sb	121.76	6.69	0.0331	4.3
52	Te	127.6	6.24	0.0295	5
53	I	126.904	4.93	0.0234	3.6
54	Xe	131.29	0.0059	0.000027	4.3
55	Cs	132.905	1.873	0.0085	20
56	Ba	137.33	3.5	0.0154	8
57	La	138.906	6.19	0.0268	15
58	Ce	140.116	6.78	0.0292	9
59	Pr	140.908	6.78	0.029	4
60	Nd	144.24	6.95	0.029	16
62	Sm	150.36	7.7	0.0309	5
	Sm_2O_3	348.7	7.43	0.0128	22.6
63	Eu	151.96	5.22	0.0207	8
	Eu_2O_3	352.0	7.42	0.0127	30.2
66	Dy	162.5	8.56	0.0317	100
	Dy_2O_3	372.92	7.81	0.0126	214
68	Er	167.26	9.16	0.033	15
69	Tm	168.934	9.35	0.0333	7
70	Yb	173.04	7.01	0.0244	12
72	Hf	178.49	13.3	0.0449	8
73	Ta	180.948	16.6	0.0553	5
74	W	183.84	19.3	0.0632	5
75	Re	186.207	21.1	0.0682	14
76	Os	190.23	22.48	0.0712	11
78	Pt	195.078	21.45	0.0662	10
79	Au	196.967	19.3	0.059	9.3

Z	Element or Compound	Atomic or Molecular Weight (g/mol)	ρ (g/cm^3)	$N \times 10^{24}$ (cm^{-3})	σ_s (b)
80	Hg	200.59	13.55	0.0407	20
81	Tl	204.383	11.85	0.0349	14
82	Pb	207.2	11.34	0.033	11
83	Bi	208.98	9.78	0.0282	9
90	Th	232.038	11.7	0.0304	12.6
92	U	238.03	19.1	0.04832	8.3
	UO_2	270.03	10.5	0.0234	16.7
94	Pu	239.0	19.74	0.0498	9.6

Index